SPRINGER
LAB MANUAL

Springer-Verlag Berlin Heidelberg GmbH

M. Pagano (Ed.)

Cell Cycle – Materials and Methods

With 18 Figures

Springer

MICHELE PAGANO, M.D.

Mitotix
One Kendall Square
Building 600
MA 02139
Cambridge
USA

ISBN 978-3-540-58066-9

1st edition 1995, corrected 2nd printing

Cataloging-in-Publication applied for

Die Deutsche Biblithek – CIP-Einheitsaufnahme
Cell Cycle : materials and methods / M. Pagano (ed). - 2. corr.ed. - Berlin ; Heidelberg ; New York ; Barcelona ; Budapest ; Hong Kong ; London ; Milan ; Paris ; Santa Clara ; Singapore ; (Springer lab manual)
ISBN 978-3-540-58066-9 ISBN 978-3-642-57783-3 (eBook)
DOI 10.1007/978-3-642-57783-3

NE: Pagano, Michele [Hrsg.]

Originally published by Springer-Verlag Berlin Heidelberg New York in 1996

Cover design: Stuve & Partner, Heidelberg
Typesetting:Camera ready by Ulrich Kunkel
SPIN 10525971 39/3137-5 4 3 2 1 0 - Printed on acid-free paper

Preface

The field of cell cycle research has advanced enormously in the past 10 years with the help of a wide range of techniques, both old and new. Investigations using several model organisms have shed new light on how progression through different phases of the cell cycle is controlled in eukaryotic cells and how these processes are deregulated in cancer cells. The excitement about these new discoveries is such that, by analogy to the "oncogene years" of the 1980s, I would define the 1990s as the "cell cycle years".

Because of this excitement, biologists from diverse subspecialties are now interested in studying cell cycle aspects of their particular systems. The aim of this handbook is to present in sufficient experimental detail fundamental methods which have been successfully used in cell cycle studies, so that readers can apply such methods to their own work.

Some of the chapters describe "pure cell cycle methods": chapters by Patrick O'Connor and Joany Jackman, and by Bob Marraccino and Peter Keng illustrate the methods of synchronization of mammalian cells, while Jim Creanor's chapter covers techniques used to synchronize yeast cells. An overview of methods used to monitor the position of individual cells within the cell cycle is presented by Marina Giunta and Carlo Pucillo. Sunny Tam and Bob Schlegel outline protocols to quantify the percentage of cells in mitosis, either normal or premature. Didier Fesquet, Jean Claude Labbé, and Jean Claude Cavadore show us how to purify cell cycle kinases. Chapters by Lisa Connell-Crowley and Wade Harper, and by Greg Hannon tell us how to express cell cycle kinases in bacteria or in insect cells, and then how to assay them in vitro. Anindya Dutta's chapter focuses on SV40-based in vitro DNA replication assays. The remaining chapters describe methods that, though not exclusively used for cell cycle research, have played an important role in the field. In other words, these chapters have been conceived from a cell cycle point of view. So that the reader can gain a greater understanding of the role of these methods in cell cycle research, papers in which these methods have been applied are cited in the Background and in the Principle and Applications sections.

To avoid repetition, we have decided not to define specific cell cycle terms in each chapter. Rather, a glossary has been compiled for easy reference.

With regard to the nomenclature, genetic loci and species' names (e.g., *cdc2, wee1, CDC28; Xenopus*) throughout the text are italicized. Proteins appear in regular typeface with the first letter capitalized (e.g., Cdc2, Wee1, Cdc28).

As a final note, I would like thank my colleagues at Mitotix for their valuable advice and Springer-Verlag for the encouragement to edit this book.

Cambridge, MA, October 1994 Michele Pagano

Contents

An Historical synopsis of Cell Cycle Terms XVII
M. Isabel Chiu and Michele Pagano

Part One
Techniques for Studying the Mammalian Cell Cycle

1 Use of Immunohistochemistry in Cell Cycle Studies 3
Jirina Bartkova, Jìri Lukas, and Jìri Bartek

2 Immunofluorescence Techniques in Cell Cycle Studies 15
M. Cristina Cardoso and Heinrich Leonhardt

3 Fluorescence in Situ Hybridization and Human Cell Cycle Genes 29
Douglas J. Demetrick

4 Techniques for Monitoring Cell Cycle Phases 46
Marina Giunta and Carlo Pucillo

5 Centrifugal Elutriation 53
Robert L. Marraccino and Peter C. Keng

6 Synchronization of Mammalian Cells 63
Patrick M. O'Connor and Joany Jackman

7 Microinjection and Electroporation of Macromolecules into Live Cells 75
Rainer Pepperkok

8 Stable and Transient Transfections in Cell Cycle Studies 87
Mark Rolfe

9 Quantification of Premature and Normal Mitosis 93
Sun W. Tam and Robert Schlegel

Part Two
Techniques for Studying the *Xenopus* Cell Cycle

10 Preparation of *Xenopus* Egg Extracts and Their Utilization in Cell Cycle Studies 103
Paul R. Clarke

Part Three
Techniques to Study the Yeast Cell Cycle

11 Cloning and Expression of Cell Cycle Genes in Schizosaccharomyces pombe 119
Guillame Cottarel

12 Preparation of Synchronous Cultures of the Fission yeast, Schizosaccharomyces Pombe 132
James Creanor

Part Four
Techniques for Studying the Cell Cycle of Any Species

13 Identification of Novel Cell Cycle Targets Using Small Molecule Ligands 145
Vivian Berlin and M. Isabel Chiu

14 Bacterial Expression and in Vitro Assay of Cell Cycle Kinases 157
Lisa Connell-Crowley and Wade Harper

15 Heterologous Expression and Secretion of Cell Cycle Proteins Using the Yeast Pichia pastoris 169
Guillaume Cottarel

16 SV40 Based in Vitro DNA Replication Assay 175
Anindya Dutta and Ellen Winchester

17 From Sequence to Function-Structural Motifs of Cell Cycle Proteins 186
Jens W. Eckstein

18 Purification of Cell Cycle Kinases 200
Didier Fesquet, Jean-Claude Labbé, and Jean-Claude Cavadore

19 Interaction-Based Cloning of Novel Proteins 211
Francesca Fiore, Lou Lamphere, and Jeno Gyuris

20 Expression of Cell Cycle Proteins
Using Recombinant Baculoviruses 231
Gregory J. Hannon

21 Cell Cycle-Regulated Transcription Factors: Bandshift Assays 243
Pidder Jansen-Dürr

22 Immunoprecipitation and Immunoblotting in Cell Cycle Studies 250
Christopher W. Jenkins and Yue Xiong

23 Protein mapping in Cell Cycle Studies 264
Emma Lees

24 From Peptide to Purified Antibody 271
Michele Pagano

Index 281

Contributors

Jìri Bartek
Danish Cancer Society
Division of Tumor Biology
Strandboulevarden 49
DK-2100 Copenhagen Ø
Denmark

Jirini Bartkova
Danish Cancer Society
Division of Tumor Biology
Strandboulevarden 49
DK-2100 Copenhagen Ø
Denmark

Vivian Berlin
Mitotix
One Kendall Square
Building 600
Cambridge, MA 02139
USA

M. Cristina Cardoso
Department of Cardiology, Children's Hospital
Harvard Medical School
300 Longwood Avenue, Enders 13
Boston, MA 02115
USA

Paul R. Clarke
School of Biological Sciences
G38 Stopford Building
University of Manchester
Oxford Road
Manchester M13 9PT
UK

Jean-Claude Cavadore
UPR 9008 CNRS – U 249 INSERM
Centre de Recherche de Biochimie Macromoléculaire
CNRS – 1919 route de Mende
BP 5051
34033 Montpellier
Cedex, France

M. Isabel Chiu
Mitotix
One Kendall Square
Building 600
Cambridge, MA 02139
USA

Lisa Connell-Crowley
Verna and Marrs McLean
Department of Biochemistry
Baylor College of Medicine
Houston, TX 77030
USA

Guillame Cottarel
Mitotix
One Kendall Square
Building 600
Cambridge, MA 02139
USA

Jim Creanor
Department of Zoology
University of Edinburgh
West Mains Road
Edinburg EH9 3JT
Scotland

Douglas J. Demetrick
Cold Spring Harbor Laboratory
P.O. Box 100
Cold Spring Harbor, NY 11724
USA

Anindya Dutta
Department of Pathology
Brigham and Women's Hospital
Harvard Medical School
75 Francis Street,
Boston, MA 02115
USA

Jens W. Eckstein
Mitotix
One Kendall Square
Building 600
Cambridge, MA 02139
USA

Didier Fesquet
UPR 9008 CNRS – U 249 INSERM
Centre de Recherche de Biochimie Macromoléculaire
CNRS – 1919 route de Mende
BP 5051
34033 Montpellier
Cedex, France

Francesca Fiore
Mitotix
One Kendall Square
Building 600
Cambridge, MA 02139
USA

Marina Giunta
Experimental Immunology Branch, NCI
NIH, bldg 10, room 4B17
9000 Rockville Pike
Bethesda, MD 20892
USA

Jeno Gyuris
Mitotix
One Kendall Square
Building 600
Cambridge, MA 02139
USA

Gregory J. Hannon
Cold Spring Harbor Laboratory
P.O. Box 100
Cold Spring Harbor, NY 11724
USA

Wade Harper
Verna and Marrs McLean
Department of Biochemistry
Baylor College of Medicine
Houston, TX 77030
USA

Joany Jackman
Laboratory of Molecular Pharmacology
Room 5C-19, Bldg 37
NCI, National Institutes of Health
Bethesda, MD 20852
USA

Pidder Jansen-Dürr
Deutsches Krebsforschungszentrum
Angewandte Tumorvirologie/Abt.0620
Im Neuenheimer Feld 242
D-69120 Heidelberg
Germany

Christopher W. Jenkins
Department of Biochemistry and Biophysics
The University of North Carolina at Chapel Hill
Chapel Hill, NC 27599-3280
USA

Peter C. Keng
Department of Radiation Oncology
University of Rochester
Rochester, NY 14627
USA

Jean-Claude Labbé
UPR 9008 CNRS – U 249 INSERM
Centre de Recherche de Biochimie Macromoléculaire
CNRS – 1919 route de Mende
BP 5051
34033 Montpellier
Cedex, France

Lou Lamphere
Mitotix
One Kendall Square
Building 600
Cambridge, MA 02139
USA

Emma Lees
DNAX Research Institute of Molecular and Cellular Biology
901 California Avenue
Palo Alto, CA 94304-1104
USA

Heinrich Leonhardt
Department of Cardiology, Children's Hospital
Harvard Medical School
300 Longwood Avenue, Enders 13
Boston, MA 02115
USA

Jìri Lukas
Danish Cancer Society
Division of Tumor Biology
Strandboulevarden 49
DK-2100 Copenhagen Ø
Denmark

Robert L. Marraccino
University of Rochester
Department of Radiation and Biophysics
Box 704, 601 Elmwood Avenue
Rochester, NY 14642
USA

Patrick O'Connor
Laboratory of Molecular Pharmacology
Room 5C-19, Bldg 37
NCI, National Institutes of Health
Bethesda, MD 20852
USA

Michele Pagano
Mitotix
One Kendall Square
Building 600
Cambridge, MA 02139
USA

Rainer Pepperkok
Department of Cellular Biology
University of Geneva
Sciences III
30, Quai Ernest-Ansermet
1205 Geneva
Switzerland

Carlo Pucillo
Dipartimento di Scienze e Tecnologie Biomediche
Sezione di Immunologia
Universita' di Udine
Via Gervasutta, 48
I-33100 Udine
Italy

Mark Rolfe
Mitotix
One Kendall Square
Building 600
Cambridge, MA 02139
USA

Robert Schlegel
Department of Molecular and Cellular Toxicology
Division of Biological Science
Harvard School of Public Health
665 Hungtinton Av.
Boston, MA 02115
USA

Sun W. Tam
Mitotix
One Kendall Square
Building 600
Cambridge, MA 02139
USA

Ellen Winchester
Department of Pathology
Brigham and Women's Hospital
Harvard Medical School
75 Francis Street
Boston, MA 02115
USA

Yue Xiong
Department of Biochemistry and Biophysics
Program in Molecular Biology and Biotechnology
Lineberger Comprehensive Cancer Center
The University of North Carolina at Chapel Hill
Chapel Hill, NC 27599-3280
USA

An Historical Synopsis of Cell Cycle Terms

M. ISABEL CHIU AND MICHELE PAGANO

Cak (Cdk-activating kinase). In 1986, Booher and Beach showed that mutations in Thr 167 inactivate *Schizosaccharomyces pombe* Cdc2 [12]. Cdc2 mutants which cannot be phosphorylated at Thr 167 (corresponding to Thr 161 in higher eukaryotes) are inactive both in vivo and in vitro [83, 131, 226] and do not bind cyclins efficiently [38, 83, 177, 194].

Solomon et al. detected an activity in *Xenopus* extracts required for the activation and phosphorylation of Cdc2, and named it Cak (Cdk-activating kinase [226]. Fesquet et al. [64], Poon et al. [200], and Solomon et al. [225] identified *Xenopus* Cak as the product of $p40^{MO15}$ kinase [221]. $p40^{MO15}$ kinase shows 40% amino acid identity to human Cdc2 [221] and, like the latter, needs a regulatory subunit to be active.

The gene encoding the human homologue of Cak, *cak1*, has been isolated [235, 255]. Several Cdks require similar activating kinases. For example, like Thr 161 of Cdc2, phosphorylation of the homologous threonine residues in Cdk2 (Thr 160) [86] and in Cdk4 (Thr 172) [124] is also essential for the activation of these two kinases. The regulatory subunit of Cak has been cloned. It encodes a 34-KDa protein with significant sequence similarity to several cyclins. For this reason (see Cdks) this regulatory subunit has been called Cyclin H and Cak renamed Cdk7 [66, 155].

Cdc(s) (Cell division cycle genes). These are genes involved in cell cycle regulation, first identified in yeast as conditional temperature-sensitive mutations which cause cell cycle arrest at the restrictive temperature.

Cdc2. See Cdk1.

Cdc25. In 1987, Russell and Nurse isolated the *Schizosaccharomyces pombe cdc25* gene [209], a positive regulator of mitosis acting in the dephosphorylation pathway of Cdc2 [84]. In 1991, Dunphy and Kumagai [43], Gautier et al. [77], and Strausfeld et al. [228] showed that Cdc25 is a tyrosine phosphatase, able to activate Cdc2 by removing the phosphate from phospho-Tyr 15 of Cdc2.

From *Xenopus*, two homologues of *cdc25* have been cloned [113]. *Xenopus* Cdc25 is phosphorylated and activated upon entry into mitosis [113, 132].

The gene encoding the human homologue of Cdc25, *cdc25C*, was isolated by Sadhu et al. [212]. Two novel members of the Cdc25 family of protein phosphatases, A and B, were identified by Galactionov and Beach [74]. Human Cdc25C is required for initiation of mitosis [166], while human Cdc25A is required for initiation of S phase [108, 117].

Cdk(s) (Cyclin-dependent kinases). This is a family of serine/threonine protein kinases that play a central role in controlling progression through the cell cycle. Each kinase is formed by a catalytic subunit (Cdk) and a regulatory subunit (Cyclin) (for a review see [197]).

Cdk1/Cdc2. Simanis and Nurse cloned the *Schizosaccharomyces pombe cdc2* gene in 1986 [222]. Subsequent studies showed that Cdc2 is required for both G1/S and G2/M transitions [179, 236]. Cdc2 associates with Cdc13, a B-type Cyclin, essential for entry into mitosis [13, 88]. It also associates with two other B-type Cyclins both of which are dispensable to the mitotic cycle, Cig1 [17] and Cig2 [18, 28] as well as to Puc1, a putative G1 Cyclin [68].

Dunphy et al. [42] and Gautier et al. [76] isolated the *Xenopus* homologue of *cdc2*. Gautier et al. [75] showed that Cdc2, in complex with a B-type Cyclin, is a component of the MPF (maturation promoting factor) required for entry into mitosis.

Draetta et al. [36] and Lee and colleagues [139, 140] isolated the human *cdc2* homologue. Draetta and Beach [35] found that Cdc2 is a protein kinase active during mitosis, that it is itself a phosphoprotein and is periodically associated with a regulatory subunit. Giordano et al. [79] and Pines and Hunter [195] showed that this regulatory subunit is Cyclin B, and that the kinase activity of the Cdc2/Cyclin B complex is maximal during mitosis. Riabowol et al. showed that the function of human Cdc2 is essential for mitosis [205]. During the Cold Spring Harbor Symposium on Cell Cycle in 1991, it was decided that members of the family of cyclin-dependent kinases would be called Cdks. Cdc2, being the first example, would be Cdk1.

Cdk2. Paris et al. [188] cloned the *Xenopus cdk2* gene, whose product controls the G1/S transition [60].

Elledge and Spottswood [51], Ninomiya-Tsuji et al. [175] and Tsai et al. [241] cloned the human *cdk2* gene. Its kinase activity begins in late G1 and culminates during S and G2 [182, 207]. Pagano et al. [184], Tsai et al. [239], van den Heuvel and Harlow [243] showed that human Cdk2 is required for the G1/S transition. The crystal structure of Cdk2 has been solved [30].

Cdk3. Meyerson et al. [163] cloned the human *cdk3* gene. Like *cdc2* and *cdk2*, it can complement *cdc28* mutants of *Saccharomyces cerevisiæ*. The regulatory subunit associated with Cdk3 has yet to be identified. This protein kinase seems to be involved in the progression through the G1 phase of the cell cycle [243].

Cdk4. Originally cloned as a gene encoding a human *cdc2*-related kinase [90], Cdk4 has since been shown to be the major catalytic partner of D-type cyclins by Matsushime et al. [159] and Xiong et al. [259]. It differs from Cdk1, 2, 3, and 5 in substrate specificity, being unable to phosphorylate histone H1. Matsushime et al. [160] showed that Cdk4 kinase activity can be detected in G1 mammalian cells using the retinoblastoma gene product as a substrate. Tam et al. showed that the association of Cdk4 with Cyclin D1 is required for Cyclin D1 function during G1 [233].

Cdk5. Hellmich et al. [100], Lew et al. [146], Meyerson et al. [163], and Xiong et al. [259] isolated the gene encoding Cdk5 from several mammalian species. In mice, Cdk5 is specifically found in postmitotic neurons [240]. In a variety of human and murine cell lines tested, however, Cdk5 is present but not functional. An interacting partner that activates Cdk5 has been cloned but its cDNA displays no striking sequence homology to any existing cyclin gene [145, 238].

Cdk6. Originally cloned as a gene encoding a human *cdc2*-related kinase [163], Cdk6 was subsequently shown by Bates et al. [6] and Meyerson and Harlow [164] to associate with D-type cyclins. In normal cells, Cdk6 appears to be specifically expressed in lymphocytes [164, 233].

Cdk7. See Cak.

Cyclin(s). Cyclins are thought to determine the subcellular localization, substrate specificity, interaction with upstream regulatory enzymes, and timing of activation of Cdks. In mammalian cells at least eight different cyclin types have been identified acting at specific stages of the cell cycle [217].

Cyclin A. Swenson et al. [231] showed that injection of clam Cyclin A into *Xenopus* oocytes induces maturation. Minshull et al. cloned *Xenopus* Cyclin A and showed that the kinase activity associated with it reaches a peak in the cell cycle before Cyclin B-associated kinase [167].

From human cells, Giordano et al. found a novel cyclin-like subunit (p60) associated with Cdc2 and with the E1A viral oncogene product [79]. Pines and Hunter [196] showed that p60 is the product of the human *cyclin A* gene, cloned by Wang et al. [246]. Human Cyclin A has subsequently been shown to be required at two different points in the cell cycle: in complex with Cdk2 [185, 207, 241] it is involved in DNA replication [80, 185, 265], and in complex with Cdc2 it is required for the G2/M transition [185].

Cyclin B. There are three B-type Cyclins in *Schizosaccharomyces pombe*. One, Cdc13, is essential for entry into mitosis [13, 88] The other two, Cig1 [17] and Cig2 [18, 28] are both dispensable for the mitotic cycle.

Gautier et al. showed that Cyclin B, in complex with Cdc2, is an essential component of *Xenopus* MPF [75].

Pines and Hunter cloned the human gene encoding Cyclin B [195]. They, along with Giordano et al. [79], also showed that the complex formed between human Cyclin B and Cdc2 is maximally active as a kinase during mitosis.

Cyclin C. Lew et al. [144] isolated the gene encoding human Cyclin C by complementation of yeast *cln* mutants. Human Cyclin C mRNA synthesis reaches a peak in mid-G1. Neither the catalytic partner nor the function of this cyclin has been identified.

Cyclin D1, 2, 3. Several groups contributed to the identification of mammalian D-type cyclins [144, 161, 171, 253, 257]. Matsushime et al. [159] and Xiong et al. [259] found that D-cyclins have multiple partners: they can form quaternary complexes with one of the several Cdks, with PCNA, and with $p21^{Cip1/Sdi1/Waf1/Pic1}$. Cyclin D1 acts as a "true" cyclin, binding to and activating Cdk4 [125, 159, 160, 247, 259]. It also acts as an "anti-cyclin", binding specifically to the dephosphorylated (and inactive) form of Cdk2 [7, 39, 125]. Baldin et al. [4] and Quelle et al. [203] demonstrated the requirement for Cyclin D1 in G1 progression. Several investigators have formally demonstrated that Cyclin D1 is the product of a proto-oncogene [10, 106, 116, 152, 153, 247]. High levels of Cyclin D1 mRNA and amplification of chromosomal regions containing the *cyclin D1* gene (11q13) are often present in human tumors and human transformed cell lines (see [170] for a review). Although Cyclin D2 and Cyclin D3 share many properties with Cyclin D1, they are not functionally equivalent [217].

Cyclin E. Koff et al. [128] and Lew et al. [144] cloned the human *cyclin E* gene. Dulic et al. [41], and Koff et al. [129] showed that Cyclin E forms an active kinase with Cdk2 during late G1 through early S phase. Ohtsubo and Roberts [180], Resnitzky et al. [204], and Wimmel et al. [252] showed that Cyclin E is involved in the G1/S transition.

Cyclin F. W. Harper and S. Elledge cloned the human *cyclin F* gene (W. Harper, pers. comm.). Cyclin F is most closely related to Cyclin A, sharing 39% amino acid identity with it within the cyclin box. However, outside the cyclin box, the sequence appears completely unrelated to other known cyclins.

Cyclin G. Tamura et al. cloned the mammalian *cyclin G* gene [234]. Cyclin G is most closely related to human Cyclin A and *Schizosaccharomyces pombe* Cig1, sharing with them 37 and 32% identity, respectively, within the cyclin box. Cyclin G mRNA is induced within a few hours after growth stimulation. The *cyclin G* gene is a transcriptional target of p53 (K. Okamoto and D. Beach, pers. comm.).

Cyclin H. Mammalian Cyclin H is the regulatory subunit of Cak [66, 155]. Cyclin H is most closely related to mammalian Cyclin C, *Schizosaccharomyces pombe* Mcs2 and *Saccharomyces cerevisiae* Ccl1.

Cdi1/Kap. Gyuris et al. [87] and Hannon et al. [93] identified a protein that interacts with some Cdks and cloned the gene encoding it from humans. Cdi1/Kap has phosphatase activity in vitro and inhibits cell cycle progression when overexpressed in vivo.

E2F. E2F is a cellular transcription factor originally shown to be required for transcription of the adenoviral E2 promoter in infected cells [130]. The E2F-dependent transcription is stimulated by the adenoviral E1A protein [2, 3, 11]. Several cellular genes involved in cellular proliferation have E2F binding sites in their promoter regions. These include c-*myc,* N-*myc* and c-*myb* [173], *E2F* [110], *cdc2* [29], *cyclin A* [101], *cyclin D1* [103, 174], and a set of genes encoding proteins involved in DNA replication such as DNA polymerase α [192], dihydrofolate reductase [9], and thymidine kinase [127].

During G1, E2F associates with and is inactivated by the retinoblastoma gene product (pRb) [89, 104, 248]. The human papillomavirus E7 protein and the adenoviral E1A protein are able to dissociate the E2F/pRb complex [23, 183]. Dissociation of pRb from E2F triggers E2F activation and subsequently cell proliferation.

E2F is also complexed with several other regulators of the cell cycle, such as cyclins [5, 143, 172], Cdks [31, 182], p107 (mainly during S phase) [20, 214, 219], and p130 (mainly during G0/1) [26]. In 1992, three labs independently cloned E2F [98, 119, 216]. The year after, Johnson et al. [118] showed that expression of E2F induces quiescent cells to enter S-phase. Another transcription factor, DP-1, has been cloned and shown to heterodimerize and cooperate with E2F [81, 99]

Mik1. The *mik1* gene in *Schizosaccharomyces pombe* encodes a functional homologue of Wee1. Like Wee1, Mik1 is a tyrosine kinase involved in the phosphorylation of Cdc2, as well as in its inactivation. Strains harboring both *wee1* and *mik1* ts mutations undergo mitotic catastrophe at the restrictive temperature [154].

MPF. In 1971, Masui and Market reported that injection of metaphase cytoplasm from unfertilized frog eggs into resting G2 oocytes induced the latter to enter into meiosis [158]. This activity is named Maturation (or M phase) Promoting Factor. Maller et al. proposed that a kinase activity constitutes an essential part of this factor [156]. Years later, several groups confirmed this prediction by demonstrating that the two subunits of a cyclin-dependent kinase, Cdc2 and Cyclin B, are the major functional components of MPF [1, 37, 42, 75, 76, 133].

Nim1/Cdr1. Russell and Nurse [210] and Feilotter et al. [63] independently identified the same gene, *nim1* or *cdr1*, which acts as a negative regulator of the Wee1 kinase. Coleman et al. [27] , Lin and Russell [149] and Parker et al. [189] showed that *S. pombe* Nim1/Cdr1 is a protein kinase which inactivates Wee1 through phosphorylation.

p13^{Suc1}. Hayles et al. [97] first isolated the *Schizosaccharomyces pombe suc1* gene as a high-copy suppressor of *cdc2* mutants. *suc1* encodes a 13-kDa protein that interacts with Cdc2 [16, 97, 168]. Cells carrying a disruption of the *suc1* gene arrest with an elongated cdc phenotype and exhibit high histone H1 kinase activity [97, 105].

Draetta et al. [36] reported the existence of a p13^{Suc1} homologue in human cells. Richardson et al. cloned the genes encoding the human homologues of p13^{Suc1} and named them *cksHs1* and *cksHs2* [206]. Parge et al. solved the crystal structure of CksHs2 [187]. How p13^{Suc1} affects the control of the cell cycle is still unclear.

p15^{Ink4B}. Human p15^{Ink4B} was identified as a protein that binds to and specifically inhibits Cdk4 in TGFß-treated cells. The gene encoding p15^{Ink4B} is adjacent to the *p16* gene at a frequent site of chromosomal abnormality in human tumors [92].

p16^{Ink4A}. Serrano et al. cloned the human *ink4A* gene encoding a cell cycle inhibitor, p16^{Ink4} [215]. p16^{Ink4A} binds to and specifically inhibits Cdk4, in contrast to p21$^{Cip1/Sdi1/Waf1/Pic1}$ which interacts with and inhibits all known Cdks. In separate studies, the human *mts1* gene, which is deleted or mutated in a large number of transformed cell lines as well as in certain uncultured tumors [19, 111, 120, 169, 181, 227], turned out to be identical to *ink4A*. These results suggest that p16^{Ink4} is a tumor suppressor gene. The levels of p16^{Ink4A} protein oscillates during the cell cycle, peaking during S phase [232].

p18^{Ink6}. Human p18^{Ink6} was identified as a protein that binds to and specifically inhibits Cdk6 and, like Cdk6, is highly expressed in T-lymphocytes (Y. Xiong, pers. comm.).

p21$^{Cip1/Sdi1/Waf1/Pic1}$. p21, also known as Cip1, Sdi1, Waf1, or Pic1, was independently identified as a protein that interacts with and inhibits the kinase function of Cdks by several groups, hence its many names [50, 86, 96, 176, 258]. El-Deiry et al. [50] and Dulic et al. [40] showed that p21$^{Cip1/Sdi1/Waf1/Pic1}$ is a key mediator of p53-dependent growth suppression. Waga et al. [244] and Flores-Rozas et al. [67] showed that p21$^{Cip1/Sdi1/Waf1/Pic1}$ is able to directly inhibit DNA replication in vitro. Zhang et al. [262] showed that low levels of p21$^{Cip1/Sdi1/Waf1/Pic1}$ stabilize the Cdk-Cyclin complexes whereas high levels inhibit them. p21$^{Cip1/Sdi1/Waf1/Pic1}$ protein levels oscillate during the cell cycle, peaking during G1 [148, 165].

p27$^{Kip1/Ick/Pic2}$. p27$^{Kip1/Ick/Pic2}$ was identified as a protein that binds to and inhibits Cdks in cells that have been growth-arrested by one of several conditions: TGFβ addition, contact inhibition [198, 223], serum starvation, lovastatin treatment [102] and cAMP treatment [123]. Firpo et al. also found p27$^{Kip1/Ick/Pic2}$ in G1-T lymphocytes [65]. Polyak et al. [199] and Toyoshima and Hunter [237] cloned the gene encoding p27$^{Kip1/Ick/Pic2}$ from mammalian species.

p53 tumor suppressor protein. p53 was first identified as a 53-kDa phosphoprotein that forms a complex with the large T tumor antigen in SV40-infected cells [135, 150]. Harlow et al. [94] cloned the human *p53* gene. p53 also associates with the adenovirus E1B protein [213] and the E6 proteins of human papillomavirus types 16 and 18 [249]. p53 is a transcription factor binding in a sequence-specific manner to cellular DNA [49, 126, 261] and can act both as positive and negative transcriptional regulator of downstream genes [61, 73, 78, 230]. p53 is able to inhibit SV40 DNA replication both in vivo and in vitro [70, 229, 245]. It has also been shown to interact with the replication protein A (RPA) holocomplex, and once bound, inhibits the ability of RPA to bind single-stranded DNA [44].

An extremely high incidence of mutations and rearrangements of the *p53* gene is observed in several rodent and human cancers, implicating it as the most significant tumor--suppressor gene to date (see [134, 202] for a review). p53-null mice develop normally but have a high incidence of tumors [33]. Overexpression of p53 causes human cells to arrest in G1 and suppresses transformation (see [193] for a review). Kastan et al. demonstrated the participation of p53 in the cell cycle checkpoint pathway during the cellular response to certain DNA damages [121, 122]. The p53-mediated G1 arrest is effected via p21$^{Cip1/Sdi1/Waf1/Pic1}$, an inhibitor of the Cdk kinases [40, 50]. Tlsty's group showed that loss of p53 increases genetic instability [151]. p53 is also involved in apoptosis (see [202] for a review). The three-dimensional structure of p53 has been solved [25].

p107. Dyson et al. [46], Ewen et al. [55], and Whyte et al. [251] identified a cellular protein of 107-kDa that specifically complexes with adenovirus E1A and SV40 large T antigen. Ewen et al. [57] and Zhu et al. [264] cloned the *p107* gene, which exhibits a high degree of similarity to pRb. p107 interacts with human Cyclin A [54, 58] and E2F during S phase [20, 214, 219]. Like pRb, p107 can inhibit E2F-mediated transcriptional activation [214, 260, 264]. It is able to bind Myc and inhibit Myc-mediated *trans*-activation of downstream genes [8, 85]. Overexpression of p107 also inhibits cellular proliferation [264].

p110/pRb. Several groups contributed to the identification of human retinoblastoma gene product (pRb) [71, 72, 137, 138]. pRb plays an important role in inhibiting cell proliferation, and is therefore a tumor-suppressor gene product. It is phosphorylated by Cyclin-dependent-kinases. pRb is dephosphorylated as cells leave mitosis, is hypophosphorylated during most of G1, and is hyperphosphorylated in late G1, through S, G2, and M (for review see [218]). pRb binds E2F in G1 and inhibits E2F-mediated *trans*-activation [89, 104, 248]; hyperphosphorylated pRb loses its ability to bind E2F thereby failing to inhibit E2F. Oncoproteins encoded by DNA tumor viruses (e.g., adenovirus E1A, human papillomavirus types 16 and 18 E7 proteins and SV40 large T antigen) specifically bind the hypophosphorylated form of pRb, preventing it from binding E2F and thus reversing its growth-suppressive function [46, 250]. pRb overexpression arrests cells in G0/G1 (see [109] for a review). Cotransfection of G1 or S phase cyclins, but not of B-type cyclins, overcomes the pRb-mediated cell cycle block [34, 56, 107], consistent with the view that Cdk-mediated phosphorylation of pRb inactivates its growth-suppressive function. Transgenic mice homozygous for the *Rb* gene-disruption allele are not viable, and those in which only one allele is disrupted experience a high incidence of pituitary tumors [24, 114, 136].

p130. Harlow et al. [95] identified a cellular protein of 130 kDa that forms a specific complex with adenovirus E1A. Hannon et al. [91], Li et al. [147] and Xavier et al. [256] cloned p130, and showed that it is homologous to pRb and p107. p130 interacts with E2F in G0/G1 [147] . p130 was also shown to interact with human Cyclin A, Cyclin E, and D-type cyclins [91, 147].

p300. Egan et al. [48] and Whyte et al. [251] identified a cellular protein of 300 kDa that forms a specific complex with the adenovirus E1A protein. Eckner et al. [47] cloned the human gene for p300.

PCNA. The Proliferating Cell Nuclear Antigen, previously known as cyclin, is the auxiliary protein of DNA polymerases ∂ and ε, required for DNA replication [14, 15, 115, 201] and repair [22, 141, 142, 220]. It is physically associated with all the known Cdk-cyclin complexes [259, 263]. During G1, PCNA seems to be kept in an inactive state by its association with cell cycle proteins, like Cyclin D1 [186] and $p21^{Cip1/Sdi1/Waf1/Pic1}$ [67, 244].

RPA. The replication protein A, previously known as either RF-A or human Ssb, is a multi-subunit complex comprising three polypeptides: RPA70, RPA34, and RPA11 [59, 82, 254]. RPA70 binds to single-stranded DNA. This binding appears to be the initial step in DNA replication and to be required for unwinding DNA replication origins. RPA has an active role in DNA synthesis (see [157] for a review). The genes encoding each of the three subunits were isolated by Kelly and colleagues [52, 53, 242]. Din et al. showed that RPA34 is phosphorylated in the S and G2 phases of the cell cycle [32]. RPA, Cdk2 and Cyclin A colocalize at sites of DNA replication both in vitro [69] and in vivo [21, 224]. Dutta and Stillman [45] showed that RPA is phosphorylated by Cdks. p53 binds RPA, and the p53/RPA complex fails to bind single-stranded DNA [44].

Weel. In 1975, Paul Nurse showed that in *S. pombe* strains lacking Weel function, cells enter mitosis at approximately half the normal size [178]. The *weel* gene was cloned [211], and was shown to be a negative regulator of mitosis [208]. Featherstone and Russell [62] and Parker et al. [190, 191] subsequently showed that *S. pombe* Weel is the major tyrosine kinase that phosphorylates Cdc2 on Tyr15, thereby inactivating it.

Igarashi et al. isolated the human homologue of *weel* [112]. Parker and Piwnica-Worms [191] and McGowan and Russell [162] demonstrated that Weel can phosphorylate Cdc2 on Tyr15 thereby inactivating it. Weel blocks the cell cycle when overexpressed in human cells [162].

Acknowledgments. We thank all our colleagues who communicated results prior to publication. We sincerely apologize to all of the authors whose excellent contributions to the cell cycle field could not be individually cited due to space limitations. In particular, because this book primarily details methods used in cell cycle studies using fission yeast, *Xenopus* and mammalian systems, we apologize for the omission of references to the relevant work done in the budding yeast *Saccharomyces cerevisiae* under each entry of the glossary.

References

[1] Arion et al. (1988) Cell 55:371
[2] Bagchi et al. (1989) Proc Natl Acad Sci USA 86:4352
[3] Bagchi et al. (1990) Cell 62:659
[4] Baldin et al. (1993) Genes Dev 7:812
[5] Bandara, La Thangue (1991) Nature 351:494
[6] Bates et al. (1994) Oncogene 9:71
[7] Bates et al. (1994) Oncogne 9:1633
[8] Beijersbergen et al. (1994) The EMBO J 13:4080
[9] Blake, Azizkhan (1989) Mol Cell Biol 9:4994
[10] Bodrug et al. (1994) EMBO J 13:2124
[11] Boeuf et al. (1990) Proc Natl Acad Sci USA 87:1782
[12] Booher, Beach (1986) Mol Cell Biol 6:3523
[13] Booher, Beach (1988) EMBO J 7:2321
[14] Bravo et al. (1987) Nature 326:515
[15] Bravo, Macdonald-Bravo (1987) J Cell Biol 105:1549
[16] Brizuela et al. (1987) EMBO J 6:3507
[17] Bueno et al. (1991) Cell 65:149
[18] Bueno, Russell (1993) Mol Cell Biol 13:2286
[19] Caldas et al. (1994) Nature Gen 8:27
[20] Cao et al. (1992) Nature 355:176
[21] Cardoso et al. (1993) Cell 74:979
[22] Celis, Madsen (1986) FEBS 209:277
[23] Chellappan et al. (1992) Proc Natl Acad Sci USA 89:4549
[24] Clarke et al. (1992) Nature 359:328
[25] Clore et al. (1994) Science 265:386
[26] Cobrinik et al. (1993) Genes Dev 7:2392
[27] Coleman et al. (1993) Cell 72:919
[28] Connolly, Beach (1994) Mol Cell Biol 14:768
[29] Dalton (1992) EMBO J 11:1797
[30] De Bondt et al. (1993) Nature 363:595
[31] Devoto et al. (1992) Cell 68:167
[32] Din et al. (1990) Genes Dev 4:968
[33] Donehower et al. (1992) Nature 356:215
[34] Dowdy et al. (1993) Cell 73:499
[35] Draetta, Beach (1988) Cell 54:17
[36] Draetta et al. (1987) Cell 50:319
[37] Draetta et al. (1989) Cell 56:829
[38] Ducommun et al. (1991) EMBO J 11:3311
[39] Dulic et al. (1993) Proc Natl Acad Sci USA 90:11034
[40] Dulic et al. (1994) Cell 76:1013
[41] Dulic et al. (1992) Science 257:1958
[42] Dunphy et al. (1988) Cell 54:423
[43] Dunphy, Kumagai (1991) Cell 67:189
[44] Dutta et al. (1993) Nature 365:79
[45] Dutta, Stillmann (1992) EMBO J 11:2189
[46] Dyson et al. (1989) Cell 58:249
[47] Eckner (1994) Genes Dev 8:869

[48] Egan et al. (1988) Mol Cell Biol 8: 3955
[49] El-Deiry et al. (1992) Nature Genetics 1:45
[50] El-Deiry et al. (1993) Cell 75:817
[51] Elledge, Spottswood (1991) EMBO J 10:2653
[52] Erdile et al. (1991) J Biol Chem 266: 12090
[53] Erdile et al. (1990) J Biol Chem 265: 3177
[54] Ewen et al. (1992) Science 255:85
[55] Ewen et al. (1989) Cell 58:257
[56] Ewen et al. (1993) Cell 73:487
[57] Ewen et al. (1991) Cell 66:1155
[58] Faha et al. (1992) Science 255:87
[59] Fairman et al. (1988) EMBO J 7: 1211
[60] Fang, Newport (1991) Cell 66:731
[61] Farmer et al. (1992) Nature 358:83
[62] Featherstone, Russell (1991) Nature 349:808
[63] Feilotter et al. (1987) Genetics 127: 309
[64] Fesquet et al. (1993) EMBO J 12:3111
[65] Firpo et al. (1994) Mol Cell Biol 14: 4889
[66] Fisher, Morgan (1994) Cell 78:713
[67] Flores-Rozas et al. (1994) Proc Natl Acad 91:8655
[68] Forsburg, Nurse (1991) Nature 351: 245
[69] Fotedar, Roberts (1991) Cold Spring Harbor Symp on Quant Biol LVI:325
[70] Friedman et al. (1990) Proc Natl Acad Sci USA 87:9275
[71] Friend et al. (1986) Nature 323:643
[72] Fung et al. (1987) Science 236:1657
[73] Funk et al. (1992) Mol Cell Biol 12: 2866
[74] Galactionov, Beach (1991) Cell 67: 1181
[75] Gautier et al. (1990) Cell 60:487
[76] Gautier et al. (1988) Cell 54:433
[77] Gautier et al. (1991) Cell 67:197
[78] Ginsberg et al. (1991) Proc Natl Acad Sci USA 88:9979
[79] Giordano et al. (1989) Cell 58:981
[80] Girard et al. (1991) Cell 67:1169
[81] Girling et al. (1993) Nature 362:83
[82] Goetz et al. (1988) J Biol Chem 263:383
[83] Gould et al. (1991) EMBO J 10: 3297
[84] Gould et al. (1990) Science 250: 1573
[85] Gu et al. (1994) Science 264:251
[86] Gu et al. (1993) Nature 366:707
[87] Gyuris et al. (1993) Cell 75:791
[88] Hagan et al. (1988) J Cell Sci 91:587
[89] Hamel et al. (1992) Molec Cell Biol 12:3431
[90] Hanks (1987) Proc Natl Acad Sci USA 84:388
[91] Hannon et al. (1993) Genes Dev 7: 2378
[92] Hannon, Beach (1994) Nature 371: 257
[93] Hannon et al. (1994) Proc Natl Acad Sci USA 91:1731
[94] Harlow et al. (1985) J Virol 39:861
[95] Harlow et al. (1986) Mol Cell Biol 6:1579
[96] Harper et al. (1993) Cell 75:805
[97] Hayles et al. (1986) EMBO J 5: 3373
[98] Helin et al. (1992) Cell 70:337
[99] Helin et al. (1993) Genes Dev 7: 1850
[100] Hellmich et al. (1992) Proc Natl Acad Sci USA 89:10867
[101] Henglein et al. (1994) Proc Natl Acad Sci USA 91:5490
[102] Hengst et al. (1994) Proc Natl Acad Sci USA 91:5291
[103] Herber et al. (1994) Oncogene 9: 1295
[104] Hiebert et al. (1992) Genes Dev 6: 177
[105] Hindley et al. (1987) Mol Cell Biol 7:504
[106] Hinds et al. (1994) Proc Natl Acad Sci USA 91:709
[107] Hinds et al. (1992) Cell 70:993
[108] Hoffmann et al. (1994) EMBO J 13: 4302
[109] Hollingsworth et al. (1993) Curr Opin in Cell Biol 5:194
[110] Hsiao et al. (1994) Genes Dev 8:1526
[111] Hussussian et al. (1994) Nature Genetics 8:15

[112] Igarashi et al. (1991) Nature 353: 80
[113] Izumi et al. (1992) Mol Biol Cell 3: 927
[114] Jacks et al. (1992) Nature 359:295
[115] Jaskulski et al. (1988) Science 240: 1544
[116] Jiang et al. (1993) Oncogene 8: 3447
[117] Jinno et al. (1994) EMBO J 13: 1549
[118] Johnson et al. (1993) Nature 365: 349
[119] Kaelin et al. (1992) Cell 70:351
[120] Kamb et al. (1994) Nature Gen 8: 22
[121] Kastan et al. (1991) Cancer Res 51: 6304
[122] Kastan et al. (1992) Cell 71:587
[123] Kato et al. (1994) Cell 79:487
[124] Kato et al. (1994) Mol Cell Biol 14: 2713
[125] Kato et al. (1993) Genes Dev 7: 331
[126] Kern et al. (1991) Science 252: 1708
[127] Kim, Lees (1991) Mol Cell Biol 11: 2296
[128] Koff et al. (1991) Cell 66:1217
[129] Koff et al. (1992) Science 257: 1689
[130] Kovesdi et al. (1986) Cell 45:219
[131] Krek, Nigg (1992) New Biol 4:323
[132] Kumagai, Dunphy (1992) Cell 70: 139
[133] Labbé et al. (1988) Nature 335:251
[134] Lane (1992) Curr Biol 2:581
[135] Lane, Crawford (1979) Nature 278: 261
[136] Lee et al. (1992) Nature 359:288
[137] Lee et al. (1986) Science 235:1394
[138] Lee et al. (1987) Nature 329:642
[139] Lee et al. (1988) Nature 333:676
[140] Lee, Nurse (1987) Nature 327:31
[141] Lee et al. (1991) J Biol Chem 266: 594
[142] Lee et al. (1992) J Biol Chem 266: 22707
[143] Lees et al. (1992) Genes Dev 6: 1874
[144] Lew et al. (1991) Cell 66:1197
[145] Lew et al. (1994) Nature 371:423
[146] Lew et al. (1992) J Biol Chem 267: 25922
[147] Li et al. (1993) Genes Dev 7:2366
[148] Li et al. (1994) Oncogene 9:2761
[149] Lin, Russell (1993) Nature 363: 738
[150] Linzer, Levine (1979) Cell 17:43
[151] Livingston et al. (1992) Cell 70: 923
[152] Lovec et al. (1994) EMBO J 13: 3487
[153] Lovec et al. (1994) Oncogene 9: 323
[154] Lundgren et al. (1991) Cell 64: 1111
[155] Makela et al. (1994) Nature 371: 254
[156] Maller et al. (1977) Dev Biol 58: 295
[157] Marraccino et al. (1990) Curr Opin in Cell Biol 2:262
[158] Masui, Markert (1971) J Exp Zool 177:129
[159] Matsushime et al. (1992) Cell 71:323
[160] Matsushime et al. (1994) Molec Cell Biol 14:2066
[161] Matsushime et al. (1991) Cell 65: 701
[162] McGowan, Russell (1993) EMBO J 12:75
[163] Meyerson et al. (1992) EMBO J 11: 2909
[164] Meyerson, Harlow (1994) Molec Cell Biol 14:2077
[165] Michieli et al. (1994) Cancer Res 54:3391
[166] Millar et al. (1991) Proc Natl Acad Sci USA 88:10500
[167] Minshull et al. (1990) EMBO J 9: 2865
[168] Moreno et al. (1989) Cell 58:361
[169] Mori et al. (1994) Cancer Res 54: 3396
[170] Motokura, Arnold (1993) Curr Opin in Genet Dev 3:5
[171] Motokura et al. (1991) Nature 350: 512
[172] Mudryj et al. (1991) Cell 65:1243
[173] Mudryj et al. (1990) EMBO J 9: 2179
[174] Müller et al. (1994) Proc Natl Acad Sci USA 91:2945

[175] Ninomiya-Tsuji et al. (1991) Proc Natl Acad Sci 88:9006
[176] Noda et al. (1994) Exp Cell Res 211:90
[177] Norbury et al. (1991) EMBO J 10: 3321
[178] Nurse (1975) Nature 256:547
[179] Nurse, Bissett (1981) Nature 292: 558
[180] Ohtsubo, Roberts (1993) Science 259:1908
[181] Okamoto et al. (1994) Proc Natl Acad Sci USA 91:11045
[182] Pagano et al. (1992) Science 255: 1144
[183] Pagano et al. (1992) Oncogene 7: 1681
[184] Pagano et al. (1993) J Cell Biol 121:101
[185] Pagano et al. (1992) EMBO J 11: 761
[186] Pagano et al. (1994) Genes Dev 8: 1627
[187] Parge et al. (1993) Science 262: 387
[188] Paris et al. (1991) Proc Natl Acad Sci USA 88:1039
[189] Parker et al. (1993) Nature 363: 736
[190] Parker et al. (1991) EMBO J 10: 1255
[191] Parker, Piwnica-Worms (1992) Science 257:1955
[192] Pearson et al. (1991) Mol Cell Biol 11:2081
[193] Perry, Levine (1993) Curr Opin in Genet Dev 3:50
[194] Pickham et al. (1992) Mol Cell Biol 12:3192
[195] Pines, Hunter (1989) Cell 58:833
[196] Pines, Hunter (1990) Nature 346: 760
[197] Pines, Hunter (1991) Trends Cell Biol 1:117
[198] Polyak et al. (1993) Genes Dev 8:9
[199] Polyak et al. (1994) Cell 79:59
[200] Poon et al. (1993) EMBO J 12: 3123
[201] Prelich et al. (1987) Nature 326: 471
[202] Prives (1993) Curr Opin in Cell Biol 5:214
[203] Quelle et al. (1993) Genes Dev 7: 1559
[204] Resnitzky et al. (1994) Mol Cell Biol 14:1669
[205] Riabowol et al. (1989) Cell 57: 393
[206] Richardson et al. (1990) Genes Dev 4:1332
[207] Rosenblatt et al. (1992) Proc Natl Acad Sci USA 89:2824
[208] Russell et al. (1989) Cell 57:295
[209] Russell, Nurse (1986) Cell 45:145
[210] Russell, Nurse (1987) Cell 49:569
[211] Russell, Nurse (1987) Cell 49:559
[212] Sadhu et al. (1990) Proc Natl Acad Sci USA 87:5139
[213] Sarnow et al. (1982) Cell 28:387
[214] Schwartz et al. (1993) EMBO J 12: 1013
[215] Serrano et al. (1993) Nature 366: 704
[216] Shan et al. (1992) Mol Cell Biol 12:5620
[217] Sherr (1994) Cell 79:551
[218] Sherr (1994) Trends in Cell Biol 4: 15
[219] Shirodikar et al. (1992) Cell 68:157
[220] Shivji et al. (1992) Cell 69:367
[221] Shuttleworth (1990) EMBO J 9: 3233
[222] Simanis, Nurse (1986) Cell 45:261
[223] Slingerland et al. (1994) Mol Cell Biol 14: 3683
[224] Sobczak-Thepot et al. (1993) Exp Cell Res 206:43
[225] Solomon et al. (1993) EMBO J 12: 3133
[226] Solomon et al. (1992) Mol Biol Cell 3:13
[227] Spruck III et al. (1994) Nature 370: 183
[228] Strausfeld et al. (1991) Nature 351: 242
[229] Sturzbecher et al. (1988) Oncogene 3:405
[230] Subler et al. (1992) J of Virol 66:4757
[231] Swenson et al. (1986) Cell 47:861
[232] Tam et al. (1994) Cancer Res 54:5816
[233] Tam et al. (1994) Oncogene 9:2663
[234] Tamura et al. (1993) Oncogene 8: 2113
[235] Tassan et al. (1994) J Cell Biol 127:467

[236] Thuriaux, Nurse (1980) Genetics 96:627
[237] Toyoshima, Hunter (1994) Cell 78: 67
[238] Tsai et al. (1994) Nature 371:419
[239] Tsai et al. (1993) Oncogene 8:1593
[240] Tsai et al. (1993) Develop 119:1029
[241] Tsai et al. (1991) Nature 353:174
[242] Umbricht et al. (1993) J Biol Chem 268:6131
[243] van den Heuvel, Harlow (1993) Science 262:2050
[244] Waga et al. (1994) Nature 369:574
[245] Wang et al. (1989) Cell 57:379
[246] Wang et al. (1990) Nature 343:555
[247] Wang et al. (1994) Nature 369:669
[248] Weintraub et al. (1992) Nature 358: 259
[249] Weness et al. (1990) Science 248: 76
[250] Whyte et al. (1988) Nature 334:124
[252] Wimmel et al. (1994) Oncogene 9: 995
[251] Whyte et al. (1989) Cell 56:67
[253] Withers et al. (1991) Mol Cell Biol 11:4846
[254] Wold, Kelly (1988) Proc Natl Acad Sci USA 85:2523
[255] Wu et al. (1994) Oncogene 9:2089
[256] Xavier et al. (1993) Oncogene 8: 2561
[257] Xiong et al. (1991) Cell 65:691
[258] Xiong et al. (1993) Nature 366:701
[259] Xiong et al. (1992) Cell 71:504
[260] Zamanian, La Thangue (1993) Mol Cell Biol 4:389
[261] Zambetti et al. (1992) Genes Dev 6: 1143
[262] Zhang et al. (1994) Genes Dev 8: 1750
[263] Zhang et al. (1993) Mol Biol Cell 4:897
[264] Zhu et al. (1993) Genes Dev 7: 1111
[265] Zindy et al. (1992) Biochem Biophys Res Comun 182:1144

Part One

Techniques for Studying the Mammalian Cell Cycle

1 Use of Immunohistochemistry in Cell Cycle Studies

JIRINA BARTKOVA, JIRI LUKAS, AND JIRI BARTEK

1 Background

The recent enormous advances in the cell cycle research have been largely made through application of cell-free and cellular in vitro model systems. On the other hand, correlation with the situation in situ, i.e., in tissues and organs, is essential to help better understand the biology of the cell cycle and to use the knowledge gained so far for an initial phase of clinical applications.

In this chapter we will outline some basic protocols for tissue processing and immunohistochemical staining applicable to cell cycle proteins. Although there is no universal recipe which would allow immunodetection of every protein on tissue sections, the procedures given below, and the suggestions regarding their possible modifications, should allow the researchers to tailor a protocol for immunohistochemical analysis of any cell cycle protein of interest.

The features which make the immunohistochemical detection of the cell cycle proteins particularly attractive include the following:

- Detection of a protein in cells in their natural tissue environment, free from possible artifacts inevitably associated with any in vitro model.
- As opposed to immunoblotting or nucleic acid analyses based on examination of cell- and tissue extracts (which can only provide a tissue-average data), immunohistochemistry permits considerably more focused analysis at the single-cell level.
- The information obtained through immunostaining of tissue sections include subcellular localization, tissue- or cell-type specificity of expression, and semiquantitative evaluation of the protein abundance.
- The availability of large collections of histological material including a variety of pathological specimens should permit large-scale analyses of any cell cycle protein in relation to various disease states.

We believe that these significant advantages of using immunohistochemistry will contribute to increasing number of applications of this technology in the fascinating field of the cell cycle research.

2 Principle and Applications

The underlying principle of tissue preparation and immunohistochemistry in cell cycle studies is to immobilize the antigens and preserve both the tissue morphology and antigenic structure of the analyzed proteins to permit the highly specific and sensitive antibody-based techniques to reveal the presence, approximate abundance, and localization of the cell cycle regulator under study, in relation to tissue architecture. While the basic principle is analogous to immunodetection of any other group of proteins (see Harlow and Lane 1988; Cattoretti et al. 1993), some characteristic properties of the cell cycle-associated proteins require certain modifications of the commonly used protocols as indicated in Section 3 of this chapter.

A major area of application of immunohistology of the cell cycle proteins has been the study of cell proliferation state of normal and (especially) pathological tissue processes (Hall et al. 1992). The driving force behind these numerous studies, mainly performed on paraffin sections with antibodies to proteins like PCNA (Fig. 1B) and Ki-67 (e.g., Cattoretti et al. 1993; Hall et al. 1990; Sawhney and Hall 1992), has been the hope that a more objective evaluation of cellular proliferation would help the clinicians to better judge the behavior of tumors and other disease states and consequently to propose individually tailored therapeutic approaches.

More recently, antibodies to Cyclins and Cyclin-dependent kinases have also been used to estimate the position of proliferating cells in normal human tissues and to map the tissue-restricted pattern of expression of the cell cycle regulatory proteins (see Fig. 1A, and Bartkova et al. 1994b; Doussis-Anagnostopoulou et al. 1994). Similarly, studies of a potential involvement of the cell cycle proteins in regulation of tissue differentiation should also greatly benefit from the advantages of immunohistochemistry.

One of the major findings emerging from the recent cell cycle studies has been the realization that abnormalities of some of the components of the cell division cycle machinery are intimately associated with cancer. Among the cell cycle proteins, there are several good candidates for being classified as either proto-oncogenes (e.g., D or E type Cyclins) or tumor suppressors (e.g., some of the inhibitors of cell cycle regulatory kinases) and immunohistochemical analyses are proving extremely useful in the search for such potentially oncogenic aberrations (see Fig. 1C, D and, e.g., Bartkova et al. 1994a, 1994b; Gillett et al. 1994; Jiang et al. 1993). The latter area of cell cycle analysis in relation to oncogenesis is therefore likely to become a major field of application of "cell cycle immunohistochemistry" in the years to come.

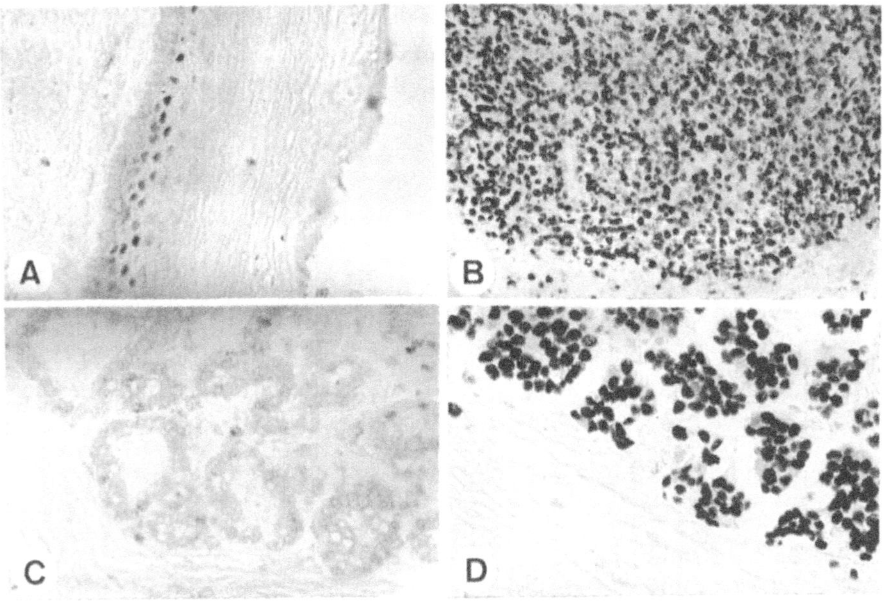

Fig. 1 A–D. Immunohistochemical detection of cell cycle proteins in human tissues. **A** Cyclin D1 in stratified epithelium of the exocervix; note the specific localization of positive nuclei in the narrow proliferative epithelial layer. **B** Expression of PCNA in a lymphocytic tissue – tonsilar germinal center, reveals a high proportion of proliferative cells. **C** Relatively low levels of cyclin D1 protein in normal human breast tissue. **D** Abnormally elevated expression of the cyclin D1 oncoprotein in an invasive breast carcinoma. Methacarn-fixed, paraffin sections (**A–D**) were probed with: the DCS-6 monoclonal antibody to cyclin D1 (**A, C,** and **D,** Bartkova et al. 1994a, b); or with the monoclonal antibody PC-10 against PCNA (**B,** Dako)

3 Methods

3.1 Tissue Preparation and Storage

Many cell cycle regulatory proteins are oscillating or labile polypeptides and, consequently, prompt processing of the tissues (preferably within minutes upon tissue resection) is essential to preserve the level of undegraded target antigens sufficient for immunohistochemical detection. The two most commonly used methods of tissue preservation are:

- freezing of specimens for preparation of cryostat sections, and
- fixation and paraffin-embedding of specimens for preparation of paraffin sections.

While the former is more gentle and therefore more likely to preserve intact target epitopes, the latter provides more histological detail and long-term durability of the samples and represents the classical approach on which the collections of archival histopathology material are usually based.

Frozen Tissue Specimens

Preparation A

1. Dissect a small piece (approximately 1 cm × 1 cm × 0.5 cm) of selected tissue, place it on a strip of card and label the card.
2. Submerge for about 1 min, the sample in isopentane (precooled to liquid nitrogen temperature) and transfer the frozen tissue to a precooled, labeled freezing vial.
3. Store in liquid nitrogen or at least at –80 °C.

Fixed, Paraffin-Embedded Tissue Specimens

Preparation B

1. Prepare the fixatives. From the wide range of fixatives in common use we describe the most frequently used one (buffered formaldehyde) and the one successfully applied in immunohistochemistry of numerous antigens including some cell cycle proteins (methacarn).
   - To prepare 1 l of a 4% solution of paraformaldehyde, add 40 g to 500 ml of water, heat to 60 °C (in a fume hood!) and add a few drops of 2 N NaOH to help dissolve. Let the solution cool to room temperature and add 500 ml 2× PBS. Store the fixative at 4 °C and use within 3 days.
   - To prepare the methacarn fixation, mix methanol, chloroform, and acetic acid 6:3:1 by volume (in a fume hood !) and keep the solution in a tightly closed bottle at room temperature for up to 2 weeks.
2. Cut small pieces of tissue approximately 1 cm × 1 cm × 0.3 cm and place immediately in the fixative.
3. Incubate overnight at 4 °C in buffered paraformaldehyde, or between 4 and 12 h at room temperature in the methacarn fixation.
4. Follow standard paraffin embedding procedures (performed in commercially available automatic apparatus, e.g., from Shandon).
5. Label the paraffin blocks and store between 10 and 20 °C.

3.2 Tissue Sections

Cryostat (Frozen) Sections

Tissue Sectioning A

1. Prior to sectioning, prepare coated glass histology slides by either dipping the clean slides for approximately 1 min into 1% gelatin dissolved in water at 50 °C; or alternatively, by incubating the slides in a solution of 1 mg/ml poly-L-lysine in distilled water for 15–20 min. Remove the coated slides and air-dry overnight at room temperature.
2. Bring the frozen tissue specimens stored at –80 °C or in liquid nitrogen to desired temperature (usually between –15 and –25 °C) in the precooled cryostat cutting chamber.
3. Cut 5–7-μm-thick sections following manufacturers' instructions for a particular cryostat model and collect the sections on the coated slides.
4. Allow the sections to air-dry at room temperature for a few minutes in a laminar-flow hood.
5. Fix the sections in a precooled mixture (–15 °C) of methanol and acetone (1:1) for 5 minutes or in acetone at room temperature for 1 min.
6. Rinse the sections in three changes of PBS, 5 min each.

The tissue sections are now ready for incubation with the primary antibody (see Sect. 3.4).

Paraffin Tissue Sections

Tissue Sectioning B

1. Coat the clean slides with poly-L-lysine as above.
2. Position the paraffin block into microtome and cut 3–4-μm-thick sections.
3. Float the sections onto slides and allow to dry overnight at room temperature.
4. Dewax the sections in two changes of xylene, 3 min each.
5. Rehydrate by passing through a series of graded alcohols: absolute ethanol (2 × 3 min), 95% ethanol (2 × 3 min), 70% ethanol (2 × 3 min).
6. Rinse in water and place in PBS.

The sections are now ready for the antigen unmasking step (see Sect. 3.3) or incubation with the primary antibody (see Sect. 3.4).

3.3 Antigen Unmasking on Paraffin Sections

Antigen unmasking

The cell cycle proteins in fixed, paraffin-embedded tissues may have been rendered inaccessible to the primary antibody by the fixation. A variety of methods help to expose ("unmask") such hidden epitopes, including mild proteolytic treatment or microwave treatment of the sections.

Trypsin treatment

1. Incubate the dewaxed, rehydrated sections in a 0.1% trypsin (e.g. from Difco), 0.1% $CaCl_2$, 20 mM Tris (pH 7.8) for 20 s–10 min (it may be necessary to optimize the digestion time for different batches of trypsin and differently fixed tissue sections).
2. Rinse the digested sections in PBS for 5 min before the primary antibody is applied.

Pronase treatment

1. Preheat the dewaxed and rehydrated sections in distilled water at 37 °C for 5 min.
2. Treat the sections with 0.02% pronase E (Sigma) in PBS at 37 °C for 2–10 min (the longer digestion times are generally required for tissues fixed for more than 24 h).
3. Stop the enzymatic treatment by rinsing the sections in PBS.

Microwave treatment

1. Place the slides with dewaxed, rehydrated sections in a jar with 0.01 M citrate buffer (use 2.1 g citric acid monohydrate to make 1 l, adjust pH to 6.0 with 2 M NaOH), cover the jar with a loosely fitting parafilm, and puncture the film.
2. Place the jar (avoid any metal!) in the center of a household or laboratory model microwave oven and heat the sections for 5 min (the buffer should boil; the sections should not dry out during the procedure and the jar may need refilling with distilled water). The number of 5-min treatments should be optimized.
3. Incubate the slides in the jar at room temperature for 15 min.
4. Rinse the sections in either PBS or TBS (0.05 M Tris/HCl, 0.15 M NaCl, pH 7.4), depending on which of the two buffers is used for washes in the subsequent immunostaining protocol.

3.4 Immunostaining Procedure: (Strept)Avidin-Biotin-Peroxidase Method

Immuno-staining

To prevent drying out of the sections, the whole staining procedure is performed in a humid chamber. The protocol as described here is aimed at high sensitivity and omits counterstaining, thus allowing detection of even weak positive reactions and documentation with black and white photography avoiding interference of commonly used nuclear counterstain (e.g., hematoxylin).

1. Rehydrated frozen or paraffin sections (the latter upon the antigen unmasking step, in case it is required) are ready for staining. Block non specific binding of primary antibodies by pretreatment of sections with a 5–20% solution of normal serum of the species in which the secondary antibody (i.e, the biotinylated antibody in this case) was raised, for 20 min at room temperature.
2. Blot excess blocking serum from sections.
3. Incubate sections with primary antibody specific for the cell cycle protein under study. Primary antibody can be a hybridoma tissue culture supernatant or polyclonal or monoclonal antibody (purified or as an ascites) appropriately diluted in PBS with added carrier protein (3% BSA or 10% serum, preferably from the same species as the blocking serum used in step 1). The suitable dilution of antibody and the optimal incubation period must be defined by preliminary staining experiments, but dilutions between 1:100 and 1:1000 for polyclonal antisera and ascites fluid, and incubations between 1 h (at 37 °C or room temperature) up to overnight (at 4 °C) are common.
4. Wash slides 3 × 5 min with PBS or TBS in a jar.
5. Incubate sections with diluted secondary biotinylated antibody (usually 1:100 to 1:200, in PBS or TBS with 2–5% blocking serum) for 1 h at room temperature.
6. Wash slides 3 × 5 min as in step 4.
7. Incubate sections with preformed streptavidin (or avidin)/biotinylated horseradish peroxidase complexes for 30 min to 1 h at room temperature. The (strept)avidin/biotinylated peroxidase reagents are available from several sources and often in a kit form (e.g., from Vector Laboratories, Dako, Amersham etc.).
8. Wash slides as in steps 4.
9. During the washing procedure (step 8), dissolve 10 mg DAB (diaminobenzidine tetrahydrochloride, Sigma) in 10 ml of PBS or TBS, add 0.1 ml of a 3% (w/v) stock solution of nickel sulfate in water and mix briefly.
10. When the slides are washed, add 10–20 µl of a 30% solution of hydrogen peroxide to the DAB/nickel sulfate solution and incubate for 1–2 min at room temperature. If a precipitate forms, filter through Whatman No. 1 filter paper.
11. Apply this substrate solution to specimens, incubate for 1–15 min (until desired stain intensity develops) at room temperature.

12. Stop reaction by washing in water. Counterstain if necessary (see Sect. Counterstaining) and air-dry the sections.

13. Mount the slides in Gelvatol (Monsanto Chemicals) or Mowiol 4-88 (Hoechst): add a small drop of the mounting media on the section, carefully place a coverslip (No. 1) on the drop, avoiding air bubbles. Leave the mounted slides on a bench to allow the mounting medium to set overnight, then store in light-tight slide boxes at room temperature until evaluation.

3.5 Optional Steps and Alternative Detection Methods

Fluorescence Detection. Although fluorochrome-labeled secondary antibodies can also be used in immunohistochemistry, these methods are not as sensitive as detection with enzyme-labeled detection systems and therefore are only rarely used for visualization of labile antigens, especially on paraffin sections.

Alternatives

Alternative Immunoenzymatic Detection Systems. While the horseradish peroxidase-based detection systems are most common in immunohistochemistry, alternative methods are also available and may be preferred (e.g., for examination of tissues with high endogenous peroxidase activity). Thus, alkaline phosphatase-labeled reagents (especially the sensitive alkaline phosphatase anti-alkaline phosphatase method), as well as β-galactosidase-labeled reagents, can be used for detection of the cell cycle proteins in tissue sections. Another possibility for a sensitive detection method is now offered by the immunogold reagents coupled with silver enhancement. All these alternative detection systems are available from a range of commercial sources.

Quenching enzyme activities

Quenching Endogenous Enzyme Activities. Endogenous enzyme activity in tissue sections (especially frozen sections) may cause problems in terms of background staining. High peroxidase activity occurs, e.g., in macrophages, monocytes, granulocytes and erythrocytes (i.e., in the spleen, bone marrow, and other tissues) and may persist even on fixed and paraffin-embedded tissue sections. Alkaline phosphatase activity can also cause problems, while there is no endogenous β-galactosidase activity in mammalian tissues under the conditions used for detection of the reagents coupled to the bacterial enzyme (unfortunately, the sensitivity of β-galactosidase-based techniques is lower than those using either peroxidase or alkaline phosphatase). Since the blocking procedures (see below) may damage some antigens, it may be better to change the detection system for cell or tissue types with high endogenous enzyme activity.

1. To block endogenous peroxidase activity, incubate the sections for 20–30 min in 0.3–0.6% hydrogen peroxide in methanol. If required, this step may be advisable to perform only after the incubation with secondary biotinylated antibody to minimize the potentially harmful effect of such treatment on the antigen under study.
2. To block endogenous alkaline phosphatase activity, include 1-tetra-amisole (levamisole, 1 mM) in the substrate solution.

Counterstaining

Counterstaining may be used to help to recognize the various cell types present in the tissue section but should be avoided if the aim is to detect even weak nuclear immunostaining, which can easily be masked by a counterstain.

1. If required, counterstaining is performed after the enzymatic reaction has been developed and the sections washed in water (i.e., after step 12 in Sect. 3.4).
2. Add a few drops of Mayer's hematoxylin (available from many suppliers) to the specimen and incubate for 20 s up to 5 min (the length of time determines the intensity of counterstaining).
3. Wash the slide gently in warm tap water (or dip the slide repeatedly into 30 mM ammonium hydroxide) until the counterstain turns blue.
4. Wash the slide gently in water and mount in Gelvatol or Mowiol (see Sect. 3.4).

3.6 Choice of Antibodies and Controls

Antibodies

High specificity of an immunoreagent for the analyzed protein is an essential prerequisite for any meaningful experiment with antibodies, but this problem is particularly apparent in the cell cycle field since both the Cyclins and Cyclin-dependent kinases represent members of large families of homologous proteins which share many epitopes. Owing to their single-epitope specificity, some monoclonal antibodies in particular are proving useful for immunohistochemical detection of the cell cycle proteins, although some polyclonal antisera (especially those made against unique peptide sequences) may also give satisfactory results (see chapter by M. Pagano, Part Four). Since the specificity of any reagent is usually verified on a panel of recombinant cell cycle proteins and/or lysates of cells cultured in vitro, it is essential to perform some control immunoblotting experiments with lysates of tissue sections to exclude the possibility that the reaction seen on sections is due to some irrelevant cross reaction. We therefore strongly recommend the following immunoblotting test, in particular when the antibody appears to react also with some nonproliferating cells on tissue sections.

1. Select a frozen piece, the sections of which show strong immunohistochemical positivity with the antibody.

2. Cut a series of some 20 consecutive cryostat sections about 20 μm thick and place the sections immediately in an Eppendorf tube containing 0.1 ml preheated 2 × electrophoresis sample (Laemmli) buffer.

3. Boil the sections in the sample buffer for 10 min, spin briefly to sediment the insoluble material, load on a SDS PAGE gel, separate the proteins, and blot onto nitrocellulose or other suitable membrane using standard procedures.

4. To probe the blot, use the antibody which showed immunohistochemical positivity on this tissue (use a sensitive detection method like ECL). Compare the mobility of the recognized band(s) with parallel lanes containing the recombinant protein and the lysates from cell lines expressing the cell cycle protein under study. If additional band(s) are recognized on the tissue lysate, this indicates either a cross reactivity with other protein(s) or, especially in tumors, potentially existence of mutant or alternatively spliced variants of the protein studied (do not confuse with degradation products present in inappropriately processed tissue!). These possibilities should be kept in mind when interpreting the immunohistochemical data.

Positive controls

1. A positive result on a known, positively reacting control tissue section obtained with the antibody to the studied cell cycle protein, confirms that the antibody is active in the dilution used, and that the detection method works satisfactorily.

2. If sections fixed for unknown periods of time and undefined in terms of fraction of proliferating cells are used for the first time, an antibody (e.g., to PCNA or Ki-67) should be included to check whether the tissue has been fixed properly and to judge the fraction of cycling cells.

Negative controls

1. The enzyme reaction should be performed on the section without the addition of any antibodies to check for potential endogenous enzyme activity.

2. The selected detection system should give negligible reactivity in the absence of the primary antibody.

3. Negative control primary antibody should be used as a crucial control on a parallel section from the same tissue. This control antibody should match the experimental antibody as to their species of origin, the form of antibody (e.g., serum versus serum, supernatant versus supernatant, ascites versus ascites, purified immunoglobulin versus purified immunoglobulin), dilution, and immunoglobulin class or subclass (if possible).

4. When available, addition of a complete detection system, including the experimental primary antibody, on a section known to lack any cells express-

ing the protein under study could also serve as an useful control for specificity of the assay.

4 Comments and Notes on Troubleshooting

- When using immunoenzymatic detection systems, the buffers used to dilute the reagents and to wash the slides should not contain any inhibitors of the enzyme (e.g., no sodium azide in case of peroxidase-based systems).
- Some of the chemicals used in immunohistochemistry are suspected carcinogens (e.g., some of the chromogenic substrates); therefore check carefully the manufacturers' recommendation as to their handling and consult your local authorities for disposal procedures.
- In case of repeatedly negative results, include the recommended positive controls (Sect. 3.6, Positive controls) and consider in particular whether the tissue was properly fixed, processed, and stored before the experiment; and whether the antibody used has been shown to work on similarly processed material using the same detection method. The procedures recommended for "unmasking" the antigenic sites on paraffin sections (see Sect. 3.3) should also be tried.
- Background problems represent probably the most frequently encountered difficulty in immunohistochemistry. The controls which omit each of the individual stages of the staining procedure should help to find out whether the high background is caused by nonspecific sticking of the reagents to tissue or by specific cross reactions.

  Nonspecific sticking can be eliminated by
  - Careful titration of the primary antibodies (avoiding "overload").
  - Pretreatment of sections with blocking proteins (see Sect. 3.4).
  - Diluting the reagents in buffers containing the blocking protein.
  - Reducing the incubation times with antibodies.
  - More extensive washes between the individual steps.

  If the background persists, try to change the detection system. Specific background problems can be due to contaminating antibodies reacting with unrelated antigens, to immunoglobulin-binding proteins expressed by the tissue, or to genuine cross-reactivity of the primary antibody with more than one protein. The tricks which may help to reduce this problem include

  - Titration of ascites or polyclonal antisera to "dilute out" the unwanted reactivity.
  - Affinity purification of the antibody.

- Substitution of the primary antibody by other antibody against the same cell cycle protein.
- Preabsorption of the antibody preparation with a mammalian acetone liver powder (available, e.g., from Sigma, helps to remove the spurious activities of the polyclonal antisera or secondary reagents).
- Preincubation of the sections with normal serum from the species in which the labeled (secondary) antibody was raised.

- A "leakage" of the normally nuclear protein to the cytoplasm is sometimes observed in immunohistochemical staining and is caused by inappropriate handling (e.g., a long period between resection and processing) or fixation of the tissue under study.
- Be patient! Optimization of the immunohistochemical staining protocol with a new immunoreagent may require several weeks of testing the individual steps and alternative procedures. The labile nature of the cell cycle proteins makes this task even more difficult and is particularly demanding with regard to tissue processing. On the other hand, a successfully established protocol is likely to yield important information not accessible by other methods.

References

Bartkova J, Lukas J, Müller H, Lützhøft, D, Strauss M, Bartek J (1994a) Cyclin D1 protein expression and function in human breast cancer. Int J Cancer 57:353–361

Bartkova J, Lukas J, Strauss M, Bartek J (1994b) Cell cycle-related variation and tissue-restricted expression of cyclin D1 protein. J Pathol 172:237–245

Cattoretti G,. Pileri S, Parravicini C, et al. (1993) Antigen unmasking on formalin-fixed, paraffin-embedded tissue sections. J Pathol 171:83–98

Doussis-Anagnostopoulou I, Tennant R, Gannon J, Gatter K (1994) Distribution of the cdc2 gene product in normal tissues: an immunological study using four new monoclonal antibodies. Histopathology (in press)

Gillett C, Fantl V, Smith R, et al. (1994) Amplification and over-expression of cyclin D1 in breast cancer detected by immunohistochemical staining. Cancer Res 54:1812–1817

Hall PA, Levison DA, Wright NA (eds) (1992) Assessment of cell proliferation in clinical practice. Springer, Berlin Heidelberg New York

Hall PA, Levison DA, Woods AL et al. (1990) Proliferating cell nuclear antigen (PCNA) immunolocalization in paraffin sections: an index of cell proliferation with evidence of deregulated expression in some neoplasms. J Pathol 162:285–294

Harlow E, Lane DP (1988) Antibodies. A laboratory manual. Cold Spring Harbor Laboratory Press, Cold Spring Harbor

Jiang W, Zhang, Y–J, Kahn SM et al. (1993) Altered expression of the cyclin D1 and retinoblastoma genes in human esophageal cancer. Proc Natl Acad Sci USA 90:9026–9030

Sawhney N, Hall P (1992) Ki-67 – structure, function, and new antibodies. J Pathol 168:161–162

2 Immunofluorescence Techniques in Cell Cycle Studies

M. CRISTINA CARDOSO AND HEINRICH LEONHARDT

1 Background

A major goal in the study of cell cycle division, as in any area of cell biology, is to define the function of each cellular component and its interplay with other factors. Immunofluorescence microscopy allows the study of the abundance and subcellular localization of any protein throughout the cell cycle, and thus provides important information towards an understanding of their role. This technique combines the sensitivity of fluorescent probes with the spatial resolution of the light microscope and the specificity of antibodies. Several different fluorescent probes can be selectively detected in a mixture of molecules due to their distinct and characteristic excitation and emission wavelengths. By choosing different color fluorophores, simultaneous detection of multiple antigens within a cell is possible providing a direct means of comparing their relative spatial and temporal distribution.

2 Principle and Applications

Multiple labeling in fluorescence microscopy is an extraordinary powerful means of sorting out the factors involved in a particular cellular process in vivo. This possibility is particularly advantageous when dealing with families of proteins that show high level redundancy in in vitro assays, as in the case of the cyclin-dependent kinase family and their cyclin counterparts. Immunofluorescence makes it possible to follow the expression and subcellular localization of different cyclins, cyclin-dependent kinases or other proteins throughout the cell cycle and to compare their distribution with other family members or potential partners. In the same samples, specific phases of the cell cycle can be identified by monitoring incorporation of nucleotide analogues into the DNA (S phase) and by counterstaining DNA with Hoechst 33258 (mitosis) (Heald and McKeon 1990; Pines and Hunter 1991; Ookata et al. 1992, 1993). The combination of microinjection or transfection with multiple antigen detection by immunofluorescence was successfully used to demonstrate a requirement of Cyclin D1 during progression through G1 phase (Baldin et al. 1993), of Cyclin A and Cdk2 for entry into S phase (Girard et al. 1991; Pagano et al. 1992; Zindy et al. 1992; Pagano et al.

1993; Tsai et al. 1993), of Cyclin B and Cdc2 for progression through mitosis (Riabowol et al. 1989; Gallant and Nigg 1992), and of Wee 1 to prevent premature mitosis without completion of DNA replication (Heald et al. 1993). Furthermore, the possibility of simultaneous detection of multiple antigens provided direct evidence that distinct cellular processes like DNA methylation, DNA replication, and cell cycle regulation, which are traditionally viewed as separate and independent processes, occur at the same subnuclear compartments and are thus temporally and spatially coordinated (Leonhardt et al. 1992; Cardoso et al. 1993). Immunofluorescence thus complements biochemical and genetic assays in that it takes into account the limited spatial and temporal availability of the factors under study and allows analysis at a single cell level, avoiding problems resulting from mixed cell populations. In addition, important in vivo interactions can escape detection by biochemical means if they are part of insoluble structures.

Using a combination of in situ detergent and salt extraction followed by immunofluorescence staining, Bravo and MacDonald-Bravo (1987) showed that, during the cell cycle, there are two populations of PCNA, one that is nucleoplasmic and extracted by nonionic detergent treatment (non-S phase cells), and another that is resistant to high salt and detergent extraction and associates with sites of ongoing DNA replication during S phase. Also, the retinoblastoma protein (pRb) was shown to be tightly bound in the G1 phase nuclei and extractable under low salt conditions at S phase, and this change in extractability correlates with its phosphorylation and inactivation as a growth suppressor protein (Mittnacht et al. 1991). Moreover, Cyclin A and Cdk2 specific association with nuclear DNA replication foci is resistant to Triton X-100 extraction (Cardoso et al. 1993 and M.C.C. and H.L., unpubl.).

Fluorescence microscopy can not only yield qualitative and semi-quantitative data, but it is also a very sensitive method to precisely quantify cellular components within the cell. Fluorescence images acquired from either conventional or confocal microscopes can be digitalized and subsequently computer processed using image analysis software. This field of electronic image acquisition and analysis is rapidly evolving, but exceeds the scope of this chapter. Fluorescence microscopy, due to its sensitivity and specificity, is a unique biophysical tool to directly study specific components inside living cells. In the cell cycle field, however, the use of this technique to probe living cells is still at an early stage and we will focus in this chapter on protocols for fluorescent staining of fixed cells. A good source of information on this exciting area of fluorescence microscopy is gathered in *Fluorescence Microscopy of Living Cells in Culture, Parts A and B* (Wang and Taylor 1989). We will also not address here technical questions pertaining to light microscopy in general, which can be found in *Light Microscopy in Biology: a Practical Approach* (Lacey 1989). The goal of this chapter is to describe basic immunofluorescence protocols typically used in cell cycle studies of mammalian cells in culture, to discuss possible variations and, most importantly, to point out common problems and necessary controls.

3 Immunofluorescence Protocols

Antibodies

Immunofluorescence requires that an antibody specifically recognizes a particular antigen in the presence of very high concentrations of other molecules. Polyclonal antiserum, although usually very good for staining cells because it contains many different antibodies that can bind to multiple epitopes on the antigen, often include high concentrations of nonspecific antibodies that give rise to high background. In most cases, simple titration of the antiserum will eliminate this problem by reducing the spurious signals below detection level. The signal strength of an antiserum is directly dependent upon the concentration of the antibody multiplied by its affinity constant: the higher the affinity, the greater the working dilution. When titration is not sufficient, the antiserum can be preadsorbed with the fixed cells prior to use. Indeed, one often obtains cleaner signals by reusing the same antibody dilution multiple times. Another method to decrease unspecific staining, which is frequently used with polyclonal antisera, is immunoaffinity purification of the antibodies (see chapter by M. Pagano, Part Four).

Monoclonal antibodies can be used either as tissue culture supernatants or as ascites fluid, and can also be pooled, which generally improves signal strength, granted that the individual antibodies bind noncompetitively to the antigen. Although monoclonal antibodies usually give clean signals, some monoclonals will not react in cell staining. This can be due to the fact that their specific epitope is either masked within the cell or destroyed during fixation.

Sample preparation

Cells for staining are usually first attached to a solid support (glass slide or coverslip, No.1 thickness) to facilitate the subsequent manipulations. In the case of adherent cultures, cells should be grown directly on previously sterilized coverslips (or, alternatively, in chamber slides).

Suspension cells can be centrifuged (approximately 1000 *g* for 10 min, at room temperature) onto glass slides using a cytospin apparatus, or can be attached to the slide with poly-L-lysine or, simply, by drying (cell smears). Slides are coated by dipping into a 1 mg/ml solution of poly-L-lysine in water for 10 min and washed with water. Slides can be dried and stored or used immediately. Cells in suspension are then spotted onto the coated slide and allowed to attach for about 10 min. Binding of cells to the positively charged poly-L-lysine covered surface, although not covalent, is usually strong enough to resist most steps of the staining procedure. Hereafter cells are processed as adherent cultures.

When dealing with cell-surface antigens, cultures can be handled in suspension either as live or as formaldehyde-fixed cells and solutions are changed by carefully spinning (at 200 g for 5 min) and resuspending the cells. The steps in the staining procedure are otherwise similar to the protocol for adherent cultures.

Fixation/ permeabilization This step converts soluble native cellular constituents into insoluble derivatives, thereby preserving cellular architecture. The most commonly used fixatives are organic solvents and cross-linking agents. The former agents (mostly alcohols, such as methanol or ethanol, and acetone, or a mixture of both) extract lipids and dehydrate the cells, precipitating the proteins. The latter agents (generally aldehydes, such as formaldehyde and glutaraldehyde) form intermolecular bridges via free amino groups creating an intracellular network of linked constituents. Stabilizing the content of biological specimens during fixation often causes denaturation of protein antigens, which is why antibodies raised against denatured proteins are often better for cell staining.

The choice of fixative can only be established empirically, so we routinely try two agents, methanol and formaldehyde, process the samples, and then evaluate the results. Often, one of the methods destroys the antigen and thus cannot be used. These tests should be run with each new antibody and prior to attempting multiple labeling procedures. With pooled monoclonals, each should first be checked individually.

Cells are first washed in phosphate-buffered saline (optionally, it can be supplemented with 1.5 mM $MgCl_2$ and 1 mM $CaCl_2$) and fixed either by incubating with −20 °C methanol for 2–5 min or with 3.7% formaldehyde (Fisher, certified ACS) in PBS for 10–15 min at room temperature. Unless otherwise stated, all manipulations are at room temperature. Methanol-fixed specimens can be air-dried and stored dehydrated at −70 °C for several months. When needed, the methanol fixed samples are brought to room temperature and rehydrated by incubating for some minutes in PBS. Formalin-fixed samples are rinsed with PBS to remove fixative and can be stored at 4 °C in PBS with 0.02% NaN_3 (to prevent microbial contamination) for several weeks.

Reactive aldehyde groups, which can increase background fluorescence, can be quenched by incubating the fixed specimen for 10 min in a solution of 50 mM glycine in water. The amino group of the glycine reacts with free aldehyde groups, thereby blocking spurious crosslinks in the rest of the procedure.

After aldehyde fixation, cultures must be still permeabilised to allow access of the antibodies inside the cells. When dealing with extracellular epitopes of membrane components, this step is generally omitted. Cells can be permeabilized either by incubation for 30 s in acetone or, more commonly, by the action of detergents. In our hands, a 10-min-incubation with 0.25% Triton X-100 in PBS works well with most cell types.

An optional procedure includes permeabilization prior to fixation. Multiple conditions can be used and are reviewed in Pachter (1992). These protocols allow one to perform in situ fractionation of the cells, which can be followed either by biochemical or cytological analyses. For the purpose of this chapter, extraction and subsequent immunofluorescent detection can provide valuable information on the association of a particular antigen to insoluble cellular components. Often, specific localization patterns are difficult to visualize by conventional microscopy due to masking by either background fluorescence or by the soluble fraction of the antigen, or both. This is specially critical when dealing with low abundance nuclear proteins and/or low affinity antibodies. Incubating the unfixed cultures for about 30 s in a physiological buffer (100 mM NaCl, 300 mM sucrose, 3 mM $MgCl_2$, 10 mM PIPES pH 6.8) containing 0.5% Triton X-100, releases approximately 70% of the cellular protein content, 75% of the tRNA, and almost all the phospholipids in the "soluble" fraction, retaining the nuclei, polyribosomes, and cytoskeleton. These Triton-extracted cells (Triton cytoskeletons) can be fixed and stained or further treated for 2 min in PBS containing different salt concentrations (NaCl from 10 mM up to 1 M). The salt-extracted cytoskeletons are then fixed and processed.

Using procedures similar to the one just described, it has been possible to distinguish during the cell cycle, populations of PCNA and pRb with different nuclear affinities (see Sect. 2, Principle and Applications).

Antibody binding and detection

Once the cells are fixed (after or before the permeabilization step), the coverslips or slides (cell-side-up) are transferred to a humidified dark chamber where all subsequent antibody incubations and washes are performed. A diagram of this setup is presented in Fig. 1. It is simply a petri dish covered with aluminum foil to prevent long exposure of the fluorochrome-conjugated

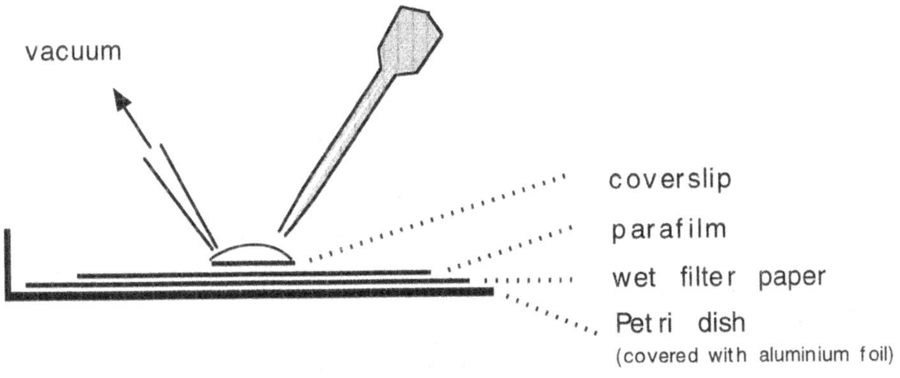

Fig. 1. Diagram of the setup for immunostaining

reagents to light with consequent fading of the signal, and with a wet filter paper covered with parafilm in the bottom of the dish to avoid drying out of the sample.

Before adding the antibody solutions, nonspecific binding should be inhibited by preincubation of the sample for about 30 min with blocking agents, a commonly used one being bovine serum albumin (BSA) at 3% in PBS (which also blocks reactive groups from aldehyde fixation). In addition, one should add the same blocking agent to all antibody dilutions. Even though BSA can be an efficient blocker and an inexpensive one, the most appropriate agent to saturate the specimen is 5% normal serum (diluted in PBS) from the same host as the labeled antibody. Normal sera, which have been lipid extracted and dialyzed, are available from a variety of commercial suppliers. For example, an anti-rabbit IgG conjugated antibody raised in goat is best blocked by the use of 5% normal goat serum in PBS. In all instances, never block a specimen with normal serum from the same host species of any of the primary antibodies, because the labeled secondary antibody would react with it.

After blocking, the antibody is added to the sample. Depending on the size of the glass coverslip, volumes as small as 10 μl can be enough to cover the cells. The hydrophobic parafilm layer will help to prevent the liquid from running off the edge of the coverslip, and it also facilitates recovering the coverslip from the dish with fine forceps. As in other immunoassays, the primary antibody can be itself labeled (in that case, the staining is referred to as direct immunofluorescence), or it can be detected with a labeled secondary antibody, which binds specifically the primary antibody (it is then called indirect immunofluorescence). Both methods are commonly used, even in combination when performing multiple stainings. The direct detection, though, requires labeling of purified primary antibody and gives less signal intensity, since the number of fluorochromes is lower. The indirect method is more sensitive because more than one second antibody molecule can react with each first antibody. The major disadvantage of the indirect method is that the background can be higher and one has to control for possible interspecies cross reaction when doing multiple antibody combinations.

Primary antibodies should be tested at concentrations between 0.1–50 μg/ml, depending on the binding affinity. As a general guideline, monoclonal antibodies are used as straight hybridoma tissue culture supernatant, ascites fluid and polyclonal antiserum should be assayed at various dilutions between 1/100–1/10 000. Incubation is carried out for 1–2 h. Since the antigen is immobilized, the time needed for antibody binding is longer than when both antigen and antibody are free in solution. With some antibodies, raising the incubation temperature to 37 °C may help improving signal strength.

Even though antibodies bind to their particular antigen with high specificity, they can also give rise to nonspecific binding. This can be particularly dangerous in cell staining since nonspecific reactions are not as easy to recognize as in other immunoassays where, for example, proteins are separated in a gel. The major causes for nonspecific binding are:

- antibodies present in the polyclonal antiserum reacting with other cellular antigens;
- cross reaction of the specific antibody with other molecules in the cell that are different but have similar or identical epitopes;
- or, when more than one molecule share the same epitope.

The first type of artifact can be easily detected by appropriate negative controls. The antibody staining should always be compared with a similar dilution of a preimmune serum of the same animal used to prepare the specific antiserum (or an antiserum preadsorbed with the original antigen), in the case of polyclonal antibodies, or to an ascites fluid or tissue culture supernatant from the same source as the specific antibodies, in the case of monoclonal antibodies. When possible, antibodies from the same class and subclass should be used as controls. In addition, if the antigen is known to be absent from some cell types, these can also be used as negative staining controls. These controls do not need to be run every time but must be performed when trying a new antibody. The second cause of nonspecific binding, i.e., cross reaction with other molecules, can generally be eliminated by further dilution of the antibody, since the affinity of the antibody for these other molecules is usually lower than to the specific antigen. The third cause is more difficult to trace and to eliminate. Testing the antibodies also in western blots of cell extracts should help addressing these types of problems.

After incubation with the first antibody, the sample is washed with several changes of 0.1% NP-40 in PBS over approximately 5 min. If background is a problem, the detergent concentration can be increased to 0.5%. When the first antibody is unlabeled, a second conjugated antibody recognizing the first one is added. This should also be previously titrated and, as a general rule, used between 0.1–10 μg/ml. Incubation should be carried out for 30–60 min, followed by extensive washes again in NP-40/PBS as above. The secondary antibody can be either fluorescently labeled or, for instance, biotinylated. In the latter case, a third incubation with fluorescently labeled streptavidin for 15–30 min has to be performed, followed by washing. Once more the reagent should be first titrated with a positive first antibody control. The first time the secondary reagent(s) is used, a negative control should be included in which the first antibody is omitted, to check for nonspecific staining by the reagent(s) used to detect the primary reagent. An additional important control that should always be performed is a positive control (i.e., an antibody previously known to react with the sample in use). If no staining is observed, this generally means that a reagent was either inadvertently omitted or the wrong reagent was used instead. Care must also be taken not to allow the specimen to dry out at any time during the procedure and to prevent evaporation of the antibody solution during incubation, since it would result in intense background and overstaining, respectively.

Mounting After the last antibody incubation, the nuclei can be counterstained with the DNA-intercalating agent Hoechst 33258 (bisbenzimide). Hoechst 33258 staining of condensed chromosomes allows identification of mitotic cells and, in more detail, of the specific stage of the mitotic process the cell is undergoing. The staining is accomplished by incubating the sample for 5 min in wash buffer (NP-40/PBS) with 0.1 μg/ml Hoechst 33258 (stock solution 1 mg/ml in H_2O), followed by washing. The samples are now ready to be mounted in an aqueous-based mountant. We use a polyvinyl alcohol-based mounting media that hardens within a few hours, forming a permanent preparation. The pH of the mountant is very important, particularly when using fluorescein conjugates, since their maximal fluorescence occurs only at pH 8.5 or higher. In addition, an anti-fading agent such as DABCO (1,4-diazobicyclo-[2.2.2]-octane) should be added to diminish bleaching, which is specially critical with fluorescein. Other anti-fading agents that can be used instead are the antioxidants *n*-propyl gallate (added to 5%) or *p*-phenylenediamine (added to 0.1%). To prepare the mounting media, add 8 g of Mowiol 4-88 (or Gelvatol, or Elvanol) to 40 ml of 0.2 M Tris-HCl pH 8.5 and dissolve by heating at 50–60 °C with occasional stirring. After cooling down, add 20 ml of glycerol and 1–2.5% DABCO. Centrifuge at 5000 rpm for 15 min and store in aliquots at –20 °C. After thawing, mount the coverslip in one drop of mounting media and let it harden at room temperature for some hours, in the dark. Usually, in 30 min the specimen can already be viewed. If this mountant is not available, the samples can be mounted in 90% glycerol in PBS (if possible with an anti-fading agent added) and the coverslips sealed to the slide with nail polish. These are not permanent mounts though. The mounted samples should be stored in the dark, at 4 °C or frozen.

Multiple labeling Fluorescence microscopy is suited for the detection of multiple antigens in the same specimen, due to the availability of different color fluorochromes with well resolved excitation and emission spectra. In the field of mammalian cell cycle, this technique has been successfully exploited to sort out the spatio-temporal localization of the several cell cycle regulators isolated thus far. Furthermore, the ability to relate subcellular localization of a specific protein to the distribution of its possible partner(s) and to cellular processes such as DNA replication and mitosis, has provided invaluable information on cellular functions in vivo. In a broader perspective, it can explain how the organization of the cytoplasm and nucleoplasm into metabolically distinct compartments may have general regulatory functions including cell cycle control. This functional organization can be best studied by multiple staining techniques.

The choice of fluorophores to use depends, obviously, on the instrument setup and the degree of color separation between the fluorophores (in order to minimize overlap between excitation and emission spectra). A good combination of fluorochromes for a triple labeling is AMCA/FITC/Texas Red. (AMCA and Texas Red are trademarks of BioCarb Chemicals and Molecular Probes, Inc., respectively.) Filter sets (i.e., an excitation filter, transmitting a portion of the emission from the light source; a dichroic mirror, reflecting the selected light to the sample; and an emission filter, blocking any residual excitation light and transmitting the longer wavelengths) that match the spectral properties of a vast array of fluorescent dyes and have good transfer efficiency, are available from microscope manufacturers and also from other companies, including Omega Optical and Chroma Burlington. The latter also provide filter sets for multiple-dye combinations, which can be very useful for colocalization studies using multiple labeling immunostaining. These filters avoid problems such as optical shift, often encountered when recording dual fluorescence images by doubly exposing photomicrographs.

To use more than one fluorochrome in the same specimen, care must be taken to ascertain oneself that each antigen is recognized by only one labeled reagent. The safest way is to use primary antibodies that are labeled with different fluorochromes. The staining can then be performed simultaneously, but control samples individually stained should be included to monitor for cross reaction. Most commonly, though, the primary antibodies are not labeled and only indirect detection is possible. In this case, the antibodies should be from different species, class, or subclass. Then, one can use labeled secondary antibodies that are specific for the individual primary antibodies. Always the specificity of the detection reagent must be checked since antibodies against one species, class or subclass may cross react with a number of other species, class, or subclass unless they have been specifically adsorbed (secondary antibodies preadsorbed with immobilized serum proteins from other species are available commercially from several manufacturers). Hence, one should try to use secondary antibodies that are derived from the same host species. For example, when using a rabbit polyclonal antibody and a mouse monoclonal antibody, one should use anti-rabbit IgG and anti-mouse IgG antibodies both produced in, for instance, goat and the blocking agent should be 5% normal goat serum. In addition, an important control to monitor species cross reactivity of the detection reagents, is to perform the immunostaining using each of the primary antibodies with the secondary antibody specific for the other primary antibody.

Multiple stainings can also be done sequentially with fixation steps in between to help minimizing cross reaction. When dealing with two antibodies, such as two mouse monoclonal antibodies that are from the same type, which happens quite often, it is better to have one directly labeled. It is also advisable to carry out the staining sequentially, with the labeled primary antibody incubated last, to avoid possible binding of the labeled secondary to this antibody.

In summary, in order to perform multiple labeling immunostaining one must take into account two important criteria:

- the fluorochromes chosen must have well resolved spectral characteristics
- the secondary antibodies available are derived from the same host so that they do not recognize one another. They must also not recognize the other primary reagents used or any other immunoglobulins present in the assay, and they should not react with the sample directly.

Monitoring S phase by BrdU incorporation

For the study of cell cycle progression it is essential to have a procedure to identify cells in S phase and in M phase. The latter is easily assayed with any DNA-intercalating dye such as Hoechst 33258, as mentioned earlier. The former is feasible by labeling the cultures with biotinylated dUTP, followed by detection with fluorescently labeled streptavidin, or, more commonly, with the thymidine analogue 5-bromo-2'-deoxyuridine (BrdU), followed by immunofluorescence staining with an antibody to BrdU (see also chapter by M. Giunta and C. Pucillo, Part One). Labeling can be carried out for several hours to allow detection of any cell that was replicating its DNA throughout the period of time prior to fixation. The continuous labeling protocol has been very useful in experiments where antibodies or antisense constructs were microinjected into the cells and the function of the inactivated products in the G1/S phase transition was being assayed. Pulse-labeling the cultures for periods of 2–10 min allows visualization of sites within the nucleus where replication occurs. These studies have shown that replication takes place in discrete patterns of nuclear foci, which undergo characteristic changes throughout S phase (Nakamura et al. 1986; Mills et al. 1989; Nakayasu and Berezney 1989; Fox et al. 1991). Furthermore, DNA synthesis proteins (Bravo and Macdonald-Bravo 1987; Hozak et al. 1993), DNA methyltransferase (Leonhardt et al. 1992) and cell cycle regulators (Cardoso et al. 1993) have been shown to specifically localize to DNA replication foci within the nucleus, providing a direct link between these different cellular processes.

Immunodetection of BrdU incorporation requires partial denaturation of the DNA to allow access of the monoclonal antibodies to the antigenic thymidine analogue. Various procedures can be used to produce single-stranded DNA exposing the halogenated pyrimidines. These include treatment with HCl, thermal denaturation in the presence of up to 50% formamide (Beisker et al. 1987) and enzymatic digestion using either restriction endonucleases alone or endonucleases followed by exonuclease III (Dolbeare and Gray 1988). We will describe two relatively mild procedures, which fall in the first and third categories, and thereby can be used in conjunction with multiple labelings.

BrdU is added to the culture media at a final concentration between 20–100 μM for the desired period of time (from a short 2–10 min pulse to several hours). Cells can be simultaneously fed with 5-fluoro-2'-deoxyuridine (0.4 μM), an inhibitor of thymidilate synthetase, which increases the amount of incorporated BrdU into replicating DNA by lowering the competition with endogenous thymidine. After fixing and permeabilizing (alternatively, the cells can be extracted prior to fixation), in the enzyme digestion-based method, the samples are treated for 30 min with EcoRI (15 units in 100 μl of the respective enzyme buffer), followed by 30 min exonuclease III digestion (30 units in 100 μl of the recommended buffer). Following this treatment, they are incubated with anti-BrdU monoclonal antibody and detected as usual (Dolbeare and Gray 1988; Fox et al. 1991). The enzyme digestion can also be done at the same time as the antibody incubation and together with other antibody binding in dual labeling experiments (Cardoso et al. 1993). A cocktail of diluted anti-BrdU monoclonal antibody with nucleases is commercially available from Amersham and Boehringer Mannheim.

In the acid denaturation method, fixed and permeabilized cells are incubated 10 min in 1.5–4 M hydrochloric acid (inclusion of Triton X-100 to

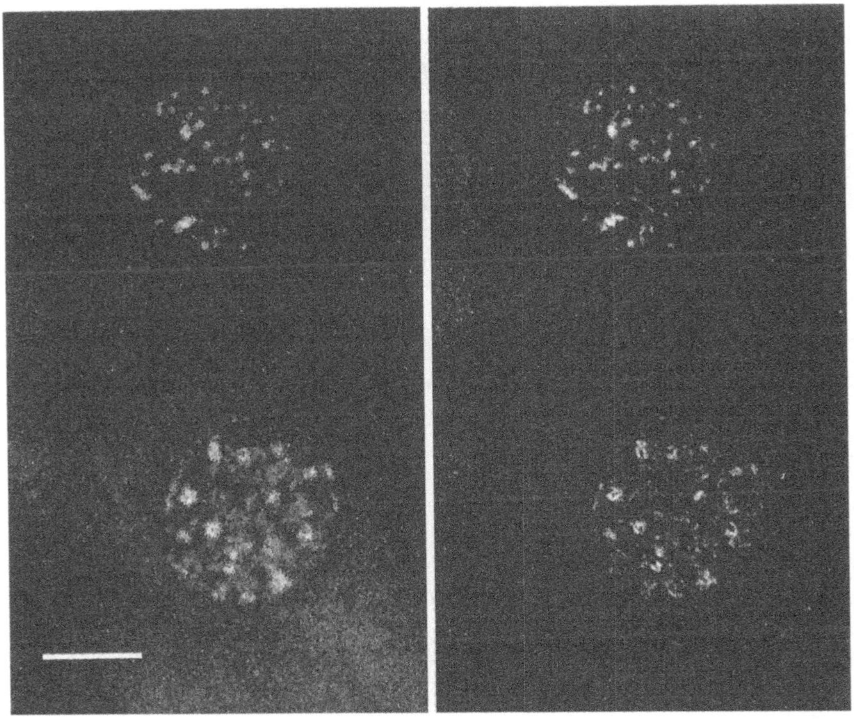

Fig. 2. Specific localization of Cyclin A (*right*) at subnuclear sites of ongoing DNA replication (*left*) as visualized by double staining with BrdU-specific monoclonal antibodies. *Bar* 10 μm

0.7% helps to extract chromatin proteins), neutralized 10 min with PBS and finally stained with anti-BrdU monoclonal antibody as described above. When performing dual stainings, the samples should first be stained for the other antibody as explained earlier, and sequentially fixed again, denatured, neutralized, and finally stained with the BrdU-specific antibodies. An example of the latter staining procedure, where Cyclin A was immunolocalized in an S phase nucleus of a mouse myoblast cell and analyzed by optical dissection with a confocal laser microscope, is presented in Fig. 2 (for specific details see Cardoso et al. 1993).

Fluorescently labeled dUTP analogues are now commercially available (Boehringer Mannheim) and make it possible to directly label and visualize sites of DNA replication in living cells.

The previous sections should have made it clear that there is no such thing as a universal and perfect protocol for cell staining. The procedure has to be optimized for each antibody and application. Possible variations and controls to be performed are described in the respective sections. The following simple and basic protocol may serve as a starting point and is only one of numerous, possible protocols.

Basic protocol for immunofluorescent staining

1. Wash cells in PBS and fix for 10 min in 3.7% formaldehyde (all solutions are made in PBS and incubations at room temperature).
2. Wash in PBS and permeabilise for 10 min in 0.25% Triton X-100.
3. Block cells for 10 to 30 min in 5% normal serum derived from the same species as the labeled antibody (blocking buffer).
4. Incubate for 60 min with primary antibody diluted in blocking buffer in a humidified chamber (see Fig. 1).
5. Wash with several changes of 0.1% NP-40 over 5 min.
6. Incubate for 30–60 min with a labeled secondary antibody diluted in blocking buffer.
7. Wash with several changes of 0.1% NP-40 over 5 min.
8. Counterstain DNA by adding Hoechst 33258 at 1 µg/ml to the second but last washing step.
9. Mount coverslip in mowiol (pH 8.5) with 2.5% DABCO.

References

Baldin V, Lukas J, Marcote MJ, Pagano M, Draetta G (1993) Cyclin D1 is a nuclear protein required for cell cycle progression in G1. Genes Dev 7:812–821

Beisker W, Dolbeare F, Gray JW (1987) An improved immunocytochemical procedure for high-sensitivity detection of incorporated bromodeoxyuridine. Cytometry 8:235–239

Bravo R, Macdonald-Bravo H (1987) Existence of two populations of cyclin/proliferating cell nuclear antigen during the cell cycle: association with DNA replication sites. J Cell Biol 105:1549–1554

Cardoso MC, Leonhardt H, Nadal-Ginard B (1993) Reversal of terminal differentiation and control of DNA replication: cyclin A and Cdk2 specifically localize at subnuclear sites of DNA replication. Cell 74:979–992

Dolbeare F, Gray JW (1988) Use of restriction endonucleases and exonuclease III to expose halogenated pyrimidines for immunochemical staining. Cytometry 9:631–635

Fox MH, Arndt-Jovin DJ, Jovin TM, Baumann PH, Robert-Nicoud M (1991) Spatial and temporal distribution of DNA replication sites localized by immunofluorescence and confocal microscopy in mouse fibroblasts. J Cell Sc 99:247–253

Gallant P, Nigg EA (1992) Cyclin B2 undergoes cell cycle-dependent nuclear translocation and, when expressed as a non-destructible mutant, causes mitotic arrest in HeLa cells. J Cell Biol 117:213–224

Girard F, Strausfeld U, Fernandez A, Lamb NJ (1991) Cyclin A is required for the onset of DNA replication in mammalian fibroblasts. Cell 67:1169–1179

Heald R, McKeon F (1990) Mutations of phosphorylation sites in lamin A that prevent nuclear lamina disassembly in mitosis. Cell 61:579–489

Heald R, McLoughlin M, McKeon F (1993) Human wee1 maintains mitotic timing by protecting the nucleus from cytoplasmically activated Cdc2 kinase. Cell 74:463–474

Hozak P, Hassan AB, Jackson DA, Cook PR (1993) Visualization of replication factories attached to nucleoskeleton. Cell 73:361–373

Lacey AJ, (ed) (1989) Light microscopy in biology: a practical approach. IRL Press, Oxford

Leonhardt H, Page AW, Weier HU, Bestor TH (1992). A targeting sequence directs DNA methyltransferase to sites of DNA replication in mammalian nuclei. Cell 71:865–873

Mills AD, Blow JJ, White JG, Amos WB, Wilcock D, Laskey RA (1989) Replication occurs at discrete foci spaced throughout nuclei replicating in vitro. J Cell Sci 94:471–477

Mittnacht S, Hinds PW, Dowdy SF, Weinberg RA (1991) Modulation of retinoblastoma protein activity during the cell cycle. Cold Spring Harbor Symp Quant Biol 56:197–209

Nakamura H, Morita T, Sato C (1986) Structural organizations of replicon domains during DNA synthetic phase in the mammalian nucleus. Exp Cell Res 165: 291–297

Nakayasu H, Berezney R (1989) Mapping replicational sites in the eukaryotic cell nucleus. J Cell Biol 108:1–11

Ookata K, Hisanaga S, Okumura E, Kishimoto T (1993) Association of p34cdc2/cyclin B complex with microtubules in starfish oocytes. J Cell Sci 105:873–881

Ookata K, Hisanaga S, Okano T, Tachibana K, Kishimoto T (1992) Relocation and distinct subcellular localization of p34cdc2-cyclin B complex at meiosis reinitiation in starfish oocytes. EMBO J 11:1763–1772

Pachter JS (1992) Association of mRNA with the cytoskeletal framework: its role in the regulation of gene expression. Crit Rev Eukaryotic Gene Expression 2:1–18

Pagano M, Pepperkok R, Verde F, Ansorge W, Draetta G (1992) Cyclin A is required at two points in the human cell cycle. EMBO J 11:961–971

Pagano M, Pepperkok R, Lukas J, Baldin V, Ansorge W, Bartek J, Draetta G (1993) Regulation of the cell cycle by the cdk2 protein kinase in cultured human fibroblasts. J Cell Biol 121:101–111

Pines J, Hunter T (1991) Human cyclins A and B1 are differentially located in the cell and undergo cell cycle-dependent nuclear transport. J Cell Biol 115:1–17

Riabowol K, Draetta G, Brizuela L, Vandre D, Beach D (1989) The cdc2 kinase is a nuclear protein that is essential for mitosis in mammalian cells. Cell 57:393–401

Tsai LH, Lees E, Faha B, Harlow E, Riabowol K (1993) The cdk2 kinase is required for the G1-to-S transition in mammalian cells. Oncogene 8:1593–1602

Wang Y-L, Taylor DL (eds) (1989) Fluorescence microscopy of living cells in culture. Vol 29, 30. Academic Press, San Diego

Zindy F, Lamas E, Chenivesse X, Sobczak J, Wang J, Fesquet D, Henglein B, Brechot C (1992) Cyclin A is required in S phase in normal epithelial cells. Biochem Biophys Res Commun 182:1144–1154

3 Fluorescence in Situ Hybridization and Human Cell Cycle Genes

DOUGLAS J. DEMETRICK

1 Background

Fluorescence in situ hybridization (FISH) is a technique of nucleic acid hybridization that permits the microscopic visualization of a nucleic acid target. FISH is primarily used to map genes to their respective chromosomal locations, but has many other potential applications to the study of human cell cycle genes.

Good discussions of FISH have already been published (Lichter et al. 1991; Trask 1991b; Barker and Schwab 1993; LeBeau 1993a, b; Price 1993). The reader is encouraged to consult these for an overview of the technique and some of the historical developments that have allowed more general applications of this technique. The following chapter contains a brief overview of FISH and concentrates on applications of the technique that are particularly useful to the cell cycle researcher.

Briefly, FISH methods include the labeling of a nucleic acid probe with a non-isotopic label and the detection of that probe with a fluorescent reagent to allow direct visualization of the target after hybridization and washing. The advantage of FISH includes the ability to visualize nucleic acids within a single cell and the sensitivity to detect one copy of the target. The target is often genomic DNA, although mRNA targets have been detected. The specimen is often a human chromosome spread, when gene mapping is the application, or cell cover slip preps, smears, or tissue sections, when gene copy number or mRNA detection is the application. If the probe is directly labeled with a fluorescent marker (direct FISH), detection can be accomplished by visualizing the specimen after washing off the unhybridized probe. If the probe is labeled with a hapten, visualization requires development with a secondary reagent(s) coupled to a fluorescent marker (indirect FISH). Direct FISH is faster and simpler to perform and potentially offers less background, but is considerably less sensitive than indirect FISH. Multiple labeled probes can be used if they can be distinguished by the color of their fluorochrome or a characteristic appearance in the specimen.

2 Principle and Applications

The primary application of this technique in human cell cycle research is to localize a cell cycle gene to a cytogenetic position. Using this approach, many human cell cycle genes have been mapped (Table 1). Mapping a cell cycle gene can reveal whether the gene is at a locus that has documented abnormalities in tumors. Mapping of *p16*^{INK4} to 9p21–22 demonstrated the potential for the gene to be involved in the numerous examples of homozygous deletions and loss of heterozygosity (LOH) at that locus and perhaps even the potential to be a tumor suppressor gene whose loss of function can confer a growth advantage onto a neoplastic cell (Okamoto et al., 1994). Numerous cell cycle genes have been mapped to sites of karyotypic abnormalities in

Table 1. Chromosomal location of human cell cycle genes

Gene	Map Position	Reference
Cdc25A	3p21	Demetrick and Beach (1993)
Cdc25B	20p13	Lane et al. (1993); Demetrick and Beach (1993)
Cdc25C	5q31	Sartor et al. (1992)
Cdc2 (*Cdk1*)	10q21	Nazarenko et al. (1991)
Cdk2	12q13	Demetrick et al. (1994)
Cdk4	12q13	Khatib et al. (1993); Demetrick et al. (1995)
Cdk5	7q35	Demetrick et al. (1994)
Cyclin A	4q26–q27	Blanquet et al. (1990)
Cyclin B1	5q12–qter	Milatovich and Franke (1992); Sartor et al. (1992)
Cyclin C	6q21	Demetrick et al. (1995)
Cyclin D1	11q13	Arnold et al. (1991)
Cyclin D2	12p13	Xiong et al. (1992)
Cyclin D3	6p21	Xiong et al. (1992)
Cyclin E	19q12–13	Demetrick et al. (1995)
Kap/Cdi1 (CDKN3)	14q22	Demetrick et al. (1995)
p16^{INK4} (CDKN2)	9p21–22	Kamb et al. (1994); Nobori et al. (1994); Okamoto et al. (1994)
p21 (CDKN1)	6p21.2	El Diery et al. (1993); Demetrick et al. (1995)
p^{15INK4B}	9p21–22	Kamb et al. (1994); Demetrick et al., unpublished
p19A	7q11.2	Demetrick et al., submitted
p19B	12p12	Demetrick et al., submitted
p45	5p13	Demetrick et al., submitted
p53	17p13	Isobe et al. (1986)
p107	20q11.2	Ewen et al. (1991)
p130	16q12–13	Hannon et al. (1993)
PCNA	20p12	Webb et al. (1990); Rao et al. (1991)
pRb	13q13	Friend et al. (1986)

Table 2. Cell cycle gene position and chromosomal anomalies in neoplasia. (Mitelman et al. 1990; Testa,1990; Solomon et al. 1991)

Cdc25A	Pleomorphic adenoma, renal and lung carcinomas, mesothelioma
Cdk2	Pleomorphic adenoma, lipoma, leiomyoma
Cdk4	Pleomorphic adenoma, lipoma, leiomyoma
Cyclin D1	Lymphoma, parathyroid adenoma, amplifications
Cyclin D3	Leiomyoma, carcinomas
Cyclin E	CLL, gliomas
$p16^{INK4}/p15^{INK4B}$	Melanoma, mesothelioma, ALL, glioma, carcinomas
$p21^{Cip1/Sdi1/Waf1/Pic1}$	Leiomyomas, carcinomas
p53	Carcinomas
pRb	Retinoblastomas, osteogenic sarcomas, carcinomas

both benign and malignant tumors (Table 2). Of these, *cyclin D1* is a candidate oncogene and $p16^{INK4}$ is a candidate tumor suppressor gene.

Aside from mapping, FISH of genomic probes may be used to establish copy numbers of cell cycle genes in interphase cells. The use of this approach to quantitate *c-erbB-2* is of potential diagnostic significance (Kallioniemi et al. 1992). FISH has been able to quantitate *cdk4* gene copy number in cell lines (Khatib et al. 1993) and has been used to quantitate copy number of *cyclin D1*, *cyclin E,* and various other cell cycle genes in both cell lines and tumor specimens (D. Demetrick, unpubl. data). FISH allows the precise quantitation of gene dosage in individual cells and may allow more accurate estimates of cellular copy number than tissue destructive methods such as Southern blotting (Kallioniemi et al. 1992; Stilgenbauer et al. 1993). Since abnormal *cyclin D1* gene dosage has been shown to correlate with the aggressiveness of some breast carcinomas (reviewed in Brison 1993), FISH testing for this gene (and/or others) of fine needle aspiration biopsies of breast tumors may allow more accurate predictions of the behavior of these tumors even before they are removed. As well, one can use FISH to examine the interphase nuclei of malignant cells to see whether two genes have been translocated next to each other (Barker and Schwab 1993). In this case, two distinct probes will merge to one position. A good illustration of this is the visualization of the *bcr/c-abl* translocation producing the Philadelphia chromosome seen in most cases of chronic myelogenous leukemia (CML) (Tkachuk et al. 1990). This test is currently being used to diagnostic advantage in the treatment of CML and reduces the necessity of classical karyotyping to identify the Philadelphia chromosome. Potentially, identification of the *cyclin D1/parathyroid hormone* translocation seen in many cases of parathyroid adenomas (Rosenberg et al. 1991) may be of diagnostic usefulness in distinguishing this lesion from parathyroid hyperplasia, which is treated quite differently.

Of course, FISH may be used to identify foreign genes in a tissue sample (Delecluse et al. 1993; Gersdorf et al. 1993). The use of this technique offers a much higher level of sensitivity than enzymatic detection methods and the advantage of cellular specificity over tissue destructive methods such as Southern blotting or PCR. Although several adaptations of PCR to in situ hybridization have been described, they are not popular and require specialized equipment (Nuovo 1992). FISH has also been utilized to detect nuclear mRNA (Lawrence 1991) and cytoplasmic mRNA in cellular samples (Raap et al. 1991) or cells in suspension (Grunwald 1993). FISH in situ mRNA detection offers the advantage of being much quicker and simpler to perform than isotopic methods but has not been used widely yet. This application of FISH may be quite useful to cell cycle scientists who need to know exactly what cell type in a tissue is expressing a certain cell cycle gene at a certain time, and who cannot gain that information from Northern blotting which produces an average of the expression of all cell types in a sample.

3 Equipment

Microscope

The only necessary equipment that a molecular biology lab is unlikely to have on hand is a good quality microscope equipped for epifluorescence; however, most labs will have access to such an item within their department or university. As the visualization of a small fluorescent probe may require detection of only 50–100 molecules of a fluorochrome, the quality of the microscope is important. For small probes (genomic <5 Kb or cDNA) a light source of 100 Watts is optimal, along with a 100 × fluorescence objective of at least 1.25 NA. Filters for each fluorochrome are mandatory as well as filters for the counterstain (i.e., DAPI or propidium iodide). A multiple bandpass filter (which allows visualization of multiple fluorochromes) is very useful for mapping, as the use of separate filters to visualize the marker fluor and the counterstain may result in registration errors and subsequent difficulty in aligning the fluorescent signal with the correct chromosomal position. A triple bandpass filter (e.g., Oncor, Chroma Technologies Inc.), for example, may allow simultaneous visualization of a Texas Red and FITC signal with a DAPI chromosome stain. A confocal microscope, although useful for subcellular fluorescent antibody staining, is generally not useful for FISH, as the resolution of the images is usually inferior to standard fluorescence and the power of the laser illumination source rapidly photobleaches the samples. The confocal microscope has been put to good use, however, in the FISH investigation of thick tissue sections (Thompson et al. 1994).

Since most chromosomal mapping is done by direct visualization and photography number assessment only requires counting of the fluorescent signal of a probe in an interphase nucleus, a good-quality fluorescent microscope is the only prerequisite for most FISH applications. If, however, publication or archiving of the results is required, photography is necessary. This can present significant problems. Black and white film, while very sensitive, does not allow for the discrimination between two colors with similar tonal characteristics. Thus, a weak fluorescent signal on a fluorescently stained chromosome may not show up on black and white film. Separate images may be collected using different filters, however, registration errors then become a problem. The best solution is the capture of the image by color film. When a mapping photo is captured through a multiple bandpass filter, different colored fluors are observed to be localized to their correct location. Even very weak tones can be discriminated by their different colors. Color films, however, have several problems of their own: lack of sensitivity and reciprocity failure. Generally, color slide film gives better results for FISH than color print film. Like black and white films (only worse), the more sensitive the film the larger the grain size. Significant enlargement of color slide film with ASA much faster than 400 may be unsatisfactory. Consequently, exposure times are required to be quite long in order to capture a weak signal (30 s to 3 min). It can be difficult to calibrate the exposure time to produce a good quality image when detecting tiny bright points of light on a dark field.

Reciprocity failure refers to a loss of the ability of the film emulsion to truthfully capture different colors of light at long exposure times. For short exposure times (<1 s) the film is calibrated to give the correct color rendition. At longer exposure times the ability of different colors to make a true exposure starts to diverge, thus it may become difficult to make an accurate true color exposure with exposure times longer than 10 s.

For many applications, a simple camera back and an accurate exposure meter will provide the desired results. For the mapping of a small probe, however, many mitoses will have to be photographed to obtain one publishable print. As well, it can be very expensive to have the color photographic enlargements and figures done commercially. An economical solution lies with a recently introduced commercial process called PhotoCD. Many photography stores and laboratories offer this inexpensive service by which photographic images including 35-mm slides, negatives and prints can be scanned and digitized at various resolutions and stored as an image on a CD-ROM disk. This allows the use of inexpensive software (e.g., Adobe Photoshop or Micrografx Picture Publisher) and a PC or Macintosh computer to enhance the image by improving contrast, color balance, tonal characteristics, and even pseudocoloring a color image to black and white. The advantage of psuedocoloring is that a weak fluor color can be assigned a certain grayscale value (such as pure white) so that it will show up on a light gray chromosome even when it would not if photographed using black and white film. The resulting image can be cropped and assembled to a publication quality figure on one's home or lab computer and printed relatively inexpensively on a commercial high-resolution printer or film recorder. This approach can reduce the cost of mapping a gene with a small probe (and having to publish in color) from approximately $ 1400 to $ 100 (excluding the cost of software and computer)

The best image capture system utilizes a cooled CCD Digital Cameras. This approach offers a great improvement in sensitivity (more than ten fold over photographic film), the ability to use infrared fluors (i.e., Cy5), potential for image analysis and enhancement, as well as convenience. Unfortunately, these systems are expensive ($ 30 000–50 000), however, if a facility is to be used exclusively for FISH, the deletion of a conventional camera and exposure system may save $ 10 000–20 000 from the cost of the system. This money could then be used to purchase a digital camera system. Like film, cooled CCD cameras come in two types, monochrome and color. For a pure FISH application, a monochrome camera is probably best (e.g., Hamamatsu or Photometrics). It will capture a high resolution grayscale image quickly and with a relatively small file size. This image can be simply pseudocolored to the color of the fluorochrome if desired. Unfortunately, like black and white film, the monochrome camera must be used with a separate filter for each fluor. Unlike black and white photos, these images can be pseudocolored to different colors then superimposed with registration correction via computer software in a process that can be automated (e.g., Oncor or IP Labs). The use of a multiple bandpass filter is still recommended to verify that the registration correction is accurate. Color cooled CCD cameras are a relatively new arrival on the scene (e.g., Oncor). They may have less resolution and sensitivity than the monochrome cameras, though still are good enough for most FISH applications. One disadvantage is that a true color image file size can be considerably larger than a pseudocolored monochrome image demanding more in the way of computing power and data storage than a monochrome camera. As well, though a color image taken through a wide bandpass filter is not given to registration error, the extra light permitted by the wide bandpass filter can make the image much less appealing. A color camera may be the device of choice in a multi-application setting such as a pathology or hematology lab where stained tissue or smear specimens can be imaged using the same equipment.

4 Methods

Most of the methods described herein are adapted, usually without major modification, from those used in the laboratory of Dr. Barbara Trask, University of Washington, and previously published (Trask 1991a). This reference is recommended if more procedural details or technical discussion is required. The methodology for the preparation of common solutions (e.g., standard citrate saline (SSC) and phosphate buffered saline (PBS) can be found in standard textbooks.

Specimens

The best gene mapping results come from using synchronized human lymphocytes to produce long chromosomes. The two methods detailed below result in preparations that can be used with different counter stains to generate two different banding patterns. Simultaneous mapping with the two preparations lessens the potential of mismapping genes due to chromosomal staining ambiguities. FISH on cell lines can be done with interphase cells or colchicine treated tumor cells. Cell lines and cultured tumor cells usually give inferior chromosomal morphology compared to normal lymphocyte preparations.

4.1 Synchronization of Human Lymphocytes

Two methods of preparing normal lymphocytes [methotrexate synchronization (Yunis 1976) for DA-DAPI (enhanced QFH) (Schwiezer 1976) banding and thymidine synchronization for PI (R) banding (Viegas-Pequignot et al. 1989)] are recommended for FISH mapping new genes.

Lymphocyte cultures

Venous blood is drawn into a heparinized Vacutainer tube from a volunteer using aseptic technique and used within 4 h; 0.60 ml of whole blood is added to 10 ml of RPMI 1640/20% fetal calf serum culture medium containing 10 mM HEPES (pH 7.4), 1 × penicillin/streptomycin/fungizone, 1 × L-glutamine, 13 units/ml porcine heparin, and 4% PHA (reconstituted Gibco 670-0576AD), and distributed into 50 ml tissue culture flasks. Usually 10–20 cultures are set up. The flasks are incubated in a humidified incubator containing 5% CO_2 at 37 °C for approximately 72 h (60–84). At this time, one of the two synchronization techniques is performed.

Methotrexate synchronization

One hundred µl of 100 × methotrexate solution [10 µM stock (4.55 µg/ml) in Hank's saline solution can be stored frozen] is added to each 10 ml flask and incubation is continued for a further 15–18 h. The contents of a flask are centrifuged at 200 *g* for 8 min in a benchtop centrifuge at room temperature and washed twice in prewarmed culture medium **without** PHA. After the second wash, resuspend the cells gently in 10 ml of prewarmed culture medium containing 10 µM thymidine (100 × –1.0 mM solution is prepared by dissolving 5 mg of thymidine in 20 ml of Hank's saline, which can be stored frozen), transfer them to a new tissue culture flask, and incubate them a further 5–7 h. To each flask add 120 µl (0.12 µg/ml) of Colcemid (Gibco) and incubate 10 min. Immediately centrifuge as before for 8 min at 200 *g*. Remove all but the last 0.3 ml of supernatant and resuspend the cells with gentle finger vortexing. Add 10 ml of prewarmed 0.075 M potassium chloride (1 ml slowly then 9 ml quickly – mix by inversion). Incubate for 20–30 min in a 37 °C water bath. Centrifuge as before.

Thymidine synchronization The steps are exactly the same as for the methotrexate synchronization, with different reagents. After the initial 72-h incubation, 100 μl of 100× thymidine is added to the 10-ml cultures (145 mg of thymidine in 5 ml of Hanks saline can be stored frozen). After 15–18 h of incubation, the cells are washed twice as before and resuspended in culture medium containing 0.1 mM of 2-bromo-deoxyuridine (BrdU) (100× – 30.6 mg of BrdU in 30 ml of Hanks saline can be stored frozen) **without** PHA. After a further 5–7 h of incubation, the cells are treated with Colcemid, washed, resuspended in warm 0.075 M KCl, and incubated as before.

4.2 Chromosome Preparations

Preparation High-quality microscope slides should be precleaned before metaphase chromosome spreads are prepared by rinsing in ethanol/ether (100:1) and drying with lint-free cloth or tissues. Methanol should be fresh (buy smaller bottle sizes and use only new bottles for spreads). After synchronization, Colcemid treatment and incubation with hypotonic KCl, the 10-ml cell cultures are removed from the water bath. 1.0 ml of fresh methanol : acetic acid (3:1) fixative is added to the KCl suspension and the contents are gently mixed by inversion.

After centrifugation (200 *g* for 8 min) the supernatant is removed except for the last 0.3 ml. The contents of the tube are resuspended gently by finger or low speed vortexing. Fresh methanol : acetic acid fixative is added **dropwise with gentle vortexing** after every drop for the first 3 ml, then after every five drops for the next 3 ml. The last 4 ml of fixative is added quickly with mixing by inversion for a total of 10 ml of fixative. The erythrocytes should lyse after the last aliquot of fix as evidenced by clearing of the turbidity of the cell solution. The solution is incubated for 20 min at room temperature, then centrifuged as usual.

Following centrifugation, the supernatant is removed except for the last 0.3 ml, and the pellet resuspended by gentle vortexing. A fresh 10 ml of fixative is added with vortexing following every 0.5 ml addition for the first 5 ml. Then, the cells are incubated a further 10 min at room temperature. The process is repeated for two more washes. After the final wash, the cell pellet is resuspended in 0.5–1.0 ml of fix, and all of the resuspended pellets are pooled into one centrifuge tube. If any of the cell prep tubes contain dark clumps, that preparation should be discarded. White lumps can be removed with a Pasteur pipette.

Chromosome spreads can be made by dropping single drops of the cell suspension from a height of 12 to 24 inches onto clean, dry glass slides. A test slide should be made and air dried. If the unmounted, dry slide is examined at 100–200× through a phase contrast microscope, numerous mitoses should be evident. The concentration of the cells can be adjusted to give the optimal mitotic density by adding more fixative or centrifuging and resuspending in less volume. If the cells are stored in fix for more than an 1–2 h, the number of mitoses is significantly lessened. If the spreads are too compact, the slides can be dipped in ice cold distilled water before dropping, to help separate the chromosomes. Usually two to three drops are applied per slide.

After air drying (can be overnight), the slides are heated at 70 °C for 4 h in plastic slide boxes (25 slides – make sure that they do not warp at that temperature). The warm slides are placed within Seal-a-Meal or similar bags with a small amount of desiccant and heat sealed after removing as much air as possible. Some investigators choose to flush the bags out with nitrogen gas before sealing. The slides can be stored for more than a year at –20 °C. They must be allowed to warm to room temperature before the bag is opened.

4.3 Probe Labeling

Labeling

Probes must be pure for the best labeling efficiency. Cesium or DNA column (e.g., Qiagen) purification adequate. While nonisotopic DNA labels can be generated in a variety of ways, nick translation is the best for FISH applications, as the resulting probe size (approximately 200 base pairs) is optimal for penetrating into condensed chromosomes (Lawrence and Singer 1985).

Purified probe (plasmid, cosmid or phage DNA, 0.2–1 µg) is labeled using a nick translation kit (Gibco 18 160-010) and a modified nucleotide (11-biotin or 11-digoxigenin dUTP). Additional DNAse I (Gibco 18 047-019) must be added to digest the probe to 200 base pairs. The first time the DNAse I is used, it should be titrated (1/1000–1/10 000 in water) in a labeling reaction and the reaction products should be run on an agarose minigel to determine the optimal concentration of DNAse I, resulting in a probe size range of 200–400 base pairs. If stored at –20 °C the DNAse stock is stable for more than a year. Reactions are incubated for 90 min in a 16 °C water bath (ice bucket filled with water and cooled with ice lasts several hours) before being separated on 1-ml spin columns in TE/0.1% SDS. Following spin column

1 × 50 µl Reaction	
Buffer and nucleotide mix[1]	5 µl
dUTP Label[2]	2 µl digoxygenin or 5 µl biotin
Target[3]	1 µg (Volume to 10 µl
Enzyme mix[4]	5 µl
DNAse I dilution (fresh)	2.5 µl
Sterile distilled water	Volume to 50 µl total

[1] As from Gibco kit buffer A4 (0.2 mM of dATP, dCTP, dGTP; 500 mM Tris pH 7.8; 50 mM $MgCl_2$; 100 mM 2-mercaptoethanol).
[2] Digoxigenin-dUTP 1 pmol/µl; Biotin-11-dUTP 0.3 pmol/µl.
[3] May use as little as 0.1 µg, but less than 10 µl volume of TE.
[4] From Gibco kit Solution C (0.4 U/ml DNA polymerase I; 40 pg/µl of DNAse I; 50 mM Tris pH 7.5; 5 mM magnesium acetate; 1 mM 2- mercaptoethanol, 0.1 mM PMSF; 50% (vol/vol) glycerol; 100 µg/ml BSA).

purification, dry sample down in a Speed Vac and resuspend in 25 µl of sterile distilled water.

4.4 In Situ Hybridization

Reagents for blocking DNA

- Human Genomic DNA. One mg of sonicated human placental DNA (Sigma) is mixed with 0.5 mg of human cot 1 DNA (Gibco) and ethanol precipitated with sodium acetate. After washing with 70% ethanol and air drying for 5 min, the DNA is resuspended in 100 µl of TE.

- Herring Genomic DNA. Ten mg per ml sonicated herring DNA in TE (herring DNA can be sonicated to an average size of <1000 bp and checked by gel electrophoresis). Store both DNA solutions at –20 °C.

- HybMix (10 ml)

Formamide (good quality)	5 ml
20 × SSC pH 7.0	1 ml
50% Dextran sulfate	2 ml

pH to 7.0, aliquot into 1-ml samples and store at –20 °C.

- Denaturation solution

Formamide (good quality)	70 ml
20 × SSC pH 7.0	10 ml
Sterile distilled water	20 ml

pH to 7.0 with 1.0 M HCl
Can be reused several times and is stable for 2 months if stored at 4 °C.

- 20 × SSC pH 7.0 is prepared dissolving 175.3 g of sodium chloride and 88.2 g of sodium citrate in 800 ml of distilled water. The pH is then adjusted to 7.0 with a few drops of a 10 M solution of NaOH. Finally, the volume is adjusted to 1 l with distilled water and the solution sterilized by autoclaving.

Equipment

- Water bath adjustable from 42 to 85 °C
- Heating block with suitable cover at 37 °C
- Sealable plastic boxes for incubating slides
- 37 °C incubator
- Clean, dry compressed air (canned if necessary)
- Rubber cement
- Cover slips (#1, size as desired)
- Coplan jars (10)

Hybridization protocol

Each slide can be covered with approximately 20 µl of hybridizing solution (or 10 µl for each lymphocyte "drop").

Samples (per 10 µl)	
1 µl	Labeled probe (20–50 ng)
1 µl	Blocking DNA (from 0–100% human)
8 µl	HybMix

The samples are mixed well and heat denatured at 85 °C for 8 min. Genomic probes are then quickly spun to concentrate the sample and incubated at 37 °C for 15–45 min to suppress the repetitive DNA. Nongenomic probes can be added immediately to denatured slides. Slides are denatured by heating in the denaturation solution for 4 min at 85 °C and then quenching immediately into ice cold 70% ethanol (no more than three to five slides per Coplan jar, as the ethanol becomes too warm). The slides are left in the cold ethanol for 3–5 min, then are dehydrated through 80, 90, and 2–100% room temperature ethanol baths. The slides can stay in the last 100% ethanol solution for several minutes before hybridization.

At the time of hybridization, the slides are quickly dried with the compressed air and placed on a 37 °C heating block or slide warmer and allowed to come to temperature. The warm hybridization solution, containing probe, is added to the slide, and immediately covered with a cover slip. The edges are sealed with rubber cement and the slides are placed in a pre-warmed, sealable container (preferably separated by a plastic grid or spacers from moist towels on the bottom of the container) and sealed. The samples are incubated overnight (at least 8 h) at 37 °C.

4.5 Detection of Label

Solutions

- Wash solutions
  - 50% Formamide

Formamide (good quality)	150 ml
20 × SSC pH 7.0	30 ml
Distilled water	120 ml

pH to 7.0 with 1.0 M HCl.
Can be reused several times and is stable for 2 months if stored at 4 °C.

  - 2 × SSC pH 7.0

- Antibody labeling solutions
  - Antibody blocking solution

0.1 M Sodium phosphate buffer pH 8.0
0.1% NP40 detergent
0.02% Sodium azide
5% nonfat dried milk

Mix and let stand several days at 4 °C. Centrifuge 15 min at 200 *g* to remove insoluble milk proteins. Store at 4 °C.

  - Antibody wash solution

2 × SSC ph 7.00
0.005% CHAPS detergent

Visualization protocol

The slides are removed from the humidified incubation box, de-coverslipped, and rinsed quickly in room temperature 2 × SSC pH 7.0 buffer. The slides are then washed for 5 min at 42 °C in three sequential baths of wash formamide followed by three sequential baths of 2 × SSC pH 7.0.

One hundred μl of room temperature blocking solution is applied and the slides are coverslipped and incubated in a humidified box for 30 min at room temperature.

Primary antibody solution is made up in blocking solution and Eppendorf centrifuged for 5 min prior to use [e.g., 1/100 dilution of donkey anti-digoxigenin (Boehringer Mannheim) or 1/200 dilution of FITC-avidin DCS (Vector)]. The blocking solution is tipped from the slide and the slide is quickly covered with 100 μl of the primary antibody solution and incubated at room temperature for 45–60 min.

After primary antibody labeling, the slides are rinsed in 2 × SSC pH 7.0 and washed for 2–5 min each in three changes of antibody wash solution. The slides are blocked for 5 min before being treated with 100 µl of the centrifuged secondary antibody diluted in blocking solution [e.g., 1/50 Cy3-sheep anti-donkey (Jackson Immunoresearch)] or 1/300 biotin-goat anti-avidin (Vector). The slides are coverslipped and incubated for 30 min at room temperature then washed as previously.

If a third step is required (e.g., 1/100 FITC-avidin) the slides are blocked for 5 min, then incubated with 100 µl of the antibody solution for 30 min as before. For dual staining, the primary antibodies can be mixed together if they are compatible (not reactive to each other), as can the secondary antibody solutions.

In general, Cy3 signals are brighter for a given probe than FITC, but photobleach more rapidly. It is important to not allow the slides to dry between steps to minimize background.

4.6 Counterstaining

Solution

- McIlvaine's solution pH 7.0

0.63 g	Citric acid
6.19 g	Sodium dibasic phosphate

To 500 ml with distilled water.

- Phosphate buffered saline (PBS).

Counterstaining

The slides are incubated in either 0.05 µg/ml DAPI in McIlvaine's buffer for 10–20 min (for FITC/Cy3 staining) or 0.1 µg/ml propidium iodide in PBS (for FITC staining of thymidine synchronized chromosome preps) for 5 min at room temperature. Keep Coplan jars dedicated to one or the other stain, as contamination with PI can lower the signal to noise ratio of Cy3 labeled preps.

The DAPI stained slides are rinsed in PBS then treated with Actinomycin D (250 µg/ml in PBS stored frozen or at 4 °C) for 15–30 min at room temperature.

After staining, the slides are rinsed in two changes of PBS and usually stored overnight in PBS (this decreases intensity of counterstain and reduces background), though they can be visualized immediately if desired. For visualization, the slides are almost completely air dried (with vigorous shaking to remove excess PBS), mounted in an appropriate antifade solution (e.g., Molecular Probes) and coverslipped with #1 coverslips.

The mounted slides can be kept at –20 °C for several weeks. Photobleaching is reduced and the signal-to-noise ratio is better for weak probes if the slide is mounted and stored overnight before visualizing.

5 Tips and Troubleshooting

- Probes may be labeled while in the vector (either plasmid or phage) but the higher the insert to vector DNA ratio, the stronger the signal. For routine mapping, I use purified phage DNA. The best signals are observed if a phage insert is subcloned into a small plasmid vector. Genomic probes as small as 1–2 kb can be easily mapped. cDNA probes are not usually good candidates for mapping. The signals are usually weaker than a proper genomic probe, the background is much higher and the potential for cross hybridization to pseudogenes is significant (Johnson et al. 1993). It is often better to spend a small amount of time on screening a genomic library and isolating a genomic probe than to spend a lot of time trying to map a cDNA and, perhaps, be misled.
- Thymidine synchronized lymphocyte chromosomes are often longer and more tangled than methotrexate synchronized chromosomes. These preps are better for fine mapping and to confirm map positions that have been obtained with the DA-DAPI banding. Cy3 signals on a DAPI stained chromosome are also easier to see for a given probe than FITC signals on a PI stained chromosome. For mapping, I usually label the probe with digoxygenin-dUTP and biotin-dUTP and hybridize the appropriate probes to both methotrexate and thymidine synchronized targets the same day. The following day the slides can be visualized and map position usually unequivocally obtained. Confirmation of the mapping can be done with an alpha-satellite probe (e.g., Oncor). Of course, the genomic probe should be confirmed by high stringency Southern blotting and/or partial sequence analysis to eliminate the possibility of mapping a pseudogene. If the sequence of a genomic probe is available, PCR analysis of a somatic cell hybrid panel (BIOS Laboratories) can be done to confirm the FISH mapping results.
- Interphase cells are labeled as detailed for gene mapping. The best signals are observed if DAPI counterstain concentration is used and the concentration is reduced to 0.01 µg/ml. Actinomycin D staining is still required to quench the bright DAPI signal.
- In my experience, RNAse pre-treatment is usually not necessary and often increases the background. It may be more useful if cDNA probes are utilized, or if interphase hybridization is attempted with cells expressing large amounts of the target gene. Proteinase treatment is generally to be

avoided, as significant numbers of target cells and chromosomes can be lost.

Acknowledgments. I am indebted to Dr. B. Trask (University of Washington) for her advice on FISH methodology and to the Medical Research Council of Canada and the Alberta Heritage Fund for Medical Research for salary support. I would also like to thank Dr. P. E. Barker and Dr. D. Esposito of Cold Spring Harbor Laboratory for their critical reading of this manuscript.

References

Arnold A, Motokura T, Bloom T, Kronenberg H, Ruderman J, Juppner H, Kim HG (1991) The putative oncogene PRAD1 encodes a novel cyclin. Cold Spring Harb Symp Quant Bio. 56:93–97

Barker PE, Schwab M. (1993) Junction mapping of translocation chromosomes by fluorescence in situ hybridization and computer image analysis in human solid tumours. Methods Mol Genet 2:129–154

Blanquet V, Wang J, Chenivesse X, Henglein B, Garreau F, Brechot C, Turleau C. (1990) Assignment of a human cyclin A gene to 4q26–q27. Genomics 8:595–597

Brison O (1993) Gene amplification and tumour progression. Biochim Biophys Acta 115525–41

Cobrinik D, Whyte P, Peeper DS, Jacks T, Weinberg RA (1993) Cell cycle-specific association of E2F with the p130 E1A-binding protein. Genes Dev 7:2392–2404

Delecluse HJ, Bartnizke S, Hammerschmidt W, Bullerdiek J, Bornkamm GW (1993) Episomal and integrated copies of Epstein-Barr virus coexist in Burkitt lymphoma cell lines. J Virol 67:1292–1299

Demetrick DJ, Beach DH (1993). Chromosomal Mapping of CDC25A and CDC25B Phosphatases. Genomics 18:144–147

Demetrick DJ, Zhang H, Beach, DH (1994) The chromosomal mapping of human cdk 2, cdk 4, and cdk 5 cell cycle kinase genes. Cytogenet Cell Genet 66:72–74

Demetrick DJ, Matsumoto S, Hannon GJ, Okamoto K Xiong Y, Zhang H, Beach DH (1995) Chromosomal mapping of the genes for the human cell cycle proteins cyclin C, cyclin E, cdkn1 (WAF-1/Sdi1), and KAP/Cdi1. Cytogene Cell Genet 69:190–192

El-Diery WS, Tokino T, Velculescu VE, Levy DB, Parsons R, Trent JM, Lin D, Mercer WE, Kinzler KW, Vogelstein B (1993) WAF1, a potential mediator of p53 tumour suppression. Cell 75:817–825

Ewen ME, Xing Y, Lawrence JB, Livingston DM (1991) Molecular cloning chromosomal mapping and expression of the cDNA for p107 a retinoblastoma gene product-related protein. Cell 66:1155–1164

Friend SH, Bernards R, Rogelj S, Weinberg RA, Rapaport JM, Albert DM, Dryja TP (1986) A human DNA segment with properties of the gene that predisposes to retinoblastoma and osteosarcoma. Nature 323:643–646

Gersdorf H, Pelz K, Gobel UB (1993) Fluorescence in situ hybridization for direct visualization of gram-negative anaerobes in subgingival plaque samples. FEMS Immunol Med Microbiol 6:109–114

Grunwald D (1993) Flow cytometry and RNA studies. Biol Cell 78:27–30

Hannon GJ, Demetrick DJ, Beach DH (1993) Isolation of the Rb-related p130 through its interaction with cdk2 and cyclins. Genes Dev 7:2378–2391

Isobe M, Emanuel BS, Givo, D, Oren M, Croce CM (1986) Localization of gene for human p53 tumour antigen to band 17p13. Nature 320:84–85

Johnson CV, Cool DE, Glaccum MB, Green N, Fischer EH, Bruskin A, Hill DE, Lawrence JB (1993) Isolation and mapping of human t-cell protein tyrosine phosphatase sequences: localization of genes and pseudogenes discriminated using fluorescence hybridization with genomic versus cDNA probes. Genomics 16:619–629

Kallioniemi O-P, Kallioniemi A, Kurisu W, Thor A, Chen L-C, Smith HS, Waldman FM, Pinkel D Gray J (1992) ERBB2 amplification in breast cancer analyzed by fluorescent in situ hybridization. Proc Natl Acad Sci USA 89:5321–5325

Kamb A, Gruis NA, Weaver-Feldhaus J, Liu Q, Harshman K, Tavtigian SV, Stockert E, Day RS, Hohnson BE, Skolnick MH (1994)A cell cycle regulator potentially involved in genesis of many tumour types. Science. 264:436–440

Khatib ZA, Matsushimie H, Valentime M, Shapiro DN, Sherr CJ, Look AT. (193) Coamplification of the CDK4 gene with MDM2 and GLI1 in human sarcomas. Cancer Res 53: 5535–5541

Lane SA, Baker E, Sutherland GR, Tonks I, Hayward N, Ellem K (1993). The human cell cycle gene cdc25B is located at 20p13. Genomics 15:693–694

Lawrence JB (1991) Fluorescent detection of nuclear RNA and DNA: Implications for genome organization. Methods Cell Biol 35:73–99

Lawrence JB, Singer RH (1985) Quantitative analysis of in situ hybridization methods for the detection of actin gene expression. Nucleic Acids Res 13:1777–1799

LeBeau MM (1993a) Detecting genetic changes in human tumour cells: have scientists gone FISHing? Blood. 81:1979–1983

LeBeau MM (1993b) Fluorescence in situ hybridization in cancer diagnosis. Important Adv Oncol 29–45

Lichter P, Boyle AL, Cremer T, Ward DC (1991). Analysis of genes and chromosomes by non-isotopic in situ hybridization. Genet Anal Techn Appl 8:24–35

Milatovich A, Franke U (1992) Human cyclin B1 gene (CCNB1) assigned to chromosome 5 (q13-qter). Somat Cell Molec Genet 18:303–307

Mitelman F, Kaneko Y, Trent JM (1990) Report of the committee on chromosome changes in neoplasia. Cytogenet Cell Genet 55:358–386

Nazarenko SA, Ostroverhova NV, Spurr NK (1991) Regional assignment of the human cell cycle control gene CDC2 to chromosome 10q21 by in situ hybridization. Hum Genet 87: 621–622

Nobori T, Miura K, Wu JD, Lois A, Takabayashi K, Carson DA (1994) Deletions of the cyclin-dependent kinase-4 inhibitor gene in multiple human cancers. Nature 368:753–756

Nuovo GJ. (1992) PCR in situ hybridization. Protocols and applications, 1st ed. Raven Press, New York

Okamoto A, Demetrick DJ, Spillare E, Hagiwara K, Hussain P, Bennett WP, Forrester K, Gerwin B, Serrano M, Beach DB, Harris CC (1994) Mutations and altered expression of $p16^{INK4}$ in human cancer. Proc Natl Acad Sc. USA 91:11 045–11 049

Price CM (1993) Fluorescence in situ hybridization. Blood Rev 7:127–134

Raap AK, Van de Rijke FM, Dirks RW, Sol CJ, Boom R, Van der Ploeg M (1991) Bicolor fluorescence in situ hybridization to intron- and exon mRNA sequences. Exp Cell Res 197:319–322

Rao VVNG, Schnittger S, Hansmann I (1991) Localization of the gene for human proliferating cell nuclear antigen (PCNA) to or close to 20p12 by in situ hybridization. Cytogenet Cell Genet 56:169–170

Rosenberg CL, Kin HG, Shows TB, Kronenberg HM, Arnold A (1991) Rearrangement and overexpression of D11S287E, a candidate oncogene on chromosome 11q13 in benign parathyroid tumours. Oncogene 6:449–453

Sambrook J, Fritsch EF, Maniatis T (ed) (1989) Molecular cloning. 2nd edition, Cold Spring Harbor Laboratory Press, New York

Sartor H, Ehlert F, Grzeschik K-H, Muller R, Adolph S (1992) Assignment of two human cell cycle genes CDC25C and CCNB1 to 5q31 and 5q12 respectively. Genomics 13:911–912

Schwiezer D (1976) Reverse fluorescent chromosome banding with chromomycin and DAPI. Chromosoma 58:307–324

Solomon E, Borrow J, Goddard AD (1991) Chromosome aberrations and cancer. Science 254:1153–1160

Stilgenbauer S, Dohner H, Bulgay-Morschel M, Weitz S, Bentz B, Lichter P (1993) High frequency of monoallelic retinoblastoma gene deletion in B-cell chronic lymphoid leukemia shown by interphase cytogenetics. Blood 81:2118–2124

Testa JR (1990) Chromosome translocations in human cancer. Cell Growth Differ 1:97–101

Thompson CT, LeBoit PE, Nederlof PM, Gray JW (1994) Thick-section fluorescence in situ hybridization on formalin-fixed, paraffin-embedded archival tissue provides a histogenetic profile. Am J Pathol 144:237–243

Tkachuk DC, Westbrook CA, Andreef M, Donlon TA, Cleary ML, Suryanarayan D, Homge M, Redner A, Gray J, Pinkel D (1990) Detection of bcr-able fusion in chronic myelogenous leukemia by in situ hybridization. Science 250:559–562

Trask BJ (1991a) DNA sequence localization in metaphase and interphase cells by fluorescence in situ hybridization. Methods Cell Biol 35:3–35

Trask BJ (1991b) Fluorescence in situ hybridization. Trends Genet 7:149–154

Viegas-Pequignot E, Dutrillaux B, Magdelenat M, Coppey-Moisan M (1989) Mapping of single copy DNA sequences on human chromosomes by in situ hybridization with biotinylated probes: enhancement of detection sensitivity by intensified-fluorescence digital-imaging microscopy. Proc Natl Acad Sci USA 86:582–586

Webb G, Parsons P, Chenevix-Trench G (1990) Localization of the gene for human proliferating cell nuclear antigen/cyclin by in situ hybridization. Hum Genet 86:84–86

Xiong Y, Menninger J, Beach D, Ward DC (1992) Molecular cloning and chromosomal mapping of CCND genes encoding human D-type cyclins. Genomics 13:575–584

Yeung RS, Bell DW, Testa JR, Mayol X, Baldi A, Grana X, Klinga-Levan K, Knudson AG, Giordano A (1993) The retinoblastoma-related gene, RB2, maps to human chromosome 16q12 and rat chromosome 19. Oncogene 8:3465–3468

Yunis JJ (1976) High resolution of human chromosomes. Science 191:1268–1270

4 Techniques for Monitoring Cell Cycle Phases

MARINA GIUNTA AND CARLO PUCILLO

1 Background

The cell division cycle is a coordinate and highly regulated series of events during which cells replicate their DNA content and then divide. Formally, the cell cycle has been divided in four phases: G1 (gap1 or presynthetic phase), S (synthetic phase), G2 (gap2 or postsynthetic phase) and M (mitosis). At the end of the mitotic events cells will choose between arrest (entering into the G0 phase) or go through a new cell cycle. Each of these stages is characterized by a typical content of DNA: in G_0 and in G_1 phases cells have a diploid (2N) content of DNA, in G_2 and M, they have a 4N DNA content, and in S phase they display an intermediate content of DNA.

2 Principle and Applications

The analysis of cellular DNA content and the possibility of monitoring DNA synthesis provide information about the grade of synchrony and the proliferative state of a cell population and about cell responsiveness to various stimuli. Analysis of the ploidy of cancerous cells, determining the DNA index (defined as the ratio between the DNA content of a malignant cell and that of a normal diploid cell), complements other prognostic factors such as histologic grade, tumor stage, etc. (Dressler et al. 1989). For these reasons, techniques monitoring the percentage of cells in the G1, S, and G2/M phases are not only essential in the study of the cell cycle, but they are also an important tool in many other fields including clinical investigation.

To monitor cell cycle phases, several quantitative methods have been developed. This chapter will discuss the basic procedures highlighting their practical applications. Some of the described protocols require the use of a flow cytometer. Although it is unlikely that a molecular biology lab has such a piece of equipment, most labs will have access to it within their department or university.

3 Methods

3.1 [^{3}H]-Thymidine Uptake

Procedure 1

Progression through the S phase of a synchronously growing cell population can be monitored using the following protocol:

1. Grow cells in several 100-mm petri dishes.
2. Synchronize the cells with the chosen method (see chapter by P. O'Connor and J. Jackman, Part One).
3. Add 20 µCi of [methyl, 1, 2, ^{3}H]-thymidine (111 Ci/mmol; Amersham #TRK 565 B62) to one 100-mm petri dish containing the cells synchronously progressing through the cell cycle and 10 ml of medium.
4. Incubate for 1 h at 37 °C.
5. Rinse the cells in PBS and harvest them.
6. Pellet and freeze.
7. Sample all the cells repeating these first steps at different times (e.g., every 4 h) during a period of 24 to 36 h.
8. Take all samples.
9. Add 0.5 ml of 0.3 M NaOH to the cell pellets, vortex, and leave on ice for 15 min.
10. Add an equal volume of 20% trichloracetic acid (TCA); vortex and leave on ice for 15 min.
11. Filter the extract through Whatman GF/A glass fiber filters and rinse with 5–10 ml cold 10% TCA (4 °C) and 5 ml 95% ethanol (–20 °C).
12. Dry the filters and quantitate by liquid scintillation counting.

Procedure 2

This protocol can be used to monitor the proliferative state of a cell population (Strong et al. 1973):

1. Dilute 1 mCi of [^{3}H]-methyl-thymidine (6.7 Ci/mmol; Du Pont NEN #NET-027 or ICN Biomedicals #24066 or Amersham #TRA.120) in 25 ml of sterile serum-free medium containing 10 mM HEPES (pH 7.4).
2. Using an Eppendorf repeater pipette, dispense 25 µl of diluted [^{3}H]-thymidine in each well of a microtiter plate in which cells have been cultured.
3. Incubate the cells at 37 °C for 6–18 additional h.

4. Harvest the cells using a semiautomatic multi-well harvester (Pharmacia LKB Biotechnology, Beckman). The harvester will transfer and lyse the cells and will then automatically wash the filter several time in order to remove the free [^{3}H]-thymidine.

5. Dry the filter using a microwave oven, transfer each single dot corresponding to a single well into a vial and then add scintillation fluid. Follow the recommendation of the harvester's manufacturer in choosing the number of washing cycle and the type and size of filter that should be used with a given machine.

Note ! It is always a good suggestion to perform the experiment in triplicate. This will allow identification of technical problems. Clumps, pipetting and dilution error can produce different results. Bacterial contamination can interfere with proliferation and the subsequent incorporation of [^{3}H]-thymidine. When differences in the triplicate values are higher than 10%, one of above problems should be considered.

3.2 Analysis of Cellular DNA Content Using Propidium Iodide

Propidium Iodide (PI) binds stoichiometrically to double stranded DNA. Under binding conditions, its fluorescence increases roughly 50-fold. This property, together with the simplicity of its use, makes this fluorochrome suitable for monitoring the percentage of cells in the G1, S, and G2/M phases of the cell cycle on the basis of the cellular DNA content (Pucillo et al. 1990). Furthermore, the PI staining can be combined with the immunostaining of cellular antigens in order to analyze the expression of a specific proteins during the cell cycle (Brons et al. 1990; Jianping et al. 1993; Bartkova et al. 1994; Gong et al. 1994). Finally, double staining with PI and antibody to Ki-67, a protein present only in cycling cells, allow a finer distinction between G0 and G1 cells (Schlüter et al. 1993).

Analysis of DNA

1. Wash the cells twice with PBS.

2. Incubate the cells with trypsin for 2–3 min (if cells are growing in suspension, this step is not necessary).

3. Resuspend carefully the cells and wash three times in PBS w/o Ca^{2+} and Mg^{2+}, avoiding the formation of clumps. If clumps are present, they can be removed by passing the cells through a 25-gauge needle or through a nylon filter.

4. Spin 5 min at 1000 rpm and remove the supernatant.

5. Vigorously add 1 ml of cold ethanol for each 1–2 × 10^6 cells/ml and leave the sample at 4 °C overnight (the overnight fixation allows a better resolution of DNA content). Cells may remain in fixative at 4 °C up to 3 weeks before stain procedure.

6. Centrifuge, discard the supernatant, and resuspend the pellet in 1 ml of PI staining solution (50 µg/ml of PI, 0.1 mg/ml of RNAse A in 0.1% NP40 and 0.1% Trisodium Citrate).

7. Incubate for 20 min at 4 °C to allow the RNAse to work (the RNAse degrades the double-stranded RNA, so that only the double-stranded DNA will be stained by PI). Samples can be stored at 4 °C in the dark.

8. Analyze the sample with a flow cytometer using a laser adjusted to emit 500 mW at 488 nm. Measure the logarithmic red PI fluorescence with a 600-nm long-pass filter.

3.3 DNA Synthesis Analysis Using 5-Bromodeoxyuridine Incorporation

5-bromodeoxyuridine (BrdU) is a thymidine analogue and is incorporated in the DNA during DNA synthesis, either replicative or repair (Giaretti et al. 1989; Lu and Lane 1993). The availability of monoclonal antibody specifically recognizing BrdU allows the identification of both cells in S phase and cells repairing their DNA (Pagano et al. 1994). The intensity of the nuclear fluorescence staining is proportional to the amount of BrdU that has been incorporated (Dolbeare et al. 1985) and varies according to the position of the cell in S phase (Fox et al. 1991). Since the antibody is able to recognize BrdU only in single stranded DNA, partial denaturation of the DNA is requested. The remaining undenatured DNA is stained with propidium iodine and gives a measure of the total DNA (Dolbeare et al. 1983). Several methods have been developed for DNA denaturation: high or low pH, high temperature, low ionic strength, etc. Each method presents some advantages and disadvantages. The procedure based on acidic denaturation is relatively mild, but low levels of incorporation of BrdU cannot be detected. In contrast, thermal denaturation increases the sensitivity for BrdU detection but has the highest amount of cell loss (up to 90% when using hematopoietic cells). In an experimental system where the number of cells is not a limiting factor, the thermal denaturation method should be preferred. For this protocol the use of a flow cytometer is requested (Ronot 1986; Jayat and Ratinaud 1993). For a protocol using indirect immunofluorescence, see the chapter by C. Cardoso and H. Leonhart (Part One).

Labeling

1. Freshly prepare a solution of BrdU in sterile distilled H_2O and protect it from direct light.

2. Label the cells at the density of 1–2 × $10^3/cm^2$ using complete culture medium supplemented with 30 μM BrdU (SIGMA Chemical Co.), 30 μM 2-deoxycytidine (SIGMA Chemical Co.) and 10 μM of 5-fluorodeoxyuridine (SIGMA Chemical Co.) at 37 °C and for the time requested by the experimental conditions (from few minutes to 24 h or more). The 5-fluorodeoxyuridine is added to inhibit the synthesis of thymidylate and 2-deoxycytidine is added to moderate the toxic effects of BrdU and 5-fluorodeoxyuridine.
3. After labeling, trypsinize the cells (unless they do not grow in suspension) and transfer them to a 15-ml conical tube.
4. Wash three times with cold PBS (between each wash, centrifuge at 500 *g* for 1 min at 4 °C).
5. Discard the supernatant, add 5 ml of 50% ethanol in PBS and vortex.

Thermal denaturation

1. Wash and resuspend 2 × 10^6 fixed cells in 1.5 ml of PBS containing 0.5 mg/ml of RNAse (SIGMA Chemical Co.).
2. Incubate 30 min at 37 °C.
3. Centrifuge at 500 g for 1 min, vortex and resuspend the pellet in 3 ml of 0.5% paraformaldehyde solution.
4. Incubate for 30 min at room temperature.
5. Centrifuge, discard the supernatant and resuspend the pellet, gently vortexing, with 1.5 ml 0.1 M HCl, 0.5% Triton-X100.
6. Incubate for 10 min at 4 °C.
7. Centrifuge, discard the supernatant and resuspend the pellet in H_2O.
8. Incubate at 95 °C for 10 min.
9. Stop the denaturation by removing from the water bath and placing the sample on ice. The time of this step should be optimized empirically for each cell type in order to avoid a large lost of cells.
10. Wash once with 3 ml of cold PBS.
11. Remove clumps by dispersing with a pipette and then filtering through nylon.

Acidic denaturation

The acidic denaturation protocol is similar with the difference that the concentration of HCl will be higher (1.5 M) and no incubation at 95 °C will be performed.

Immunostaining

1. Resuspend the cells in 50 µl of anti-BrdU monoclonal antibody (Becton Dickinson, Caltag, Dako) diluted (according to the manufacturer's directions) with 1% BSA (or 1% gelatin) and 0.5% Tween-20 in PBS.
2. Incubate for 30 min at 4 °C.
3. Wash three times with 0.5% Tween-20 in PBS.
4. Drain the pellet and add the FITC-conjugated secondary antibody (Becton Dickinson, Cappel etc.) for 30 min at 4 °C.
5. Wash three times and add 1.5 ml of PI solution (see Sect. 3.2).
6. Analyze with a flow cytometer using a laser adjusted to emit 500 mW at 488 nm. Measure the logarithmic green BrdU fluorescence using a 514-nm bandpass filter and the logarithmic red PI fluorescence using a 600 nm long-pass filter.

References

Bartkova J, Lukas J, Strauss M, Bartek J (1994) Cell cycle-related variation and tissue-restricted expression of cyclin D1 protein. J Pathol, 172:237–245

Brons PP, Pennings AH, Haanen C, Wessels HM, Boezman JB (1990) Simultaneous measurement of DNA content and cell surface immunofluorescence of human bone-marrow cell using a single laser cytometer. Cytometry 11:837–844

Dolbeare F, Beisker W, Pallavicini MG, Vanderlaan M, Gray JW (1985) Cytochemistry for bromodeoxyuridine/DNA analysis: stoichiometry and sensitivity. Cytometry 6:521–530

Dolbeare F, Gratzner H, Pallavicini MG, Gray JW (1983) Flow cytometry measurement of total DNA content and incorporated bromodeoxyuridine. Proc Natl Acad USA 88: 5573–5577

Dressler LG, Bartow MA, Bartow SA (1989) DNA flow cytometry in solid tumors: practical aspects and clinical applications. Semin Diagnostic Pathol 6:55–82

Fox MHT, Arndt-Jovin MP, Baumann H, Robert-Nicoud M (1991) Spatial and temporal distribution of DNA replication sites localized by immunofluorescence and confocal microscopy in mouse fibroblasts. J Cell Sci 99: 247–253

Giaretti W, Nusse M, Brno S, Di Vinci A, Geido E (1989) A new method to discriminate G1, S, G2, M and G1 postmitotic cells. Exp Cell Res 182:290–295

Gong J, Traganos F, Darzynkiewicz Z (1994) Staurosporine blocks cell progression through G1 between the Cyclin D1 and Cyclin E restriction point. Cancer Res 54: 3136–3139

Jayat C, Ratinaud MH (1993) Cell Cycle analysis by flow cytometry: principles and applications. Biol Cell 78:15–25

Jianping G, Traganos F, Darzynkiewicz Z (1993) Expression of Cyclins B and E in individual MOLT-4 cells and in stimulated human lymphocytes during their progression through the cell cycle. Int J Oncol 3:1037–1042

Lu X, Lane PD (1993) Differential induction of transcriptionally active p53 following UV or ionizing radiation: defects in chromosome instability syndromes? Cell 75:765–778

Pagano M, Theodoras AM, Tam SW, Draetta G (1994) Cyclin D1-mediated inhibition of repair and replicative DNA synthesis in human fibroblasts. Genes Dev 8:1627–1639

Pucillo C, Salzano S, Pepe S, Vitale M, Formisano S, Rossi G (1990) Regulation of expression of low-affinity IgE receptor (FceRII) in the human monocyte-like cell line (U937) by phorbol esters and TPA. Int Arch Allergy Appl Immunol 93:330–337

Ronot X (1986) G2 arrest, binucleation, and single parameter DNA flow cytometer analysis. Cytometry 7:286–290

Schlüter C, Duchrow M, Wohlenberg C, Becker M, Key G, Flad H, Gerdes J (1993) The cell proliferation-associated antigen of antibody Ki-67: a very large, ubiquitous nuclear protein with numerous repeated elements, representing a new kind of cell cycle-maintaining proteins. J Cell Biol 123:513–522

Strong DM, Ahmed AA, Thurman GB, Sell KW (1973) In vitro stimulation of murine spleen cells using a microculture system and a multiple automated sample harvester. J Immunol Methods 2:279–287

5 Centrifugal Elutriation

ROBERT L. MARRACCINO AND PETER C. KENG

1 Background

Centrifugal elutriation is a velocity sedimentation method that separates cells on the basis of size, shape, and density. Lindahl first used this technique to separate cell subpopulations from a variety of biological systems (Lindahl 1956). Since cell volumes increase from G1 through G2/M phases of the cell cycle (Anderson and Petersen 1964; Steen and Lindmo 1978), synchronous mammalian cell populations in each phase of the cell cycle can be isolated by this separation method (Meistrich et al. 1977a, b; Keng et al. 1980). Several advantages are provided by centrifugal elutriation in obtaining synchronous cells. Using approximately 10^9 cells, between 10^7 and 10^8 synchronized G1, S and G2/M cells can be obtained by this method in a short period of time (less than 1 h). There is a high degree of homogeneity in the cell populations obtained by this method, usually 90–95% for G1 phase cells, 75–85% for S phase cells, and 60–75% for G2/M phase cells (Liu et al. 1989). Finally, centrifugal elutriation does not require an arrest of the cell cycle to obtain synchronized cells, and the entire separation procedure is performed in tissue culture medium. Since there is little or no perturbation to cell growth, this method can be used to establish the temporal sequence of cell cycle events. The utility of this method can be exemplified in the study of the roles of cyclin A (Marraccino et al. 1992; Pagano et al. 1992) and cyclin E (Koff et al. 1992) in the human cell cycle.

A number of other procedures are developed to obtain synchronous populations of mammalian cells, but involve specific cell cycle blocking agents or nutritional deprivation (Bhuyan and Groppi 1989; LaLande 1990). These methods which arrest the cell cycle are subject to the criticism that the synchronization process may introduce distortions into the normal sequence of events that occur during the cell cycle. Mitotic selection is another synchronization method that is often used, and does not perturb the cell cycle (Terasima and Tolmach 1963). This method is applicable only to cells that grow in monolayer culture. Upon entering mitosis, cells are not as firmly attached to the substrate, and can be collected in the growth medium after agitation. However, the yield of cells obtained from mitotic selection is frequently low, particularly for cells with a doubling time longer than 24 h. Cells that are tightly adherent to the surface of the culture vessel or to each other, and cells that grow in suspension culture cannot be synchronized by mitotic selection.

2 Principle and Applications

The principle of cell separation by centrifugal elutriation has been described in detail (Lindahl 1956; Sanderson et al. 1976; Meistrich et al. 1977; Pretlow and Pretlow 1979; Keng et al. 1981). Briefly, cells suspended in the separation chamber of the elutriator system are subjected to two opposite forces: the centrifugal force generated by the rotation of the rotor in an outward direction and the fluid force pumped into the separation chamber in an inward or centripetal direction. The special shape of the separation chamber allows a gradient of flow velocity to establish in the chamber. This gradient decreases toward the center of rotation. Cells are sedimented to a position in the separation chamber according to their sedimentation velocity that is determined by the size, density, and shape of the cell. The cells remain in the chamber as long as the two opposite forces are in equilibrium. By increasing the flow rate of the fluid or by decreasing the centrifugal force, homogeneous populations of cells with a specific size are eluted out of the rotor. Since each cell must sediment to a position in the separation chamber, any clumping of the cells will adversely affect the homogeneity of the eluted populations.

In order to obtain the optimum conditions for separation experiments, the relationship between the elutriated cell volume (V; μM^3) and the collection rotor angular velocity (ω) or the fluid flow velocity (v) must be known. This can be established by equilibrating the three forces; Fc (centrifugal force), Fb (buoyant force), and Fd (fluid drag force); on cells suspended in the separation chamber (Keng et al. 1981).

$$Fc = (\pi/6)d^3\rho\omega^2R \quad (1)$$

$$Fb = -(\pi/6)d^3\rho_o\omega^2R \quad (2)$$

$$Fd = -3\pi\eta dv, \quad (3)$$

where d is the diameter of the cell, ω is the rotor angular velocity, R is the distance from the cell to the center of the rotor, ρ_o is the fluid density, ρ is the cell density, η is the fluid viscosity, and v is the local fluid velocity. At an equilibrium point when the net force on a cell is zero, the cell stays at a fixed position in the chamber. This is represented by the following equation.

$$Fc + Fb + Fd = 0. \quad (4)$$

By rearranging the parameters in Eq. (4), the relationships between the size of the cell and the rotor speed or flow rate used to maintain the cell's position in the chamber are represented in Eqs. (5) and (6).

$$V = k\,\omega^{-3}. \quad (5)$$

k is an "elutriation constant" that relates the rotor speed to the cell volume (V) when a specific flow rate is used for cell separation.

$$V = k' \, v^{1.5}. \qquad (6)$$

k' is a different "elutriation constant" that relates the flow rate to the cell volume when a given rotor speed is used for cell separation.

Homogeneous populations of cells with different sizes are eluted out of the separation chamber when the equilibrium is interrupted by increasing the fluid force (flow rate) or decreasing the centrifugal force (rotor speed). The relationships between the cell volume and rotor speed or flow rate are shown in Fig. 1. The elutriation constant k shown in Fig. 1 is pertinent to most mammalian cells. These constants are useful in determining the specific rotor speed and flow rate for cell synchronization experiments. For example, S-phase CHO cells with a median cell volume of 950 μM^3 can be isolated at a rotor speed of 2690 rpm and a flow rate of 30 ml/min as calculated from the constant $k = 1.85 \times 10^{13}$ μM^3 rpm^3. The constant for the large capacity chamber (50 ml) is different and the value is $k = 8.7 \times 10^{11}$ μM^3 rpm^3 at a flow rate of 50 ml/min.

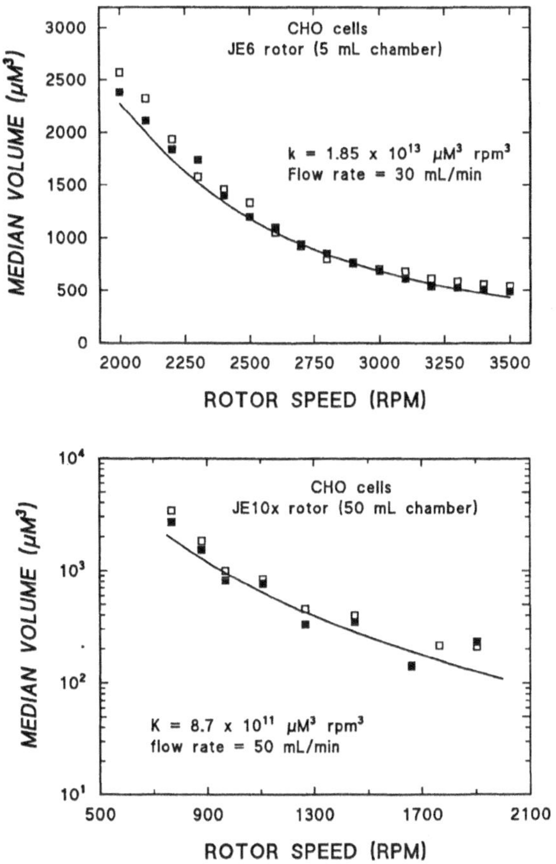

Fig. 1. The median volume of elutriated, CHO cells as a function of rotor speed. An asynchronous CHO cell population is separated as described in Table 1. Two different Beckman rotors are used in separate elutriations: the JE6 rotor (top panel) and JE10x rotor (bottom panel). At each specific rotor speed, two 40-ml fractions are collected. The median cell volume in each of these two fractions (□, ■) is determined with a Coulter Counter and Channelyzer system, and graphed. The solid lines in both graphs represent the theoretical calculations using the elutriation constant, k, that is derived from empirical data and values calculated from Eq. 5. There is a significant correlation between the theoretical relationship expressed in Eq. 5 and the empirical data, but there are slight deviations at the extremes of rotor speed

3 Materials and Methods

A step-by-step procedure for isolating G1, S, and G2/M cells from exponentially growing CHO cells is provided as an example in this chapter. Procedures for other mammalian cells are similar to those of CHO cells except that the relationship between cell volume and cell cycle position may be different for other cell lines.

3.1 Equipment

Centrifuge

Cell separation is performed with the Beckman JE-6 (5-ml chamber) or JE-10× (50-ml chamber) rotor driven by a Beckman J6 M centrifuge. The rotor speed is controlled by a digital board (Beckman Instrument Inc., Palo Alto, CA) to allow the speed to be accurate within ±10 revolutions per minute (rpm). Rotor speed is measured with an electronic detector connected to the driver motor. Fluid flow force is generated by a Multiperpex pump equipped with a fine velocity control (LKB Instruments, Baltimore, MD). The flow rate is continuously monitored with an in-line flow meter (VWR Scientific Co., Rochester NY). The flow rate during the separation process is maintained within ±0.1 ml/min.

Rotor Assembly: Several types of rotor are available for the Beckman elutriator system. A slightly different procedure is required for each rotor. Make certain that all the connections, O-rings, and inlet-outlet tubings are installed and aligned correctly. Any failure in rotor assembly can result in fluid leaks that may lead to a variety of problems in cell separation.

Check for Fluid Leaking: Prior to elutriation, the rotor is checked for any external or internal leakage caused by defective seals or O-rings. No experiment is performed until any leakage problems are corrected.

Rotor check

1. A 50-ml syringe filled with 30–40 ml distilled water is used for this procedure.
2. The separation chamber is removed during this procedure to expose the loading conduit (the off-center hole on the shaft) and exit conduit (center hole on the shaft). Please refer to the instruction manual for details.
3. The syringe is connected to the inlet feed fitting on the top of the seal housing with a silicon rubber tubing. After the water is forced through the rotor by pressing the plunger of the syringe, the only place that water should exit is the loading conduit.

4. If water is seen coming out of the outlet feed fitting of the seal housing (top of the rotor) or out of the exit conduit (center hole on the shaft), either the rotor is incorrectly assembled or the rubber O-rings and seals are worn out and need to be replaced.

5. Reinstall the separation chamber after the rotor passes the leakage test. Reconnect the syringe to the inlet feed fitting and fill the separation chamber with distilled water.

6. Block the outlet feed fitting with clamped tubing.

7. Apply approximately 15 psi pressure through the syringe plunger and determine that water is not leaking from the rotor.

Elutriator system

Elutriator System Assembly. The entire elutriator system consists of a fluid reservoir, a fluid flow meter, a loading tube, an LKB 2115 Multiperpex peristaltic pump, a pressure gauge, a bubble trap chamber, an elutriator rotor, and fifty-sixty conical centrifuge tubes (50-ml) for sample collection. The sequence of assembly of these components is shown in Fig. 2. The fluid reservoir (2 l) is placed inside an ice bucket to keep the elutriation fluid (tissue culture medium with 1–10% serum) at 0–4 °C during the separation. The percentage of serum in the elutriation fluid is determined by the nutri-

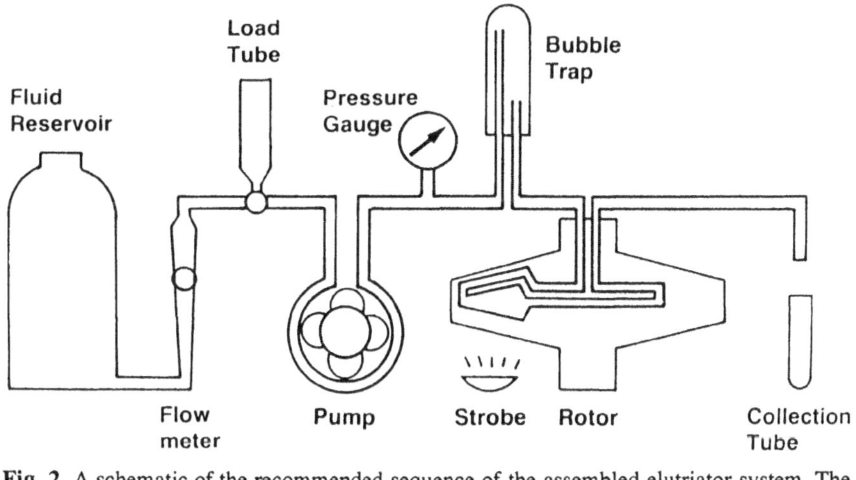

Fig. 2. A schematic of the recommended sequence of the assembled elutriator system. The cell sample passes through the peristaltic pump. An in-line flow meter and pressure gauge are necessary for the monitoring conditions throughout the elutriation. The use of three-way valves (not depicted) are attached to the loading tube and bubble trap, and control the flow of media and cells through the system. When loading cells, the valves are turned so that the cells are pumped out of the loading tube, and directly into the elutriator rotor. When nearly all of the cell suspension is pumped, the valves are set so that the flow of media is directed from the fluid reservoir, through the bubble trap and into the rotor, and the loading tube is bypassed. Extreme care is required to avoid the injection of bubbles into the system especially during the loading procedure

tional requirements of the cells and economical considerations. The serum in the fluid that is routinely used in the elutriation of mammalian cells ranges between 1–10%, but fluid with 10% serum is used throughout the separation of CHO cells as described below. A ring stand with three clamps is used to secure the positions of flow meter, loading tube, and bubble trap chamber.

When the separated G1, S, and G2/M cells are used for experiments that require aseptic conditions such as cell survival analysis, the entire elutriator system should be sterilized within 12–24 h. prior to the separation experiment to avoid bacteria and yeast contamination.

Sterilization of elutriator system

1. At the end of each separation experiment, the entire elutriator system is flushed with 500–750 ml of distilled water using the LKB peristaltic pump.
2. Afterward, the elutriator system is disassembled and the fluidics are divided into two groups. The first group includes the fluid reservoir, flow meter, and loading connector with silicon rubber tubing. The second group is the bubble trap collection bell with its connecting tubing.
3. The fluidics are wrapped with aluminum foil and sterilized by a steam autoclave with a dry cycle for at least 35 min.
4. The elutriator system is assembled with the sterilized fluidics. Make certain that every connector is sprayed with 75% ethanol to prevent contamination during the assembly.
5. The inlet and outlet feed fitting of the elutriator rotor are also sterilized with 75% ethanol before connecting them to the loading and collecting tubing.
6. Once the system is completely assembled, the fluid reservoir is filled with 500 ml 75% ethanol.
7. A total of 300 ml 75% ethanol is pumped through the system at a flow rate of 30 ml/min. During this procedure, the air bubble trap chamber is turned upside down several times until the entire chamber is filled with alcohol.
8. The rotor is turned by hand to remove bubbles trapped in the separation chamber.
9. The entire system is filled with 75% ethanol and allowed to sit overnight.
10. At 1 h before the separation experiment, the pump is turned on to remove the ethanol from the system.
11. After the ethanol is completely removed from the reservoir, the remaining ethanol in the system is washed off by 500 ml sterile distilled water. Make certain that no air bubbles enter the system during this procedure.
12. After the sterile distilled water is pumped through the system, the fluid reservoir is filled with tissue culture medium for cell separation.

3.2 Cells and Solutions

Cells

- Chinese hamster ovary (CHO) cells are maintained in culture with F-10 medium supplemented with 10% (v/v) fetal bovine serum (Grand Island Biological Co., Grand Island, NY) as described previously (Liu et al. 1989).
- Cells that are growing exponentially are removed from flasks by trypsin and resuspended in 20 ml (5-ml separation chamber) or 100 ml (50-ml separation chamber) ice-cold F-10 tissue culture medium before elutriation.
- The optimum number of cells for separation with the 5-ml chamber is 4–6 $\times 10^7$ cells (8–12 75 cm^2 flasks). The optimum cell number for the 50-ml separation chamber is 4–6 $\times 10^8$ cells.

Medium

- Three to five liters of tissue culture medium with 10% serum is required for a complete separation (G1, S, and G2/M cells).

3.3 Collection procedure

The selection of optimum rotor speed and flow rate for isolating synchronous G1, S, and G2/M CHO cells is determined by Eqs. (5) and (6). The cell volume distribution of a CHO population in exponential growth is shown in Fig. 3. The median cell volume of unseparated cells is 870 μM^3, and median cell volumes of separated G1, S, and G2/M cells are 620 μM^3, 950 μM^3, and 1260 μM^3, respectively (Fig. 3). The corresponding rotor speeds for

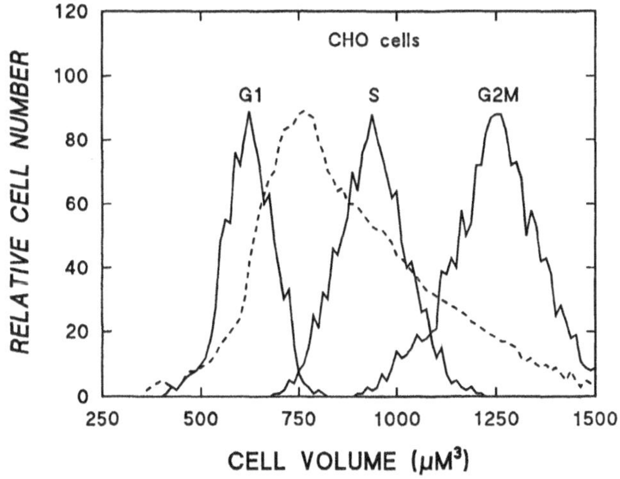

Fig. 3. Volume distribution of CHO cells as determined with a Coulter Counter and Channelyzer system. The frequency distributions of cell volumes are graphed in an asynchronous, exponential CHO population (-----) and separated CHO cells in G1, S, and G2/M (—) after elutriation which is described in Table 1

Table 1. Elutriation protocols for separating CHO cells

		JE6 rotor (5-ml chamber) $k = 1.85 \times 10^{13}$ μM³ rpm³ Flow rate = 30 ml/min	JE10X rotor (50-ml chamber) $k = 8.71 \times 10^{11}$ μM³ rpm³ Flow rate = 50 ml/min
Fraction Number		Rotor speed (rpm)	Rotor speed (rpm)
1		3300	1203
2		3230	1165
3	G1	3136	1132
4	G1	3053	1102
5	G1	2979	1075
6		2911	1051
7		2849	1029
8		2792	1008
9	S	2739	989
10	S	2690	971
11	S	2645	955
12		2602	939
13		2562	925
14		2524	911
15	G2/M	2489	898
16	G2/M	2455	886
17	G2/M	2423	875
18		2393	864
19		2364	853
20		0	0

collecting these separated cells are 3053 rpm, 2690 rpm, 2455 rpm, at a flow rate of 30 ml/min ($k = 1.85 \times 10^{13}$ μM³ rpm³). However, the removal of cells from the separation chamber is a step-by-step process with the smallest (G1 phase) cells coming out of the rotor first, followed by medium (S phase) and large (G2/M phase) sized cells. To isolate very homogeneous cell populations (size or cell cycle position), a procedure with a gradual change in rotor speed or flow rate is required (Keng et al. 1981). Table 1 shows the separation procedure routinely used to collect synchronous CHO cells.

Loading and collection

1. The CHO cells are elutriated in ice-cold, complete tissue culture medium with serum (10%) that is chilled in an ice bath.
2. Bring the rotor speed and flow rate to the initial loading condition as shown in Table 1. The initial condition is chosen so that the resulting centrifugal and fluid forces retain the cells in the separation chamber.
3. Load the sample with the peristaltic pump and do not allow the sample to enter the air bubble trap.

4. Immediately after the loading, two 40-ml fractions (5-ml separation chamber) or four 40-ml fractions (50-ml separation chamber) of eluted fluid are collected at each rotor speed. The rotor speed is selected so that eluted cells have a constant increase in volume at each collection [Eq. (5) or (6)].
5. During the separation procedure, the cell number and cell size distribution of each separated fraction are analyzed using a Coulter Counter and Channelyzer system.
6. For each fraction, the median cell volume is determined from the volume distribution with a calibration constant. The calibration constant is derived from latex spheres of uniform size and density.
7. Cells at different phases of the cell cycle are selected initially based on their median cell volume for further cell cycle studies. The homogeneity of each fraction is confirmed by flow cytometry.

Flow cytometry

The percentage of G1, S, or G2/M phase cells in separated fractions is determined from the DNA histograms measured by flow cytometry.

1. Cells are first removed from the medium by gentle centrifugation (150 *g*).
2. Remove the supernatant and fix the cells (10^6) in 75% ethanol overnight 3 ml.
3. Prior to flow cytometric analysis, cells are washed in PBS to remove ethanol, treated with RNAse (1 mg/ml) at room temperature for 30 min (10^6 cells per ml), and then stained with propidium iodide (10 mg/ml) for at least 15 min (10^6 cells per ml).
4. A flow cytometer that is equipped with an argon laser operated at an excitation wavelength of 488 nm is used to monitor cellular DNA content. The fraction of cells in the G1, S, and G2/M phases of the cell cycle is determined by computer analysis of the DNA histograms using mathematical models described by Dean and Jett (1974) and Fried (1976).

References

Anderson EC, Petersen DF (1964) Synchronized mammalian cells: an experimental model of synchrony decay. Exp Cell Res 36:423–426.

Bhuyan BD, Groppi VE (1989) A critical evaluation of methods using cell cycle specific inhibitors. Pharmac Ther. 42:307–348

Dean P, Jett JH (1974) Mathematical analysis of DNA distributions derived from microfluorometry. J. Cell Biol 60:523–527

Fried J (1976) Method for quantitative evaluation of data from flow microfluorometry. Comput Biomed Res 9:263–276

Keng PC, Li CKN, Wheeler, KT (1980) Synchronization of 9L rat brain tumor cells by centrifugal elutriation. Cell Biophys 2:191–206

Keng PC, Li CKN, Wheeler KT (1981) The characterization of the separation properties of the Beckman elutriator system. Cell Biophys 3:31–56

Koff A, Giordano A, Desai D, Yamashita K, Harper JW, Elledge S, Nishimoto T, Morgan D, Franza BR, Roberts JM (1992) The formation and activation of a cyclin E-Cdk2 complex during the G1 phase of the cell cycle. Science 257:1689–1694

LaLande M, (1990) Mimosine reversibly blocks the mammalian cell cycle in late G1 phase. Exp Cell Res 186:332–339

Lindahl PE (1956) On counter-streaming centrifugation in the separation of cells and cell fragments. Biochem Biophys Acta 21:411–415

Liu Y, Marraccino RL, Keng PC, Bambara RA, Lord EM, Chou W, Zain, SB (1989) The requirement for proliferating cell nuclear antigen (PCNA) expression during the cell cycle. Biochem 28:2967–2974

Marraccino, RL, Firpo EJ, Roberts JM (1992) Activation of p34/Cdc2 protein kinase at the start of S phase in the human cell cycle. Mol Biol Cell 3:389–401

Meistrich, ML, Meyn RE, Barlogie B (1977b) Synchronization of mouse L-P59 cells by centrifugal elutriation. Exp Cell Res 105:169–177

Meistrich ML, Grdina KJ, Meyn RE, Barlogie B (1977a) Separation of cells from mouse solid tumors by centrifugal elutriation. Cancer Res 37:4291–4269

Pagano M, Pepperkok R, Verde F, Ansorge W, Draetta G. (1992) Cyclin A is required at two points in the human cell cycle. EMBO J 11:961–971

Pretlow TG., Pretlow TP (1979. Centrifugal elutriation of cells. Cell Biophys 1:195–210

Sanderson RJ, Bird KE, Palmer NF, Brenman J (1976) Design principles for a counter-streaming centrifugation cell separation chamber. Anal Biochem 71:615–622

Steen, HB, Lindmo T (1978) Cellular and nuclear volume distribution during the cell cycle of NHIK 3025 cells. Cell Tissue Kinet 11:69–81

Terasima T, Tolmach LJ (1963) Growth and nucleic acid synthesis in synchronous HeLa cell populations. Exp Cell Res 30:344–362

6 Synchronization of Mammalian Cells

PATRICK M. O'CONNOR AND JOANY JACKMAN

1 Background

An essential and probably primary task in the study of cell cycle related processes is the synchronization of cells into specific cell cycle phases. Cell synchronization improves conditions by which an actual process under scrutiny can be studied and helps clarify linkage of the process to a particular cell cycle phase transition. In this chapter we will provide methods that have proven effective in synchronizing mammalian cells into defined cell cycle phases. We will discuss the advantages and disadvantages of each procedure and point to some key situations where such procedures have been employed successfully in the biochemical dissection of cell cycle control processes. Our focus here will be on the use of chemicals or growth restriction conditions to synchronize exponentially growing cells. We will present methodology to arrest cells in G0/G1, G1/S, G2 and mitosis-*like* states (Fig. 1). We will discuss the use of (1) serum starvation, isoleucine-withdrawal and chemicals (lovastatin, mimosine) for arrest of cells in a G0-*like* or G1-*like* state, (2) S phase blocking agents (aphidicolin, hydroxyurea, thymidine), (3) DNA damaging agents to produce G1 and/or G2 arrested cells, and (4) nocodazole to arrest cells in a mitotic state. Alternative nonchemical strategies using centrifugal elutriation as a basis for cell synchronization are discussed in the chapter by Marracino and Keng (Part One). Monitoring procedures to validate the success and selectivity of the arrest procedure, and subsequent cell cycle progression of the synchronize cohort will be discussed in the chapter by Giunta and Pucillo (Part One). Taken together, these chapters will appraise the currently available techniques that are commonly used for mammalian cell synchronization and provide the reader with methodology applicable to the study of a wide variety of cell cycle processes.

2 Principle and Applications

The principle element of cell synchronization is to create a completely synchronized population of cells that progress uniformly through the cell cycle. Achieving this task requires (1) a complete arrest at one defined point in the cell cycle, (2) the arrest procedure be completely reversible, and (3) the re-

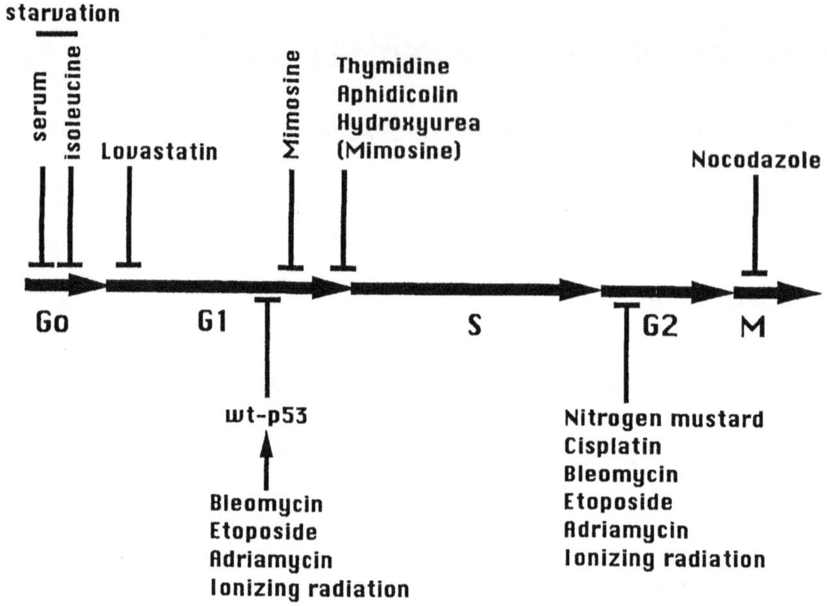

Fig. 1. Agents and conditions that arrest cell cycle progression. A variety of techniques can be used to synchronize cells into particular phases of the cell cycle. Reversible arrest of cells in G0 can be accomplished by serum or isoleucine starvation. Reversible arrest of cells in G1 can be accomplished using the chemical agents, lovastatin or mimosine, and G1/S phase synchronized cells can be obtained using DNA synthesis inhibitors, aphidicolin, hydroxyurea, or excess thymidine. The precise point of arrest for mimosine is still unclear. The microtubule inhibitor, nocodazole, can be used to synchronize cells in mitosis. Arrest of cell cycle progression in G1 and/or G2 of the cell cycle is dependent upon the DNA damaging agent employed and the status of the p53 tumor suppressor gene in the cells. G1 arrest can be accomplished in wild-type p53 but not mutant p53 cell lines with DNA double strand breaking agents (bleomycin, etoposide, adriamycin, or ionizing radiation). Arrest of cells in G2 of the cell cycle can be elicited by both DNA strand-breaking and DNA crosslinking agents

leased cells recover and progress uniformly through the cell cycle. In the practical sense, although the procedures currently available for cell synchronization aim towards these requirements, they do not entirely meet them. Problems arise in the degree of arrest obtainable and the uniformity with which the released population progresses through the cell cycle. For a cell line not yet characterized for synchronization one may have to experiment with several different procedures to obtain the most suitable technique.

Prior to selecting a method of synchrony there are several factors to consider: (1) which cell cycle phase will be the focus of your studies; (2) what is the level of synchronization required; and (3) what is the number of cells required for analysis. Most of the synchrony protocols discussed here are recommended for cells with doubling times less than 30 h. While all synchrony methods result in a population of cells enriched in a particular cell cycle phase, cell synchronization is transient producing synchronous populations for a single cycle. Generally, synchrony diminishes each time cells pass through G1 phase. Additionally, the tightness of the cycle block can influence the method of synchrony chosen. Some techniques lend themselves to the production of large volumes of synchronized cells but require special equipment or lengthy preparation. Other techniques produce only small numbers of synchronized cells but are both rapid and inexpensive. Chemicals which inhibit cell progression are advantageous for synchronization because they are effective in a large variety of cell types, require no special equipment and lend themselves to applications which require large numbers of cells. Indeed, chemical inhibitors are often the primary method of synchronizing cells at a particular stage of the cell cycle. The disadvantages of chemical synchronization are possible disruption of normal cell cycle regulatory processes (Schimke et al. 1991) and, for some chemicals like mimosine, the difficulty in elucidating the exact cellular target/s.

3 Methods

3.1 Synchronization of Cells in G0/G1 Phase

Two of the most commonly used methods for synchronizing cells in a G0/G1-like state are serum starvation and isoleucine deprivation. Both procedures are described below. Before using either technique, it is important to have determined both the doubling time of the exponentially growing culture and the confluent/maximum cell density the cells will achieve (cells/ml or cells/cm^2). The first method relies upon serum starving the cells (Campisi et al. 1984). It may take 24 to 48 h for an entire population to respond to serum withdrawal (Pardee 1989). Serum withdrawal requires optimization of both the amount of serum withdrawn and the length of withdrawal. Often withdrawal of serum for extended periods will actually reduce the synchronicity of the restimulated population. For some cell types, serum starvation is not a viable method of synchronization either because cells permanently arrest in G0, do not arrest, or undergo apoptosis. The method described below is for suspension cultures, although the procedure can be adapted to monolayer cultures.

Protocol 1, G0/G1

1. Exponentially growing cells are centrifuged (600 *g*, 5 min), washed twice in sterile-filtered phosphate buffered saline, pH 7.4 or medium lacking serum at 37 °C and then resuspended at 30% of their confluent density in medium that contains 0 to 0.5% serum depending on the cell type. For example, REF-52 cells will stably arrest in serum-free medium (Girard et al.

1991), while NIH-3T3 cells (Morla et al. 1989) can be rendered quiescent by culture in medium containing 0.5% serum.

2. Cells are serum starved for 24–48 h to sequester cells into a G0-like state. Cells which have passed the restriction point (Pardee 1974, 1989) will progress through the remaining cell cycle and divide. Upon reentry into G1 of the next cell cycle, the cells will ideally arrest.

3. Cells are stimulated to reenter cycle by resuspension in medium containing the appropriate amount of serum used for exponential growth conditions. Cells will progress into S phase approximately 12 h after the stimulation; however, this time may vary between 6 and 20 h depending on the cell type.

The procedure has proven applicable in many cell cycle related studies including analysis of the cell cycle dependent reversible tyrosine phosphorylation of Cdc2 (Morla et al. 1989) and regulation of Cyclin A (Girard et al. 1991). Such studies should prove useful for additional information.

Protocol 2, G0/G1 Preparation of a large number of synchronous G0-cells can also be obtained by amino acid starvation (Tobey and Crissman 1972). Omission of isoleucine, an essential amino acid from the media causes cells to sequester in G0/G1-like state. However, not all cell types will respond to this synchronization procedure. The method described below is for the human Burkitt's lymphoma cell line, CA46.

1. Exponentially growing cells are centrifuged (600 *g*, 5 min), the medium discarded, and the cells washed twice in sterile-filtered phosphate buffered saline, pH 7.4, or medium lacking serum and isoleucine at 37 °C.

2. Cells are then resuspended at 35–40% of their confluent density in isoleucine-free Minimal Essential Medium prepared with serum that has been dialyzed to remove any residual free isoleucine from the serum.
Medium packs in which individual lyophilized amino acids can be removed before rehydration are sold by several companies. We purchase our medium kits and dialyzed serum from Gibco Life Technologies Inc. Be careful to remove the isoleucine vial from the preparation area before reconstituting the medium.

3. Cells are cultured in the isoleucine-free medium for up to 30–42 h, or one to two cell cycles, which ever is less. Prolonged culture under these conditions will lead to increased toxicity and in some cases irreversibility of the arrest.

4. After incubation in isoleucine-free medium, the cells are induced to reenter the cell cycle by addition of fresh complete medium containing serum. Avoid rough or unnecessary handling of the cells following the star-

vation procedure, i.e., suspension cells should be pelleted at low speed (500 *g*, 5 min) prior to resuspension in the appropriate growth media as they may be more fragile than exponentially growing cells. Attached cells should be treated cautiously as they may not be as tightly attached to the plate, and therefore may be dislodged by the removal of the isoleucine-free medium or addition of the fresh complete medium.

In general, synchrony does not continue for more than one cell cycle with either of the two starvation methods described above. Depending on the cell type, cells will enter S phase as a parasynchronous wave within 8–12 h. Starvation methods are useful when combined with other types of synchronizing agents, in particular release from starvation into the presence of aphidicolin or mimosine for G1/S cells (Heintz et al. 1982; Burhans et al. 1990; Dijkwel and Hamlin 1992).

3.2 Synchronization of Cells in G1 Phase

There are two general reversible inhibitors of G1 progression, lovastatin and mimosine. Each of these agents causes cells to block reversibly at different stages of G1. The cellular target of the first inhibitor is known, but the direct target of the second is as yet unclear. The first inhibitor, lovastatin, causes cells to block very early in G1 rather than G0, as defined by the presence of the proliferation specific antigen Ki-67 (Keyomarsi et al. 1991). Lovastatin inhibits HMG-CoA reductase (3-hydroxy-3-methylglutaryl-coenzyme A reductase), resulting in the depletion of mevalonate an essential precursor of cholesterol synthesis and a requirement for isoprenylation of substrate molecules including $p21^{ras}$. Addition of lovastatin to cells results in the sequestration of cells primarily in G1, although in some cells types, there is also a partial arrest of cells in G2/M (unpubl. observ.). Described below is a general method by which lovastatin can be used to synchronize cells in G1 phase.

Protocol 1, G1

1. Lovastatin (Merck, Sharp and Dohme Research Pharmaceuticals, Rahway, NJ) must be activated before addition to cell cultures. To activate lovastatin, the drug is diluted at a concentration of 50 mg/ml in 95% ethanol. The pH of the solution is raised by addition of 813 µl 1N NaOH and the solution subsequently neutralized with 1 N HCl . The activated solution is then diluted in sterile-filtered deionized water to a concentration of 4 mg/ml (10 mM) and stored at –20 °C for future use.
2. Lovastatin must be included in culture medium for at least one doubling time. Cell density should be reduced to 30–35% of maximum to prevent growth arrest due to confluence. While lovastatin concentrations must be

optimized for each cell type, effective concentrations ranging between 10–60 μM have been reported (Keyomarsi et al. 1991).

3. Reversal of the lovastatin block is accomplished by washing the cells twice in phosphate buffered saline or lovastatin-free medium at 37 °C. Cells are then resuspended at 35–40% of their confluent density in media supplemented with mevalonate. The mevalonate concentration should be 100-fold in excess of the concentration of lovastatin used. It is important to note that recovery from lovastatin does prolong the G1 phase of many cell types (Keyomarsi et al. 1991).

Protocol 2, G1 Mimosine, a plant amino acid whose cellular target is still unclear, is thought to block cells either in late G1 (Lalande 1990) or at the G1/S boundary (Mosca et al. 1992). The discrepancy appears to be based on the relevance of the recovery period to the timing of initial ^{3}H-thymidine incorporation. Nonetheless, when used alone or in combination with a prior isoleucine synchronization, mimosine is an effective synchronizing agent (Dijkwel and Hamlin 1992; Mosca et al. 1992). Described below is a simple procedure for the synchronization of CA46 cells.

1. Mimosine can be purchased from Sigma Chemical Company. Mimosine is dissolved at a final concentration of 100 mM in filter-sterile phosphate buffered saline pH 7.4. This solution can be stored at 4 °C before use.
2. Exponentially growing cells or cells synchronized by a prior isoleucine-deprivation are resuspended in media containing 100–400 μM mimosine. Cells are cultured for 16–24 h (exponentially growing cells) or 14 h (released from isoleucine block) in the presence of mimosine.
3. The mimosine-blocked cells are released into cycle by washing cells twice in sterile-filtered phosphate buffered saline or complete medium and then resuspending the washed cells in complete medium at a cell density equivalent to 35–40% of their confluent density.

3.3 Synchronization of Cells at the G1/S Border

The procedures utilized to synchronize cells at the G1/S phase border rely primarily upon an inhibition of DNA synthesis. This can be performed with a variety of chemical inhibitors, including aphidicolin, hydroxyurea, or excess thymidine. Historically, synchronization of cells with excess thymidine proved to be the first reliable method to gain early widespread use (Bootsma et al. 1964; Stein and Borun 1972). The method is most useful to monitor S phase and particularly G2/M phase events. The procedure is good for one cell cycle passage, thereafter the cells lose synchrony rapidly. The method involves two sequential exposures to high levels of thymidine, with each exposure separated by an interval determined from knowledge of the normal cell cycle distribution of the exponentially growing cells and the recovery time from the thymidine blocks. The procedure described below is for the synchronization of HeLa cells based on a doubling time of 24 h. Knowledge of the normal cell cycle distribution pattern of the exponentially growing cells can be gained from flow cytometry (see chapter by Giunta and Pucillo, Part One). In our example, the human cervical carcinoma cell line, HeLa S3 which can be grown as a suspension, have 35% of cells in G1, 50% of cells in S phase and 15% of cells in G2/M phases. Mitosis is probably about 45–60 min long, giving a G2 period of about 2.5–3 h. All handling of the cells-washing, diluting, centrifuging should be performed at 37 °C.

Protocol, G1/S

1. The thymidine (Sigma Chemical Company) can be prepared as a 100 mM stock solution in phosphate buffered saline, pH 7.4, sterile filtered (or autoclaved) and stored at –20 or –80 °C for at least 3 months.
   Exponentially growing cells should be diluted to 2.5×10^5/ml with fresh medium supplemented with 2 mM thymidine for 12 h. During this period, the G2/M cells will progress into G1 and then with the original G1 population acquire a biochemical state equivalent to a G1/S phase border cell. (G2M = 3.6 h + G1 = 8.4 h, Total = 12 h). Any cells in S phase upon addition of thymidine will become blocked in S phase.
2. Release of the cells from the first thymidine block is performed by centrifuging suspension cells (600 *g*, 5 min), discarding the thymidine-medium and then washing twice in an equal volume of complete medium. For monolayer cultures, the thymidine can be removed by pouring off the thymidine medium and then pouring fresh medium over the monolayer. The fresh medium should be used to rinse the flask before discarding. Repeat with fresh medium two more times for three washes in total.
3. Cells should then be incubated in fresh medium for 16 h. During this period the cells will recover from the thymidine block (approximately 1–2 h) and progress through the cell cycle, divide and enter G1 of the next cell cycle.
   Entry into G1 of the next cell will commence with the cells that were blocked at the end of S phase (the leading-edge cells). This will take about 5–6 h following release from thymidine block (1–2 h recovery + 4 h progression through G2/M). Entry into G1 of the next cell cycle will end with the lagging-edge cells (those arrested at G1/S) progressing through the cell cycle and dividing (between 16 and 18 h). The leading-edge cells will be at the G1/S phase border of the next cell cycle at the end of the 16 h. The lagging edge of cells will be in G2/M at the end of the 16 h. Counting suspension cells prior to and after the 16 h-incubation gives an indication of the degree of progression by the number of cells that have divided. The real trick to this procedure is to prevent the leading-edge cells from enter-

ing S phase of the next cell cycle while ensuring the lagging-edge cells have left S phase of the previous cell cycle. Timing is of the essence!

4. At the end of the 16-h release period, cells are diluted to 2.5×10^5/ml and reincubated with 2 mM thymidine containing medium for 12–14 h. Cells in G2/M or G1 would progress and arrest at the G1/S phase border. Flow cytometric determination of cell cycle position at 8–10 h following the readdition of thymidine will ensure that the population of cells is sufficiently synchronized before the more complex cell cycle study is commenced.

5. Release of cells from the second thymidine block follows essentially the washing procedures laid out in Step 2.

Modification Variations on the double-thymidine block strategy have evolved with time. A commonly used variant technique involves a first block with thymidine and a second block with aphidicolin (Heintz et al. 1983), or alternatively a double-aphidicolin block (O'Connor et al. 1993a). Essentially, the same basic principles as described above for the double-thymidine block apply to the use of aphidicolin. Both Heintz et al. (1983) and O'Connor et al. (1993a) detail the use of aphidicolin as a replacement for one or both of the thymidine-block steps. The use of thymidine and aphidicolin procedures for synchronizing cells at the G1/S phase border proved useful in determining the regulation of various cell cycle related processes including the regulation of histone gene expression (Heintz et al. 1983), regulation of cyclin B (Pines and Hunter 1989) and effect of DNA damage on the formation and activation of cyclin A and cyclin B-dependent kinases (O'Connor et al. 1993a).

3.4 Synchronization of Cells by Mitotic "Shake-Off"

Collection of pure populations of mitotic cells allows one to measure cell cycle related events during the mitosis-G1-S phase transitions. The cells then rapidly lose synchrony as they progress through S phase. Thus, the mitotic shake-off procedure is not useful for assessment of the G2 to M phase transition. The procedure relies upon the characteristic that monolayer cells round-up as they progress through mitosis. This leaves the cells' anchorage with the surface of the flask weakened and ideal for dislodging the mitotic cells into suspension. Thereafter, the suspension cells can be recovered by centrifugation. This procedure has the advantage that chemical agents, which may perturb the cell biochemistry, are not used. However, only a small proportion of the cells may be in mitosis at any one time, therefore a limitation of the procedure is the low yield of mitotic cells obtained from exponentially growing cultures. To enrich for mitotic cells, one can carry out short incubations with mitotic inhibitors, and/or release cells from a G1/S phase synchronization procedure and capture cells by shake-off as they progress into mitosis. The procedure described below is applicable to almost all monolayer cells. Our description is based on the human colon carcinoma cell line, HT-29.

Protocol, Mitotic cells

1. Semi-confluent cells (70–80%) are trypsinized from 162 cm^2 plates by first removing the growth medium, washing once with serum-free medium and then incubating the cells in 3 ml of trypsin solution for 5 min at 37 °C. At the end of the incubation, tap the flasks on the bench to shake off any loosely attached cells, add 20 ml of complete medium and centrifuge the cells (600 *g*, 5 min). Discard the supernatant and then add 20 ml of complete medium to wash the cells. Re-centrifuge the cells and then resuspend them in 20 ml of complete medium. Count the cells and then dilute to 2.5×10^5/ml.
   The trypsin solution can be purchased lyophilized form various sources; we use Gibco Life Technologies Inc. For mitotic selections we rehydrate with 40 ml (twice the manufacturer's recommended volume) of deionized sterile filtered water into a lyophilized powder of 5 g Trypsin, 2 g EDTA.4 Na, 8.5 g NaCl/l. The solution can be stored at –20 °C for at least 3 months.

2. Add back 20–40 ml of cells at 2.5×10^5/ml to 162 cm^2 flasks and leave for 6 h at 37 °C. This allows the cells to reattach to the flask surface. Remove any unattached cells by first shaking the flask and swirling the medium over the flask surface before pipetting off the medium. Repeat two more times with fresh medium before incubating the cells in fresh medium for 10 additional h.

3. At the 10-h time point, the flasks are gently shaken or tapped on the bench. The medium in the flask is swirled around the flask surface and the medium then removed to recover the mitotic cells. The medium from several flasks is pooled and then centrifuged (600 *g*, 5 min) to pellet the mitotic cells. The cells are then resuspended at 5×10^5/ml and analyzed by flow cytometry, giemsa staining, and/or phase contrast microscopy.

Enrichment of mitotic cells can be accomplished by addition of chemical agents to the mitotic shake-off procedure. For example, during the 10-h incubation mentioned at the end of step 2, a mitotic inhibitor, such as nocodazole (40 ng/ml, Aldrich) can be added. Preference is given to the addition of nocodazole for the last 6-h period to enrich for viable mitotic cells. Step 3 is performed essentially as described except an additional centrifugation/wash step is included to ensure removal of nocodazole from the cells. An alternative approach employs a one-step thymidine or aphidicolin block for 12-h prior to step 1 to enrich for cells at G1/S. When these cells are released into cycle at step 2 to step 3, the cells will progress across the cell cycle and more cells will be in mitosis at the time of the mitotic shake off. Each cell type will require some characterization to ensure the most suitable times at which cells should be harvested.

Pure populations of G1 cells can be gained using the mitotic shake off technique by placing the mitotic cells in a flask and incubating them for 2 h at 37 °C. The mitotic cells will re-attach to the flasks and flatten out as they enter G1 phase. Mitotic laggards can be removed from the flask by gently swirling the medium over the attached cells and then decanting the medium-off. The remaining attached cells will consist of primarily G1 cells and this can be confirmed by flow cytometry.

The use of mitotic shake-off procedures for synchronizing cells at the metaphase-anaphase border and enriching for G1 cells has proved useful in determining the regulation of various cell cycle related processes including the regulation of the Cdc2 kinase (Morla et al. 1989), Cyclin A (Pagano et al. 1992), and Cyclin B (Pines and Hunter 1989) proteins.

3.5 Use of DNA Damaging Agents to Arrest Cell in G1 and/or G2 Phases

A rapidly developing area of mammalian cell cycle research is the molecular dissection of checkpoint control systems which arrest cell cycle progression following DNA damage. Arrest can occur in G1 and/or G2 phases of the cell cycle and the arrest is dependent both on the status of the *p53* tumor suppressor gene in the cells and the type of DNA damaging agent employed. Arrest of cells in G1 phase requires that the cell line contain wild type *p53* genes and an intact *p53* control system (O'Connor and Kohn 1992). Normal human primary cells or cancer cell lines that still retain wild type *p53* activity can be utilized for this purpose. Some examples of the latter include the breast carcinoma cell line MCF-7, the renal carcinoma cell line ACHN, and the lung carcinoma cell line A-549. Some mutant-*p53* cancer cell lines include MDA-MB-435 (breast carcinoma), CA46 (lymphoma) and HT-29 (colon carcinoma), all of which are available through the American Type Tissue Culture Collection. Arrest of the cells in G1 can be induced with DNA double strand breaking agents such as ionizing radiation (O'Connor et al. 1993b) or anticancer drugs: bleomycin, Etoposide, Adriamycin (unpubl. observ.). DNA crosslinking agents such as nitrogen mustard or cisplatin produce much weaker G1 arrest than the DNA strand breaking agents (unpubl. observ.). Arrest of cells in G2 of the cell cycle can be elicited by both DNA strand-breaking and DNA crosslinking agents (O'Connor and Kohn 1992). Synchronization of cells at G1/S and then release before DNA damage is most commonly used for biochemical assessment of the G2 arrest phenomenon. G1 arrest can be studied in cells released from G0/G1 induced by serum starvation or the isoleucine block procedures which arrest cells in G1 before the *p53*-dependent arrest point.

References

Bootsma D, Budke L, Vos O (1964) Studies on synchronized divisions of tissue culture cells initiated by excess thymidine. Expt Cell Res 33:301–304

Burhans WC, Vasselev LT, Caddle MS, Heintz NH, DePamphilis MD (1990) Identification of an origin of bi-directional replication in mammalian chromosomes. Cell 62: 955–965

Campisi J, Morreo G, Pardee AB (1984) Kinetics of G1 transit following brief starvation for serum factors. Expt Cell Res 152:459–462

Dijkwel PA, Hamlin JL (1992) Initiation of DNA replication in the DHFR locus is confined to the early S period in CHO cells synchronized with the plant amino acid mimosine. Mol Cell Biol 12:3715–3722

Girard F, Strausfeld U, Fernandez A, Lamb NJC (1991) Cyclin A is required for the onset of DNA replication in mammalian fibroblasts. Cell 67:1169–1179

Heintz NH, Hamlin JL (1982) An amplified chromosomal sequence that includes the gene for DHFR initiates replication within specific restriction fragments. Proc Natl Acad Sci USA 79:4803–4087

Heintz N, Sive HL, Roeder RG (1983) Regulation of human histone gene expression: Kinetics of accumulation and changes in the rate of synthesis and the half-lives of individual histone mRNAs during the HeLa cell cycle. Mol Cell Biol 3:539–550

Keyomarsi K, Sandoval L, Band V, Pardee, AB (1991) Synchronization of tumor and normal cells from G1 to multiple cell cycles of lovastatin. Cancer Res 51:3602–3609

Lalande M (1990) A reversible arrest point in late G1 phase of the mammalian cell cycle. Exp Cell Res 186:332–339

Morla AO, Draetta G, Beach D, Wang JYJ Reversible tyrosine phosphorylation of cdc2: dephosphorylation accompanies activation during entry into mitosis. Cell 58:193–203

Mosca PJ, Dijkwel PA, Hamlin JL (1992) The plant amino acid mimosine may inhibit initiation of origins of replicaiton in CHO cells. Mol Cell Biol 12:4375–4383

O'Connor PM, Kohn, KW (1992) A fundamental role for cell cycle control in the chemosensitivity of cancer cells. Sem Cancer Biol 3:409–416

O'Connor PM, Ferris DK, Pagano M, Draetta G, Pines J, Hunter T, Longo DL, Kohn KW (1993a) G2 delay induced by nitrogen mustard in human cells affects cyclin A/cdk2 and cyclin B1/cdc2 complexes differently. J Biol Chem 268:8298–8308

O'Connor PM, Jackman J, Jondle D, Bhatia K, Magrath I, Kohn KW (1993b) Role of the p53 tumor suppressor gene in cell cycle arrest and radiosensitivity of Burkitt's lymphoma cell lines. Cancer Res 53:4776–4780

Pagano M, Pepperkok R, Verde F, Ansorge W, Draetta G (1992) Cyclin A is required at two points in the human cell cycle. EMBO J 11:961–971

Pardee AB (1974) A restriction point for control of normal animal cell proliferation. Proc Natl Acad Sci USA 71:1286–1290

Pardee AB (1989) G1 events and regulation of cell proliferation. Science 246:603–608

Pines J, Hunter T (1989) Isolation of a human cyclin cDNA: Evidence for cyclin mRNA and protein regulation in the cell cycle and for interaction with p34cdc2. Cell 58: 833–846

Schimke RT, Kung AL, Rush DF, Sherwood SW (1991) Differences in mitotic control among mammalian cells. Cold Spring Harbor Symp Quant Biol 56:417–425

Stein GS, Borun TW (1972) The synthesis of acidic chromosomal proteins during the cell cycle of HeLa cells. J Cell Biol 52:292–296

Tobey RA, Crissman HA (1972) Preparation of large quantities of synchronized cells in late G1 in the pre-DNA replicative phase of the cell cycle. Expt Cell Res. 75:460–464

7 Microinjection and Electroporation of Macromolecules into Live Cells

RAINER PEPPERKOK

1 Background

1.1 Microinjection

Capillary microinjection is one of the most efficient methods for introducing macromolecules into living mammalian cells (Proctor 1992). Samples can be delivered directly into either the nucleus or the cytoplasm. In synchronized cell cultures, the molecules can be transferred at well defined stages of the cell cycle and modifications of culture conditions are possible, before, during, or after injection. Therefore, microinjection of target specific inhibitors (e.g., antibodies, synthetic peptides, antisense mRNAs) or activators (e.g., proteins, peptides, cDNAs under the control of a constitutive promoter) has been a useful tool to study the in vivo function of proteins in cell proliferation and cell cycle control.

Critical aspects of the technique have been recently considerably improved. Firstly, the pulling and handling of glass micropipettes has been simplified by the development of micropipette pullers with built-in computers, allowing the highest degree of reproducibility in tip shape and diameter. Secondly, the number of cells which can be injected per experiment was significantly increased by computer automation of the technique (Ansorge and Pepperkok 1988). More than 1500 cells can be injected per hour with such an automated injection system (AIS) allowing biochemical analysis of microinjected cells on a routine basis (Pepperkok et al. 1993a). The upgrading of the AIS with technology for low light level fluorescence microscopy and image analysis allows fully automated evaluation of microinjection experiments on an integrated system (Pepperkok et al. 1993b).

1.2 Electroporation

Electroporation is an alternative method for the introduction of molecules into living cells. Membrane pores are generated by exposure of cells to short pulses of sufficiently high electrical fields. Molecules to be introduced into the cells are transported through these pores by diffusion, electrical drift and electro-osmosis. The technique has successfully been used in many different applications (Chang et al. 1992; Potter 1993; Weaver 1993) as the transient and stable transfection of bacteria, plant and mammalian cells with exogenous DNA (electrotransfection) or the introduction of enzymes, antibodies, or particles into living mammalian cells.

As a consequence of all these different applications, a large variety of procedures and protocols exists. Therefore, this chapter will give only general outlines for the selection of electroporation parameters and tested protocols as examples.

2 Principle and Applications

2.1 Microinjection

Ultrafine glass micropipettes (tip diameter <1 μm) are used to introduce the macromolecules into living cells by microinjection. Under microscopic control the operator penetrates the cells to be injected with the micropipette tip and delivers the sample into the cells by pressure applied from the open rear end of the pipette. Typically, injected volumes range from 0.05 pl for nuclear to 0.2 pl for cytoplasmic injections, representing 5 to 10% of the cell volume (Pepperkok et al. 1988). Thus, upon microinjection into cells, the sample concentration is diluted by a factor of 10 to 20. Therefore, in microinjection experiments sample concentrations should be at least 20 times higher compared to working concentrations used in in vitro assays (e.g., in immunofluorescence).

The injection technique described here is most efficient for applications using mammalian monolayer cells with an average size above 5 μm. However, modifications which allow cell attachment to a transparent support suitable for phase contrast microscopy will also permit microinjection of oocytes or suspension cells. Human diploid cells (IMR-90, WI-38, breast ductal epithelial cells) and rodent immortalized fibroblasts (REF-52 and NIH3T3) have been successfully used for microinjection experiments in cell cycle studies (Pepperkok et al. 1988; Sorrentino et al. 1990; Lane et al. 1993; Pagano et al. 1993; Twamley-Stein et al. 1993; Pepperkok et al. 1994; Tam et al. 1994). They can be easily microinjected and reversibly synchronized in the G1 phase of the cell cycle by serum deprivation (see chapter by O'Connor and Jackman, Part One).

Since the number of cells which can be injected per experiment is limited, analysis of microinjected cells is commonly performed on a single cell basis. Immunofluorescence is the method of choice for the simultaneous detection of microinjected molecules and protein expression in single cells (Sorrentino et al. 1990; Pagano et al. 1993; Pepperkok et al. 1994; see also chapters by Cardoso and Leonhart and by Bartkova et al., Part One). Unspecific antibodies not interfering with cell cycle progression and cell function may be used as coinjection markers in case the injected molecules of interest cannot be directly visualized by immunofluorescence (Baldin et al. 1993). Low light level fluorescence microscopy and computerized image analysis allow the automated and precise quantification of immunofluorescence stainings and enhance the speed, quality and statistical significance of the evaluation of results in microinjection experiments (Pepperkok et al. 1988; Lorenz et al. 1993; Pepperkok et al. 1993b; Todorov et al. 1994).

Biochemical analysis or electron microscopy of microinjected cells is usually more difficult to perform, and achievement of reproducible results requires a high degree of experience in microinjection technique. However, in rare cases, the signals obtained from 200 to 300 microinjected cells which had been labeled with ^{35}S methionine or ^{32}P are already sufficient to carry out SDS-PAGE followed by autoradiography analysis (Gautier-Rouviere et al. 1990; Lane et al. 1993). The high injection rate which can be achieved with a computer automated microinjection system (AIS) has significantly facilitated such biochemical analysis (Pepperkok et al. 1993a).

2.2 Electroporation

Important parameters affecting the efficiency in electroporation experiments with mammalian cells are:

- **Electrical field (voltage applied to the cell chamber).** The voltages used vary from 50 V to several thousand volts. The optimum range depends on type and size of the cells to be electroporated. If the applied voltage is too high, the survival rate of the cells is dramatically decreased. Too low voltages result in unsatisfactory electroporation efficiencies.
- **Duration of the pulses applied.** The duration of the applied electrical pulse is typically varied between a few microseconds and several milliseconds. Low voltages usually require long pulses and vice versa.
- **Concentration of cells and molecules to be introduced.** Typically, cells are prepared at concentrations of 10^6–10^7 per ml. The risk for cell fusion during electroporation is rather high for cell densities above 10^7 per ml. DNA concentrations of 5 μg/ml already give good transformation efficiencies, higher concentrations improve the results, but concentrations above 300 μg/ml are toxic for the cells. For the electroporation of proteins or antibodies, concentrations are between 0.1 mg/ml and 10 mg/ml, depending on their activity and the specific requirements of the experiments. As for microinjection, a dilution of the sample concentration (10 to 20 times) occurs upon electroporation and has to be taken into consideration.
- **Culture conditions and electroporation buffer.** Cells are cultured under normal conditions. The electroporation buffer should be of low ionic

strength and should not contain serum, because even small amounts of it will lower the electroporation efficiency. PBS and HBS are preferred for mammalian cells.

- **Size of molecules to be electroporated.** Because introduction of molecules into cells by electroporation is based on the diffusion of the sample molecules through the membrane pores generated by the electrical pulse, the size of the sample molecules affects electroporation efficiency. Obviously, high molecular weight molecules are less efficiently electroporated than low molecular weight molecules.

Compared to microinjection, electroporation has the advantage that the molecules of interest can be introduced into many cells (10^7 cells per ml) at the same time. Therefore, biochemical analysis of electroporated cells is easily possible. When monolayer cells are used for the experiments they have to be trypsinized for electroporation and subsequently replated onto culture dishes or glass coverslips. Because cells need several hours to settle down and attach on their support, a lag phase occurs between introduction of the molecules into the cells and their analysis. In some cases (e.g., studies on signal transduction) this lag phase will make cell cycle studies by electroporation impossible. However, electroporation may successfully be used for studies on cell cycle progression when this lag phase is not relevant for the analysis of the results (Lukas et al. 1994).

3 Methods

Chemicals and reagents

^{35}S methionine and ^{32}P are both obtained from NEN (DuPont, Cat. No. NEG-009A or NEX-053C, respectively), purified luciferase (cat. no. L5256) is bought from Sigma (Detroit, USA), cell culture petri dishes 35 mm Ø (cat. no. 153066) are from NUNC (Roskilde, DK), glass coverslips are from Menzel (Braunschweig, FRG), fluorescein or rhodamine conjugated dextrans with different molecular weights are available from Sigma (Detroit, USA).

Microinjection equipment

Raw glass capillaries are from Clark Electromedical Instruments (Reading, UK; cat. no. GC120TF-10); the pipette puller type P-87 is from Sutter Instruments (Novato, USA); the Microinjector model 5242 for controlling the pressure applied to the micropipette, and the microloaders for filling micropipettes (cat. no. 0030001.22) are from Eppendorf (Hamburg, FRG).

The automated microinjection system (AIS) including microscope, motorized scanning stage and micromanipulator, and the control computer can be obtained from Zeiss (Oberkochen, FRG).

Electroporation equipment

For the tested protocols described here, a GenePulser (cat. no. 165–2977) and electroporation cuvettes with 0.4 cm electrode gap (cat. no. 165–2086), both from Biorad, were used.

Pulling of micropipettes

1. No special treatment of the glass capillaries before pulling is required.
2. Insert raw capillary carefully into the puller without touching the heating filament (square shape).
3. Choose the pulling program recommended by the supplier for pulling injection micropipettes.
4. Adjust the value of the current through the heating filament to your individual needs (usually this value should not vary more than 10% of the value recommended by the supplier) and pull the micropipette. The tip diameter of the micropipette decreases with increasing current and vice versa. Capillaries with outer tip diameter between 0.2 and 1.0 μm are easily available with the used puller.
5. Two micropipettes will be obtained after 20–30 s Both can be used for microinjection because they are identical with respect to their physical parameters.

Characterization of micropipettes by bubble pressure measurement

The protocol below describes characterization and classification of the pulled micropipettes by bubble pressure measurements (Mittman et al. 1987). This is a nondestructive method for estimating the tip diameter. It has the advantage that after the measurement the micropipettes can still be used for microinjection experiments without any restrictions. This is not the case when micropipettes have been characterized by electron microscopy.

1. Insert the micropipette into the microinjector capillary holder.
2. Dip the tip into a beaker filled with 100% ethanol (filtered through a 0.22 μm syringe filter).
3. Increase the pressure of the microinjector until air bubbles from the micropipette tip appear, indicating that it is open. Pressure values of micropipettes, used in our laboratory for nuclear and cytoplasmic injections, are between 2000 and 4000 hPa. Lower values indicate a tip size above 1 μm of diameter and the corresponding micropipettes should not be used for microinjection into mammalian cells. The tip diameter is rather small for pressure values higher than 4000 hPa and the risk of clogging the micropipette during microinjection is rather high.
4. Store the micropipettes in a dust-free and dry environment until use.

Filling micropipettes with injection sample

1. To avoid micropipette clogging during microinjection, the sample should always be centrifuged for at least 10 min at 10 000 *g* at 4 °C.
2. Deliver the sample through the rear open end of the micropipette down near to the tip using a sterile microloader from Eppendorf. Typically loaded volumes are 0.2–1 µl.
3. Insert the filled micropipette quickly into the capillary holder of the microinjector, mount the holder onto the micromanipulator and lower the micropipette rapidly until the tip reaches the medium of the cell culture dish.

The speed of step 3 is important because the sample in the micropipette tip may easily dry out if it remains for too long in the air and micropipette clogging occurs.

Centering the micropipette in the microscope field

The following procedure is very useful to prevent breakage of the micropipette tip during its centering in the microscope field. It is directly applicable when using the Zeiss automated injection system (AIS) but can also be used for the manual injection technique (using, e.g., micromanipulators from Eppendorf or Leitz).

All steps should be performed in the capillary position "Cell" (AIS program menu).

1. Center the micropipette above the culture dish roughly by eye.
2. Lower the micropipette (using the AIS program menu "Adjust") until it touches the culture medium.
3. Change to the lowest magnification available on the microscope (e.g, 45 ×).
4. Focus on the cells.
5. Center the micropipette in the field of view in the microscope and lower it carefully, but not all the way to its focus (breakage of micropipette!)
6. Change to the next higher magnification.
7. Repeat steps 4–6 until you reach the working magnification (320 ×). Now the cells should be in focus and the micropipette should be centered, but not in focus (breakage of micropipette!).
8. Lower the micropipette carefully until the tip comes into focus. Both cells and tip are on the same focal plane.
9. Use the computer mouse to mark the tip position (AIS menu program "Mark Tip").
10. Proceed directly with automated microinjection.

Steps 8 to 10 are only necessary when the AIS is used and are omitted for manual microinjection.

Automated Microinjection

For the following, the micropipette should be centered in the field of view. The pressure applied to the micropipette remains constant during automated injection. It is adjusted with the microinjector at the beginning of the session according to the cell type and injection sample used.

1. Switch the capillary position to "Wait" (AIS program menu).
2. Choose the program menu "Position OK."
3. Choose the program menu "Mark Next."
4. Mark all the cells to be injected by pointing with the computer mouse to the exact place of injection on the cells (displayed on the video monitor). Nuclear injections should be performed in the center of the cell nucleus. Cytoplasmic injections are performed just next to the nucleus.
5. Choose the program menu "Inject". All the marked cells in the field are injected automatically under computer control and their stage coordinates are stored on computer disk.
6. Switch to a neighboring field containing noninjected cells and repeat steps 3 to 5.

Quantification of microinjected sample volume

1. Add purified luciferase (Sigma, final concentration 2 mg/ml) to the sample solution to be injected in biological experiments.
2. Microinject exact number of cells (50 cells will be enough).
3. Lyse cells and determine luciferase activity of the total cell extract in a luminometer as described (DeWet et al. 1987).
4. Dilute the sample solution containing luciferase (2 mg/ml) in steps of one order of magnitude or less, and determine the luciferase activity of the dilutions. A calibration curve which correlates the different amounts of sample solution with their luciferase activities will be obtained.
5. Using this calibration curve and the luciferase activity of the total cell extract, as measured in step 3, the total volume of sample solution microinjected into all cells (e.g., 50) can be derived.
6. Divide the volume calculated in step 5 by the number of injected cells (e.g., 50) to obtain the average volume injected per cell.

The average volume microinjected per cell is varied by the pressure applied to the micropipette and the time (controlled by computer when the AIS is used) the tip stays inside the cell. It depends also on the cell type and injection solution used. Typical values are 0.1–0.5 pl for cytoplasmic and 0.01–0.05 pl for nuclear injections.

Retrieval of microinjected cells

When the AIS is used, every injected cell can be retrieved by its stage coordinates stored on computer disk during microinjection. For this purpose, cells should be grown on glass coverslips which contain two distinguishable orientation marks. They may be easily scratched onto the glass coverslips using a diamond pen. With the cursor mouse, the position of these orientation marks must be entered into the computer twice, at the beginning of the microinjection session and when the "Find Cells" AIS program menu is used to retrieve injected cells.

For manual microinjection, coinjection markers (e.g., purified IgGs or fluorescently labeled dextrans), not interfering with cell functions, may be used to identify microinjected cells. They should be easily detectable (e.g., immunofluorescence) and compatible with the assay system (e.g., immunofluorescence) used to analyze microinjected cells.

Microinjecting the same cells twice

With the AIS, the method to retrieve injected cells on glass coverslips is also used to microinject the same cells twice. Microinjected cells are first retrieved by their stage coordinates stored on the computer disk (see above) and then microinjected a second time as described in the paragraph *Automated microinjection* above.

With manual microinjection systems double injections are more difficult to perform. Fluorescently labeled dextrans have been used as coinjection markers to trace injected cells (Pepperkok et al. 1988). They have to be mixed with the sample solution for the first injection (a final concentration of 3 µg/µl for fluorescein dextran is typically used). Microinjected cells are first detected by fluorescence microscopy (excitation of fluorescein with 488 nm) and then microinjected a second time. However, because exposure to the excitation light and bleaching of the fluorophores (e.g., fluorescein) during observation may be harmful to the cells, such an approach is limited.

Biochemical analysis of microinjected cells

Speed and efficiency of the automated injection system allow biochemical analysis of microinjected cells. Between 500 and 1000 cells which have been labeled with ^{35}S methionine or ^{32}P will be sufficient to analyze labeled proteins by gel electrophoresis and autoradiography. The following protocol is routinely used in our laboratory for this purpose. It may have to be modified according to the specific requirements of the experiments.

1. Plate 250 cells in 5 µl droplets in the center of a glass coverslip (10 × 10 ml).

2. Place coverslips into a humid chamber and incubate at 37 °C until cells have attached to the glass (takes usually 6–8 h).

3. Transfer coverslips into 3.5-cm Petri dishes containing 2 ml of culture medium and let cells grow for 2 days at 37 °C. Usually between 500 and 1000 cells will be in the center of the coverslip after this time.
4. Microinject all cells on the coverslip with the AIS.
5. Incubate cells for 15 min in methionine free culture medium (labeling medium) to deplete cellular stores. (Skip this step for ^{32}P labeling and proceed directly with step 6.)
6. Label for the appropriate time with ^{35}S methionine or ^{32}P in 40 μl labeling medium. Typical concentrations are: 1 μCi/μl ^{35}S methionine or 10 μCi/μl ^{32}P.
7. Wash cells three times with chase medium (complete medium containing normal concentrations of methionine or phosphate).
8. Incubate cells for the appropriate time (depends on particular experiment) with 2 ml chase medium.
9. Lyse cells in 50 μl lysis buffer containing nonradioactively labeled cell lysate at a protein concentration of 1 mg/ml. The presence of these "carrier proteins" in the lysis buffer is extremely important for the success of subsequent steps of the experiment (e.g., immunoprecipitation and analysis of specific proteins).
10. Proceed with biochemical analysis of lysed proteins.
    Alternatively, steps 5–7 can be performed before microinjection (step 4).

Determination of electroporation efficiency/ optimization of parameters

Optimization of electroporation parameters is most easily and rapidly performed with test molecules of appropriate size. Fluorescein or rhodamine conjugated dextrans are convenient for this purpose. They are commercially available (Sigma) in a wide range of molecular weights and are easy to detect by fluorescence microscopy. For the tests, all parameters are kept constant except the one under investigation, which is varied in different experiments. Starting conditions which give reasonable electroporation efficiencies (tested for HeLa, NIH3T3, IMR-90, WI-38, REF-52) with the Biorad Gene Pulser and 0.4 cm electrode gap are:

voltage = 275 V; resistance = infinitive; capacitor = 125 μF; 50 μl of cell suspension per experiment (see also chapter by M. Rolfe, Part One).

1. Grow cells to 70–80% confluency.
2. Remove medium and wash 2 × with PBS.
3. Add 1 ml trypsin EDTA and incubate until cells detach from the plastic.
4. Add 10 ml of culture medium to stop trypsinization and count cells.

5. Spin down cells for 10 min at 180 g in a cooled centrifuge.
6. Aspirate supernatant with a Pasteur pipette and wash cell pellet with 10 ml of cold PBS.
7. Spin down cells as in step 5.
8. Aspirate supernatant and resuspend pellet in cold fluorescein dextran solution (final concentration of 5 µg/µl in PBS) to a final concentration of 10^7 cells/ml.
9. Transfer 50 µl of the cell suspension into a precooled cuvette and apply a single pulse (duration of pulse is approx. 10 ms for the settings described above).
10. Leave transfected cells on ice for 15 min.
11. Replate cells onto 6 cm Ø culture dishes containing glass coverslips and prewarmed culture medium.
12. Culture electroporated cells for 16 h.
13. Fix cells for 15 min in paraformaldehyde (final concentration 3% in PBS) at room temperature.
14. Stain cells with Hoechst dye 33258 (from Serva, final concentration 1 µg/ml in PBS) for 5 min at room temperature.
15. Wash cells with PBS for 5 min and mount cells on a glass slide using mowiol or any other suitable mounting medium used for immunofluorescence.
16. Analyze cells by fluorescence microscopy. For each field of view determine the number of fluorescein dextran positive cells (homogeneous green cellular fluorescence) and Hoechst dye positive cells (all cells in the field contain a bright nuclear Hoechst dye fluorescence).
    Calculate the percentage of successfully electroporated cells (a good electroporation condition should yield to about 30–50% of fluorescein dextran positive cells).

Modification This protocol (with optimized parameters) has been used in our laboratory with slight modifications (steps 13 to 16 have to be changed according to the biological questions of the particular experiment) for electrotransfection of DNA (2 µg of DNA per experiment), electroporation of RNA (final concentration of 5 µg/µl), antibodies (final concentration of 1–10 µg/µl), synthetic oligodeoxynucleotides (final concentration of 10 µg/µl) and synthetic peptides (final concentration of 1–10 µg/µl).

Acknowledgments. I thank M. Lowe and D. Masson for critically reading the manuscript.

References

Ansorge W, Pepperkok R (1988) Performance of an automated system for capillary microinjection into living cells. J Biochem Biophys Meth 16:283–292

Baldin V, Lukas J, Marcote MJ, Pagano M, Draetta G (1993) Cyclin D1 is a nuclear protein required for cell cycle progression in G1. Genes Dev 7:812–821

Chang DC, Chassy B, Saunders JA, Jordon CA (eds) (1992) Handbook of electroporation and electrofusion. Academic Press, San Diego

DeWet JR, Wood K, DeLuca M, Helsinki D, Subramini S (1987) Firefly luciferase gene: structure and expression in mammalian cells. Mol Cell Biol 7:725–737

Gautier-Rouviere C, Fernandez A, Lamb NJC (1990) Ras-induced c-fos expression and proliferation in living rat fibroblasts involves C-kinase activation and the serum response element pathway. EMBO J 9:171–180

Lane HA, Fernandez A, Lamb NJC, Thomas G (1993) $p70^{sk6}$ function is essential for G1 progression. Nature 363:170–172

Lorenz P, Pepperkok R, Ansorge W, Pyerin W (1993) Cell biological studies with monoclonal and polyclonal antibodies against human casein kinase II subunit ß demonstrate participation of the kinase in mitogenic stimulation. J Biol Chem 268:2733–2739

Lukas J, Pagano M, Staskova Z, Draetta G, Bartek J. (1994) Cyclin D1 protein oscillates and is essential for cell cycle progression in human tumor cell lines. Oncogene 9: 707–718

Mittman S, Flaming DG, Copenhagen DR, Belgum JH (1987). Buble pressure measurement of micropipette tip outer diameter.. Neurosci Meth 22:161–166

Pagano M, Pepperkok R, Lukas J, Baldin V, Ansorge W, Bartek J, Draetta, G (1993) Regulation of the cell cycle by the cdk2 protein kinase in cultured human fibroblasts. J Cell Biol 121 101–111

Pepperkok R, Zanetti M, King R, Delia D, Ansorge W, Philipson L, Schneider C (1988) Automatic microinjection system facilitates detection of growth inhibitory mRNA. Proc Natl Acad Sci USA 85:6748–6752

Pepperkok R, Scheel J, Horstman H, Hauri HP, Griffiths G, Kreis TE (1993a). ß-COP is essential for biosynthetic membrane transport from the endoplasmic reticulum to the Golgi complex in vivo. Cell 74:71–82

Pepperkok R, Herr S, Lorenz P, Pyerin W, Ansorge W (1993b) System for quantitation of gene expression in single cells by computerized micro-imaging: application to *c-fos* expression after microinjection of anti-casein kinase II (CKII) antibody. Exp Cell Res 204:278–285

Pepperkok R, Lorenz P, Ansorge W, Pyerin W (1994) Casein kinase II is required for G0/G1, early G1 and G1/S phase of cell cycle. J. Biol. Chem. 269:6986–6991

Potter H (1993). Application of electroporation in recombinant DNA technology. Methods Enzymol 217:461–478

Proctor GN (1992) Microinjection of DNA into mammalian cells in culture: theory and practise. Meth Mol Cell Biol 3:209–231

Sorrentino V, Pepperkok R, Davis RL, Ansorge W, Philipson L (1990) Cell proliferation inhibited by *MyoD1* independently of myogenic differentiation. Nature 345:813–815

Tam SW, Theodoras AM, Shay JW, Draetta GF, Pagano M (1994) Differential expression and regulation of cyclin D1 protein in normal and tumor human cells: association with Cdk4 is required for cyclin D1 function in G1 progression. Oncogene 9:2663–2674

Todorov IT, Pepperkok R, Philipova RN, Kearsy SE, Ansorge W, Werner, D (1994) A human nuclear protein with sequence homology to a family of early S phase proteins is required for entry into S phase and for cell division. J Cell Sci 107:253–265

Twamley-Stein GM, Pepperkok R, Ansorge W, Courtneidge SA (1993) The src family tyrosine kinases are required for platelet-derived growth factor-mediated signal transduction in NIH3T3 cells. Proc Natl Acad Sci USA 90:7696–7700

Weaver JC (1993) A general phenomenon for manipulating cells and tissue. J Cell Biochem 51:426–435

8 Stable and Transient Transfections in Cell Cycle Studies

MARK ROLFE

1 Background

The ability to transfer DNA into living cells has revolutionized the study of biology. Transfection of DNA into mammalian cells is now a widely used technique which, inevitably, has evolved an equally wide number of technical variants. This chapter will describe some of the easiest and most reproducible methods for transferring DNA into a variety of mammalian cells, both those that grow as adherent monolayers and those that grow in suspension.

2 Principle and Application

Transient Transfection. Transient transfection, as the name implies, involves transferring DNA into cells in a short lived experiment that can last anywhere from 6 to 72 h. The seminal work on transient transfection was done in the cos cell system, which utilises a fibroblast-like cell line established from CV-1 African Green Monkey kidney cells. In these cells plasmids that contain an SV40 origin of replication are able to replicate to extremely high copy numbers owing to the presence of a constitutively expressed T-antigen gene present in the cos cell genome (Gluzman 1981; Aruffo 1992). More recently transient transfections have been done in many other cell types – in these cases the ability to detect expression of the transfected gene has resulted from a combination of factors including the use of strong viral promoters to achieve high levels of expression and the development of highly sensitive reporter assays.

In studies of cell cycle regulatory proteins, transient transfections have been of most use in confirming that cloned cDNAs do encode for proteins of the expected size and antibody cross-reactivity (Hinds et al. 1992). Transient expression of dominant mutant Cdk kinases has been used to address the requirement for kinases of this family in cell cycle progression (van den Heuvel and Harlow 1993). Transient transfection experiments have also been used to determine whether point-mutated Cdks retain kinase activity (M. Rolfe, unpubl.) and whether affinity tagged proteins (e.g., GST-Cdc2) retain their enzymatic activities and ability to bind to Cyclins (M. Rolfe, unpubl.). These types of experiments also make it possible to screen Cdk/Cyclin deletion mutants to map domains that are important for Cdk/Cyclin interaction. Finally, transient transfection by means of microinjection of DNA into living cells has been used to assess the function of many cell cycle proteins (see chapter by R. Pepperkok, Part One).

Stable Transfection. In stable transfections, the gene (or genes) under study is transferred into a mammalian cell along with a selectable marker which usually encodes a drug-resistance gene (see Bebbington and Rolfe 1992). After a short period of drug-free growth, the cells are exposed to the selective agent. Only those cells that retain a copy of the drug-resistance gene stably integrated into the host genome are able to survive continued exposure to the selective agent. After a 14–21-day period of selective growth, most cells have died, but those that do express the drug resistance gene at high enough levels have grown to form drug resistant colonies that can be cloned or pooled for further analysis.

Many groups cotransfect their gene of interest along with a selectable marker that is carried on a separate plasmid, but it is becoming more common to construct plasmids where the gene (or genes) under study can be engineered to be expressed from the same plasmid as the drug resistance gene (Bebbington and Rolfe 1992).

In the last 3 years there have been several reports from different groups that have created engineered cell lines expressing moderate to high levels of cell cycle regulatory proteins. Livingston's and Weinberg's groups demonstrated that ectopic expression of human G1 Cyclins could regulate the function of the retinoblastoma gene product, probably by promoting its phosphorylation (Hinds et al. 1992; Dowdy et al. 1993; Ewen et al. 1993a). Ohtsubo and Roberts (1993) and Quelle et al. (1993) showed that constitutive expression of either Cyclin E or Cyclin D1 has the ability to shorten the G1 phase of the cell cycle. Resnitzky et al. (1994) and Wimmel et al. (1994) obtained similar results by expression of Cyclin D1 or Cyclin E with a tetracycline-inducible system. Jiang et al. (1993) established some Rat-6 cell lines that overexpressed human Cyclin D1. Such cell lines had a shortened G1 and were tumorigenic when injected into nude mice. Similarly, Hinds et al. (1994) and Lovec et al. (1994) reported that overexpression of Cyclin D1 induces transformation in primary rat cells in cooperation with other oncogenes. Ewen et al. (1993b) showed that constitutive expression of Cdk4, but not of Cdk2, reduces the serum requirement in mink epithelial cells. Work from my laboratory has demonstrated that stably transfected Rat-1 fibroblasts can be established that over express individual Cdks or Cyclins and pairwise combinations of Cdks and Cyclins. Some of these transfected cells have a more transformed phenotype than the parental cells in that they detach readily from Petri dishes and some can form colonies in soft agar (M. Rolfe, unpubl.).

3 Methods

Solutions

1 × HBS: 8 g NaCl, 6 g Hepes, 0.2 g Na_2HPO_4 (anhydrous). Dissolve in ~800 ml water and pH to 7.1 with 1 M NaOH. Make vol 1000 ml, filter sterilize, and store at 4 °C.

1 × PBS: 8 g NaCl, 0.2 g KCl, 1.44 g Na_2HPO_4, 0.24 g KH_2PO_4. Dissolve in ~800 ml water and pH to 7.4 with HCl. Make vol 1000 ml, autoclave, and store at room temperature.

1 × TBS: 8 g NaCl, 0.2g KCl, 3 g Tris base. Dissolve in ~800 ml water and pH to 7.4 with HCl. Make vol. 1000 ml, autoclave, and store at room temperature.

3.1 Transient Transfection

I shall describe three common techniques for transiently transfecting a variety of cell types. A further way to obtain transient transfection is to microinject DNA directly into living cells. For this technique see the chapter by R. Pepperkok (Part One).

$CaPO_4$ transfection

This method is used almost exclusively for adherent, fibroblastic cells. The protocol is for transfection in a 24-well plate, but can be scaled up easily.

DAY 1

1. Seed 5 × 10^5 cells/well. Leave in medium containing fetal calf serum (FCS).

DAY 2

2. Re-feed cells with 1 ml fresh medium containing FCS.
3. Prepare DNA precipitate as follows: – 47.5 µl 1 × TE, 20 µl DNA (500 µg/ml), 7.5 µl 2.5 M CaCl2
4. Add above mix to 75 µl 2 × HBS. Mix.
5. Add precipitate directly to medium on cells.
6. Incubate 37 °C 3 h.
7. Remove medium.
8. Add 1 ml 15% glycerol in PBS/well. Leave 1 min.
9. Remove glycerol. Take great care not to dislodge cells – some cell types become less well attached after glycerol shock. I use a P1000 Gilson pipette. I do not use an aspirator.
10. Wash cells with PBS or serum-free medium.
11. Add 1 ml growth medium and leave at 37 °C for 6–72 h.

DEAE dextran method

This is the method of choice for cos cells and can work well for some suspension cells (e.g., mouse B cells).

DAY 1

1. seed cells 2–3 × 10^5/well in 1 ml medium containing FCS.

DAY 2

2. Replace medium with 1 ml serum-free medium.
3. Prepare DMEM/Tris (20 ml DMEM + 5 ml 0.25 M Tris pH 7.5).
4. Set up tubes for transfection: 0.8 ml DMEM/Tris + 1 μg plasmid DNA.
5. Wash cells with serum-free medium.
6. Add 0.2 ml 1 mg/ml DEAE-dextran in TBS to transfection tubes.
7. Remove medium from cells and add DEAE-dextran/DNA.
8. Leave 6–8 h at 37 °C.
9. Remove complex.
10. Add 1 ml 10% DMSO in HBS and leave for 2–5 min (time is not critical as long as it is >2 min).
11. Carefully remove DMSO.
12. Wash cells with serum-free medium.
13. Incubate in medium + FCS for 6–72 h.

Electroporation

Electroporation is best known for creating stable cell lines (see below) but it can be used for transient transfection when the above two methods have failed (see also chapter by R. Pepperkok, Part One). Conditions for electroporation tend to fall into two classes:

- For adherent cells, a very high voltage with low capacitance seems to be best, whereas suspension cells tend to the reverse.
- A medium voltage with very high capacitance. I have used the following two sets of parameters with good success:
  - For adherent cells 1500 V, 25 μF, two pulses.
  - For suspension cells 250 V, 960 μF, one pulse.
  - In each case use 5 × 10^6 cells /transfection. Collect cells and wash in PBS. Resuspend in 1 ml PBS + 10 μg plasmid DNA. Electroporate. Incubate on ice for ~5 min. Replate cells in 5–10 ml fresh growth medium. Collect and analyze 6–72 h later.

3.2 Stable Transfection

I shall describe two robust protocols for isolating stably transfected cells. The basis of these methods is the same as for transient transfections. $CaPO_4$ transfection works well for adherent cells. DEAE-dextran cannot be used to create stable cell lines. Electroporation works very well for both adherent and suspension cells.

$CaPO_4$ transfection

1. For transfection in 9 cm^2 plates seed 1–2 × 10^6 cells/plate.
2. Follow recipes for transient transfection but use 3 ml growth medium and 0.5 ml precipitate.
3. After 24 h nonselective growth add selective medium.
4. Re-feed cells every 7 days with fresh selective medium.
5. After 14–21 days colonies should be visible. These can be cloned using cloning cylinders or pooled for further analysis.

Electroporation

Linearized DNA must be used to create stable cell lines by electroporation. I usually linearize 10–40 μg DNA in a reaction volume of 100 μl and check 1/100th on an agarose gel prior to electroporation. Once the DNA is linearized, I add it directly to the 1 ml of cells in PBS.

1. Use same parameters as for transient transfection.
2. After the pulse dilute the 1 ml of cells into 25 ml growth medium.
3. Plate out 2 × 96 well plates with 100 ml cells/well.
4. To the remaining ~5 ml cells add 20 ml growth medium.
5. Repeat plating out in 2 × 96 well plates.
6. Dilute remaining ~5 ml cells with 20 ml growth medium.
7. Plate out final 2 × 96 well plates.
8. Incubate all six 96 well plates at 37 ° for 24 h.
9. Add 100 μl 2 ˣ selective medium to each well.
10. After 14–21 days colonies should be visi.ble in some of the plates.

This serial dilution should give one pair of 96 well plates where there is >> 1 colony/well. You can be very confident that such colonies are clonal. The colonies can be harvested (with trypsin if necessary) and expanded as cell lines very easily. If the transfected cells are growing in suspension, take great care not to disturb the plates during selective growth as this can make identification of wells containing single colonies very difficult.

3.3 Lipofection

Using cationic lipids such as DOTAP (N-[1-(2,3-dioleoyloxy)propyl]-N,N,N-trimethyl ammonium methyl sulfate), negatively charged macromolecules spontaneously form liposomes which can be used to transfer the macromolecules into living cells. Several "lipofection" kits are commercially available and can be used for transient and stable transfection studies in a wide variety of cell types. Because most researchers who do lipofection use the commercial kits, protocols are not included here; simply following the manufacturer's instructions should give satisfactory results.

References

Aruffo A (1992) In: Ausubel et al. (eds) Current protocols in molecular biology. Chapt 16: Expression of proteins in mammalian cells. pp 16.13.1–16.13.7

Bebbington CR, Rolfe M (1992) In: Ausubel et al. (eds) Current protocols in molecular biology. Chapt16: Expression of proteins in mammalian cells. pp 16.14.7–16.14.13

Dowdy SF, Hinds PW, Louie K, Reed SI, Arnold A, Weinberg RA (1993) Physical interaction of the retinoblastima protein with human D cyclins. Cell 73:499–511

Ewen M, Sluss H, Whitehouse L, Livingston D (1993a) Functional interaction of the retinoblastoma protein with mammalian d-type cyclins. Cell 74:1009–1020

Ewen ME, Sluss HK, Sherr CJ, Matsushime H, Kato JY, Livingston DM (1993b) TGFß inhibition of cdk4 synthesis is linked to cell cycle arrest. Cell 73:487–497

Gluzman Y (1981) SV40 transformed simian cells support the replication of early SV40 mutants. Cell 23:175–182

Hinds PW, Mittnacht S, Dulic V, Arnold A, Reed SI, Weinberg RA. (1992) Regulation of retinoblastoma protein functions by ectopic expression of human Cyclins. Cell 70:993–1006

Hinds P, Dowdy S, Ng-Eaton E, Arnold A, Weinberg R (1994) Function of a human cyclin as an oncogene. Proc Natl Acad Sci USA 91:709–713

Jiang W, Kahn SM, Zhou P, Zhang Y-J, Cacace AM, Infante AS, Doi S, Santella RM, Weinstein IB,. (1993) Overexpression of Cyclin D1 in rat fibroblasts causes abnormalities in growth control, cell cycle progression and gene expression. Oncogene 8: 3447–3457

Lovec H, Sewing A, Lucibello F, Mueller R, Moroy T (1994) Oncogenic activity of Cyclin D1 revealed through cooperation with Ha-ras: link between cell cycle control and malignant transformation. Oncogene 9:323–326

Ohtsubo M, Roberts JM (1993) Cyclin-dependent regulation of G1 in mammalian fibroblasts. Science 259:1908–1912

Quelle DE, Ashmun RA, Shurtleff SA, Kato J-Y, Bar-Sagi D, Roussel MF and Sherr CJ (1993) Overexpression of mouse D-type Cyclins accelerates G1 phase in rodent fibroblasts. Genes Dev 7:1559–1571

Resnitzky D, Gossen M, Bujard H, Reed S (1994). Acceleration of the G1/S phase transition by expression of Cyclin D1 and E with an inducible system. Mol Cell Biol 14: 1669–1679

van den Heuvel S, Harlow E (1993) Distinct roles for Cyclin –dependent kinases in cell cycle control. Science 262:2050–2054

Wimmel A, Lucibello F, Sewing A, Adolph, S, Müller R (1994) Inducible acceleration of G1 progression thorough tetracycline-regulated expression of human Cyclin E. Oncogene 9:995–997

9 Quantification of Premature and Normal Mitosis

SUN W. TAM AND ROBERT SCHLEGEL

1 Background

Cellular chromosomes are typically visible only during mitosis or meiosis. However, premature chromosome condensation (PCC) of the G1 nucleus has been observed naturally in the pollen mother cells of *Triticum* x *Secale* hybrids (Bleier 1930), in an *Avena* hybrid (Holden and Mota 1956), and in *Haemanthus* endosperm (Ostergren and Bajer 1961). Viruses that can induce cell fusion, such as myxoviruses, herpes viruses, and adenoviruses, have also been shown to induce chromosome pulverization (Nichols et al. 1967). Pulverized chromosomes result from the premature condensation of chromosomes during DNA replication. Similarly, G1, S, and G2 chromosomes can be condensed by fusing interphase cells with mitotic cells (Johnson and Rao 1970).

Chemical compounds such as caffeine (Schlegel and Pardee 1986), the protein phosphatase inhibitor okadaic acid (Yamashita et al. 1990; Steinmann et al. 1991), and the protein kinase inhibitors 2-aminopurine, 6-dimethylaminopurine, and staurosporine (Schlegel et al. 1990; Yamashita et al. 1990; Tam and Schlegel 1992) have been shown to induce PCC in hamster cells arrested in S-phase by DNA synthesis inhibitors. The mechanism by which these chemicals bypass mitotic checkpoint control involves the activation of maturation-promoting factor (MPF) (Yamashita et al. 1990; Steinmann et al. 1991; Tam and Schlegel 1992), which consists of $p34^{cdc2}$ and cyclin B (see Norbury and Nurse 1992 for review). Newport and colleagues have proposed that both caffeine and okadaic acid promote the dephosphorylation of $p34^{cdc2}$ by inhibiting wee1 and mik1-like kinases (Symthe and Newport 1992), enzymes that phosphorylate $p34^{cdc2}$ and inhibit its activity (see Solomon 1993 for review).

More detailed studies into the molecular mechanisms of premature mitosis have reinforced the importance of MPF in this process. Mutation of Tyr-15 to Phe-15 in $p34^{cdc2}$ of *S. pombe*, and combined mutations of Thr-14 to Ala-14 and Tyr-15 to Phe-15 in human $p34^{cdc2}$ introduced into *X. laevis* egg extracts allow mitosis to occur in the presence of DNA synthesis inhibitors (Gould and Nurse 1989; Norbury et al. 1991). Similarly, overexpression of cdc25 in *S. pombe*, overexpression of the Drosophila cdc25 homologue *string* in *X. laevis* egg extracts, or the loss of mik1 and wee1 in *S. pombe* lead to premature mitosis in the presence of DNA synthesis inhibitors (Enoch and Nurse 1990; Kumagai and Dunphy 1991; Lundgren et al. 1991). Cdc25, a $p34^{cdc2}$ phosphatase that antagonizes the activities of weel or mikl, can dephosphorylate p34cdc2 and activate MPF (see Solomon 1993 for review). Hence, both the inhibitory and stimulatory pathways which regulate the activity of MPF can be manipulated to induce premature mitosis.

2 Principle and Applications

In this chapter, we present a procedure for quantification of normal and premature mitosis. We describe a protocol of inducing mammalian cells that are arrested in S phase to undergo premature mitosis by treatment with caffeine. These premature mitotic cells are collected in the presence of the microtubule polymerization inhibitor, nocodazole. By omitting all drug treatment, the same technique can be used to quantify the percentage of mitotic cells in an asynchronously growing population and to examine condensed chromosomes during normal mitosis. Normally, eukaryotic cells do not initiate mitosis when DNA replication is blocked (for review see Hartwell and Weinert 1989). This cell cycle checkpoint can be bypassed in some cells by caffeine treatment. S-phase-arrested Chinese and Syrian hamster cells such as CHEF18, CHO and BHK cells can undergo PCC, nuclear envelope disassembly, and spindle formation when exposed to caffeine, while human cells, such as HeLa or HT1080 cells, show little response under the same conditions (Brinkley et al. 1988; Schlegel and Craig 1991; Steinmann et al. 1991). Hence, only certain mammalian cell types can be used in this PCC procedure.

In order to show definitively that cells have undergone PCC, the cells must be swelled in a hypotonic solution and fixed in an acetic acid/methanol solution prior to microscopic examination. Without swelling, it is impossible to distinguish between cells undergoing apoptosis and cells undergoing PCC (see Fig. 1). Apoptotic cells display brightly staining condensed chromatin indistinguishable from PCC when cells are simply methanol-fixed on the tissue culture dishes (Meikrantz et al. 1994). After swelling, apoptotic cells can be seen to have an intact nuclear envelope and a diffuse homogeneously staining nucleus that often contains several "holes" (Arends et al. 1990). Cells undergoing PCC, however, show chromatin pulverization and nuclear envelope breakdown.

The ability to experimentally induce PCC has been applied to both basic and clinical research. For instance, chemically induced PCC has been used to investigate biochemical pathways involved in checkpoint control (e.g., Symthe and Newport, 1992; Schonthal and Feramisco 1993). This technique can also be used to screen for novel therapeutics that alter cell cycle regulatory pathways. Fluorescence in situ hybridization in conjunction with PCC has

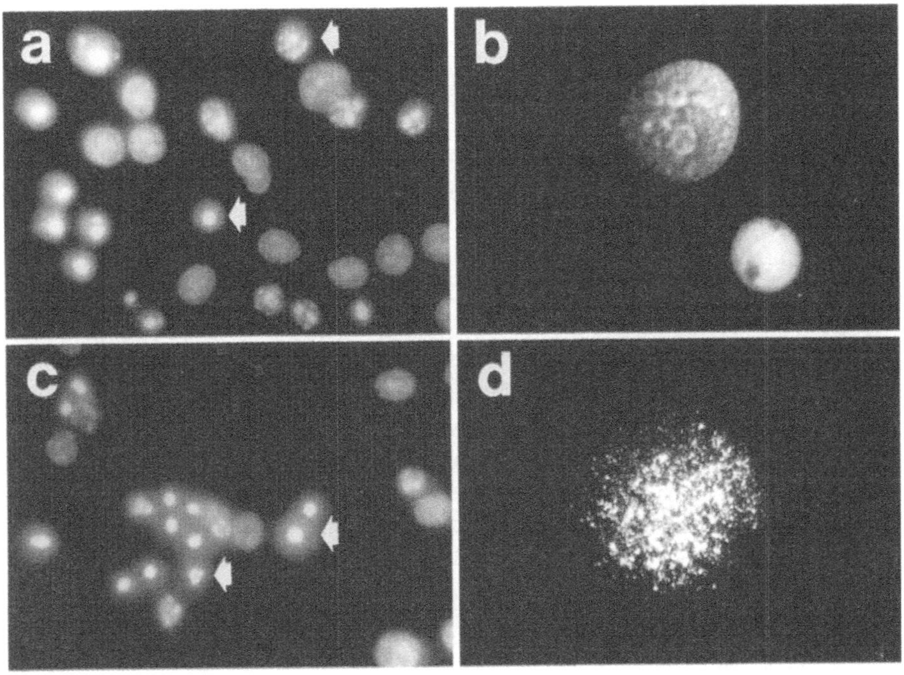

Fig. 1a–d. Fluorescence micrographs of staurosporine-induced apoptosis and premature chromosome condensation (PCC). – Apoptosis: HeLa cells were arrested in S phase with hydroxyurea (2.5 mM) for 5 h followed by treatment with staurosporine (20 ng/ml) for 8 additional h. **a** Arrows depict representative apoptotic nuclei, methanol-fixed in situ and stained with Hoechst 33258 (400×). **b** Chromosome spread showing a normal nucleus (*upper*) and an apoptotic nucleus (*lower*), stained with Hoechst 33258 (1000×). – PCC: BHK cells were arrested in S phase with hydroxyurea (2.5 mM) for 3 h followed by treatment with staurosporine (50 ng/ml) and nocodazole (0.25 μg/ml) for an additional 8 h. **c** Arrows depict representative cells undergoing PCC, methanol-fixed in situ and stained with Hoechst 33258 (400×). **d** Chromosome spread stained with Hoechst 33258 (1000×) showing chromosome pulverization and nuclear envelope breakdown, indicative of S-phase PCC. Caffeine, as well as all other PCC-inducing chemicals, produce morphological changes identical to those shown in panels **c** and **d**

been used to examine radiation-induced damage and repair of specific human chromosomes (Kovacs et al. 1994). Analysis of such chromosomes has proven to be more sensitive than the analysis of mitotic chromosomes for detecting radiation- and clastogen-induced DNA damage. The PCC technique can also be used to estimate the proliferation potential index of bone marrow, which may predict the recurrence of leukemia (Rao 1990). Many other useful applications are likely in the future.

3 Methods

3.1 Preparation of Chromosome Spreads

Reagents

- Phosphate buffered saline (PBS), consisting of 140 mM sodium chloride; 2.7 mM potassium chloride; 8 mM disodium phosphate, dibasic; 1.5 mM; potassium phosphate, monobasic; adjusted to pH 7.5.
- 0.5 mM ethylenediaminetetraacetic acid (EDTA) dissolved in PBS.
- 0.1 mg/ml nocodazole (Sigma) dissolved in dimethylsulfoxide (DMSO), stored at 4 °C.
- 100 mM hydroxyurea (Serva) dissolved in PBS; filter to sterilize and store at 4 °C. Discard after 1 month.
- 100 mM caffeine (Sigma) dissolved in PBS; filter to sterilize and store at room temperature. Discard after 2 months.
- 0.075 M potassium chloride (KCl).
- Carnoy's fixative, consisting of 1:3 (v/v) glacial acetic acid:methanol, prepared fresh for each use.
- Hoechst 33258 (Sigma).
- 0.1M N-propylgallate in 80% glycerol.
- 0.8% Na citrate.
- Giemsa.

Preparation

1. Grow mammalian cells to 60% confluency in 75 cm^2 tissue culture flasks. At least 1×10^6 cells are needed for each chromosome spread.
2. Add 250 µl of 100 mM hydroxyurea for every 10 ml of media (2.5 mM final concentration) to arrest cells in S phase. The cells are incubated at 37 °C for 3 h or until all mitotic cells have disappeared, indicating that G2 and M cells have exited from mitosis and, due to the presence of hydroxyurea, no other cells have progressed through S and G2 phases to enter mitosis.
3. Add 500 µl of 100 mM caffeine (5 mM final concentration) and 25 µl of 0.1 mg/ml nocodazole (final concentration of 0.25 µg/ml) per 10 ml of media. A control flask without caffeine addition is included for each experiment. The cells are then incubated for an additional 8 h.
4. Wash cells twice in 5 ml PBS containing 2.5 mM hydroxyurea, 5 mM caffeine, and 0.25 µg/ml nocodazole. The initial media and the PBS washes are retained.
5. Add 5 ml of 0.5 mM EDTA containing 2.5 mM hydroxyurea, 5 mM caffeine, and 0.25 µg/ml nocodazole if adherent cells, instead of suspension cells, are to be examined; otherwise proceed to step 7

6. Incubate cells for 7 to 10 min at 37 °C or until cells have detached from the flask. Hitting the flask several times with your hand should dislodge all cells.

7. Mix the initial media, the PBS washes, and the EDTA solution together in a transparent 50 ml plastic tube and spin at room temperature for 3 min at 1000 *g* to pellet the cells.

8. Discard the supernatant and add 10 ml of 0.075 M KCl and incubate the cells at room temperature for 7 to 10 min. If cells are swelled for too long, they will burst.

9. Pellet the cells at room temperature for 3 min at 1000 *g* and aspirate all but 0.5 ml of the supernatant.

10. Vortex the tube at the lowest speed while adding 10 ml of Carnoy's fixative dropwise.

11. Incubate the cells at room temperature for 10 min. At this point, cells can be stored at 4 °C overnight or processed further.

12. Spin the cells at room temperature for 3 min at 1000 *g*, and discard the supernatant.

13. Add 0.5 ml of Carnoy's fixative to resuspend cells.

14. Submerge a clean glass slide into ice-cold distilled water and drop the cell suspension with a Pasteur pipette dropwise at arm's length onto the wet slide. One to two drops are sufficient for each spread.

15. Air dry the slide for 5 min at room temperature.

16. Stain the slide with Hoechst 33258 (1 µg/ml for 7 min). Mount a cover slip over the fixed cells with 0.1M N-propylgallate in 80% glycerol to prevent fading of the Hoechst stain.

17. Examine cells by fluorescence microscopy at ×400 magnification to determine the fraction of cells having undergone premature mitosis. Staining cells with Giemsa, rather than Hoechst 33258, would allow the use of a light microscope for examination. Nuclear envelope breakdown and chromatin pulverization (see Fig. 1) are used as indicators of cells in S phase that are forced into premature chromatin condensation (Johnson and Rao 1970).

3.2 Counting Mitotic Figures of Mammalian Cells

This simplified protocol can be used to quantify the percentage of normal mitotic cells in any asynchronously growing population.

Counting

1. If adherent cells growing on glass coverslip are used, gently remove medium and wash cells twice in 5 ml PBS (vigorous washes will dislodge loosely attached mitotic cells). Immediately fix cells with Carnoy's fixative for 10 min with gentle rocking before proceeding to step 10. For suspension cells (at least 1×10^5 cells), wash twice in 5 ml PBS before proceeding to step 2.
2. Resuspend cells in 100 µl PBS (in an Eppendorf tube).
3. Add 1 ml 0.8 % Na citrate and incubate on ice for 1 min.
4. Spin gently, leave 100 µl of solution to resuspend the cells.
5. Add dropwise 1 ml Carnoy's fixative.
6. Spin again, leave 100 µl of solution to resuspend the cells.
7. Add 1 ml Carnoy's fixative.
8. Spin and resuspend cells in 20 to 100 µl Carnoy's fixative.
9. Submerge a clean glass slide into ice-cold distilled water and drop the cell suspension with a Pasteur pipette dropwise at arm's length onto the wet slide. One to two drops are sufficient for each spread.
10. Air dry the sample for 5 min at room temperature.
11. Stain with Giemsa and count mitotic figures under a light microscope.

4 References

Arends MJ, Morris RG, Wyllie AH (1990) Apoptosis: The role of the endonuclease. Am J Pathology 136:593–608

Bleier H (1930) Verhalten der Verschiedenen Kernkomponeten bei der Reduktionsteilung von Bastarden. Cellule 40:87–111

Brinkley BR, Zinkowski RP, Mollon WL, Davis FM Pisegna MA, Pershouse M, Rao PN (1988) Movement and segregation of kinetochores experimentally detached from mammalian chromosomes. Nature. 336:251–254

Enoch T, Nurse P (1990) Mutation of fission yeast cell cycle control genes abolishes dependence of mitosis on DNA replication. Cell 60:665–673

Gould KL, Nurse P (1989) Tyrosine phosphorylation of the fission yeast $cdc2^+$ protein kinase regulates entry into mitosis. Nature 342:39–45

Hartwell LH, Weinert TA (1989) Checkpoints: controls that ensure the order of cell cycle events. Science 246:629–634

Holden J, Mota M (1956) Non-synchronized meiosis in binucleate pollen mother cells of an *Avena* hybrid. Heredity 10:109–117

Johnson RT, Rao PN (1970) Mammalian cell fusion: induction of premature chromosome condensation in interphase nuclei. Nature 226:717–722

Kovacs MS, Evans JW, Johnstone IM, Brown JM (1994) Radiation-induced damage, repair and exchange formation in different chromosomes of human fibroblasts determined by fluorescence in situ hybridization. Radiat Res 137:34–43

Kumagai A, Dunphy WG (1991) The cdc25 protein controls tyrosine dephosphorylation of the cdc2 protein in a cell-free system. Cell 64:903–914

Lundgren K, Walworth N, Booher R, Dembski M, Kirschner M, Beach D (1991) mik1 and wee1 cooperate in the inhibitory tyrosine phosphorylation of cdc2. Cell 64:1111–1122

Meikrantz W, Gisselbrecht S, Tam SW, Schlegel R (1994) Activation of cyclin A-dependent protein kinases during apoptosis. Proc Natl Acad Sci USA 91:3754–3758

Nichols WW, Aula P, Levan A, Haneen W, Norrby E (1967) Radioautography with tritiated thymidine in measles and Sendai virus-induced chromosome pulverization. J Cell Biol 35:257–262

Norbury C, Nurse P (1992) Animal cell cycles and their control. Annu Rev Biochem 61: 441–470

Norbury C, Blow J, Nurse P (1991) Regulatory phosphorylation of the $p34^{cdc2}$ protein kinase in vertebrates. EMBO J 10:3321–3329

Ostergren G, Bajer A (1961) Mitosis with undivided chromosomes. I. A study on living material. Chromosoma (Berl) 12:72–79

Rao PN (1990) The discovery (or rediscovery?) of the phenomenon of premature chromosome condensation. BioEssays 12:193–197

Schlegel R, Pardee AB (1986) Caffeine-induced uncoupling of mitosis from the completion of DNA replication in mammalian cells. Science 232:1264–1266

Schlegel R, Belinsky GS, Harris MO (1990) Premature mitosis induced in mammalian cells by the protein kinase inhibitors 2-aminopurine and 6-dimethylaminopurine. Cell Growth Differ 1:171–178

Schlegel R, Craig RW (1991) Mitosis: normal control mechanisms and consequences of aberrant regulation in mammalian cells. In: Campisi J, Cunningham DD, Inouye M, Riley M (eds) Perspectives on cellular regulation: from bacteria to cancer. Wiley-Liss, New York, pp 235–249

Schonthal A, Feramisco JR. (1993) Inhibition of histone H1 kinase expression, retinoblastoma protein phosphorylation, and cell proliferation by the phosphatase inhibitor okadaic acid. Oncogene 8:433–441

Smythe C, Newport JW (1992) Coupling of mitosis to the completion of S phase in *Xenopus* occurs via modulation of the tyrosine kinase that phosphorylates $p34^{cdc2}$. Cell 68:787–797

Solomon MJ (1993) Activation of the various cyclin/cdc2 protein kinases. Cur Opin Cell Biol 5: 180–186

Steinmann KE, Belinsky GS, Lee D, Schlegel R (1991) Chemically induced premature mitosis: differential response in rodent and human cells and the relationship to cyclin B synthesis and $p34^{cdc2}$/cyclin B complex formation. Proc Natl Acad Sci USA 88: 6843–6847

Tam SW, Schlegel R (1992) Staurosporine overrides checkpoints for mitotic onset in BHK cells. Cell Growth Differ 3:811–817

Yamashita K, Yasuda H, Pines J, Yasumoto K, Nishitani H, Ohtsubo M, Hunter T, Sugimura T, Nishimoto T (1990) Okadaic acid, a potent inhibitor of type 1 and type 2A protein phosphatases, activates cdc2/H1 kinase and transiently induces a premature mitosis-like state in BHK21 cells. EMBO J 9:4331–4338

Part Two

Techniques for Studying the *Xenopus* Cell Cycle

10 Preparation of *Xenopus* Egg Extracts and Their Utilization in Cell Cycle Studies

PAUL R. CLARKE

1 Background, Principle and Applications

Early embryos, in particular those of the South African clawed frog, *Xenopus laevis*, have played important roles in the study of cell cycle processes. The cell cycle in early embryos is somewhat simplified, with a rapid alternation between mitosis and interphase, when DNA replication occurs. The cell cycles lack gap phases (G1 and G2) and many of the checkpoint controls that monitor the progression of the cell cycle are absent. No cell growth or transcription occurs, the embryos containing stores of almost all of the proteins required until the mid-blastula transition. These properties have made the early embryo of *Xenopus* a useful system for dissecting the underlying molecular mechanism controlling the progression of the cell cycle.

The large eggs of *Xenopus* can be easily microinjected and sufficient amounts can be prepared for biochemical analyses. However, observation of intracellular events in living embryos is not possible because of the opacity of the egg surface and yolk granules in the cytoplasm. Concentrated extracts made by gently crushing the eggs in a centrifuge have provided an effective cell-free system in which to study cell cycle processes. The concentration of intracellular components is maintained by avoiding dilution. In particular, the relative activities of enzymes such as regulatory protein phosphatases and kinases is not disturbed greatly. Such extracts have proved very useful in the investigation of processes such as membrane dynamics (Tuomikoski et al. 1989), nuclear assembly and transport (Newmeyer and Wilson 1991), DNA replication (Hutchison 1993; Leno and Laskey 1991), microtubule dynamics (Belmont et al. 1990; Verde et al. 1990), and the regulation of the molecular mechanism that controls and coordinates these events (Félix et al. 1993; Murray 1991).

Observations that the early embryonic cell cycles of *Xenopus* continue in the absence of DNA replication or spindle assembly (Hara et al. 1980) indicated that a cytoplasmic clock controls progression of the cell cycle. It is now known that the oscillation between interphase and mitosis is controlled by the activity of maturation-promoting factor (MPF), a protein kinase complex consisting of a *Cdc2* catalytic subunit and a B-type Cyclin (Nurse 1990). *Xenopus* oocytes are arrested in interphase until induced to mature by progesterone stimulation within the female frog or artificially by injection of active MPF into isolated oocytes. In addition, oocyte maturation can be produced by a variety of substances that activate the store of inactive preMPF already present in the oocyte. Maturation consists of the breakdown of the nucleus (the germinal vesicle) and progression through meiosis I, followed by arrest in meiosis II with MPF maintained at a high level by cytostatic factor (CSF), the composition of which is not completely defined. When a laid egg is fertilized, penetration of a sperm induces a calcium wave that inactivates CSF, releases the metaphase block and MPF activity rapidly declines to an interphase state. Thereafter, the first mitotic metaphase occurs at about 70 min, followed in rapid succession by the next 11 mitotic cell cycles, each lasting about 30 min (at 20 °C). A synchronous release from CSF-arrest can be artificially induced in a population of eggs by "activating" the eggs with a brief electric pulse or Ca^{2+} ionophore treatment. Activated eggs enter the first mitotic cell cycle, undergo periodic surface contraction waves and oscillation of MPF activity, although cell division does not occur.

Extracts that recapitulate cell cycle events were first produced using *Rana pipiens* eggs by Lohka and Masui (Lohka and Masui 1983), who crushed the eggs by centrifugation, separating the cortex, pigment granules and yolk from the cytoplasm. Subsequently, *Xenopus* extracts that would undergo single or multiple cell cycles in vitro were made by several groups using slightly different protocols (Blow and Laskey 1986; Hutchison et al. 1987; Félix et al. 1989; Murray and Kirschner 1989). When *Xenopus* sperm heads are added to such extracts, the chromatin decondenses, vesicles accumulate around it to form a nuclear envelope and DNA replication is initiated. Following the activation of MPF, nuclear envelope breakdown and chromatin condensation occurs. So the nuclear events of the entire cell cycle can be recapitulated in vitro. This system has numerous advantages for the study of cell cycle processes: components can be removed by immunoprecipitation or supplemented with recombinant proteins into which mutations can be introduced. Chemical inhibitors are also easily added. Such extracts have facilitated dissection of the complex interactions of molecules controlling cell cycle processes that are disrupted when diluted cell extracts are prepared from tissues or cultured cells. However, *Xenopus* is perhaps less useful for the identification of novel components, since genetic methods that have been so successful in uncovering the regulation of the cell cycle in budding yeast, fission yeast and *Drosophila* are not applicable.

The activation and inactivation of MPF has been studied extensively in *Xenopus* egg extracts. The ability of extracts to perform multiple cell cycles depends upon continued protein synthesis in the extracts. Murray and Kirschner (1989) demonstrated that this requirement was due solely to the synthesis of Cyclin, which occurs steadily during interphase from maternal mRNA transcripts. The mitotic Cyclins identified in the early embryonic cell cycles of *Xenopus* are B1, B2 and A (Minshull et al. 1989; Minshull et al. 1990); the complex of Cyclin B1 and Cdc2 is the predominant active kinase that induces the events of metaphase and was purified as MPF (Dunphy et al. 1988; Gautier et al. 1988; Gautier et al. 1990). As Cyclin B accumulates, it associates with Cdc2, which is present in a surplus in the early embryo. However, the kinase does not become immediately active, but is regulated post-translationally. When Cyclin B binds to Cdc2, phosphorylation of Cdc2 is induced at three sites: one is required for subsequent activity (Thr161), but the other two are dominant inhibitory sites in the ATP binding site of the kinase (Thr14 and Tyr15). Final activation of the kinase requires dephosphorylation of the inhibitory sites, catalyzed by the Cdc25 phosphatase. The timing of this transition and the entry into mitosis is controlled by a network of protein phosphatases and kinases that is still being defined (reviewed in Clarke 1994). Kinase activation initiates the process of Cyclin degradation (Félix et al. 1990), resulting in the inactivation of Cdc2/Cyclin B and return to interphase.

In this chapter, I will describe a method based upon that of Félix et al. (Félix et al. 1989) for preparing egg extracts that can be stored frozen. The activation of Cdc2 protein kinase by the addition of recombinant Cyclins and an assay of Cdc2 activity will also be described. Finally, the formation of functional nuclei that replicate their DNA and the reconstitution of a coupling mechanism between DNA replication and Cdc2 activation will be discussed. Several other protocols for preparing egg extracts specifically to study DNA replication, nuclear import and assembly, chromatin assembly and cell cycle regulation have also been published recently (Leno and Laskey 1991; Murray 1991; Newmeyer and Wilson 1991; Félix et al. 1993; Hutchison 1993).

2 Methods

2.1 The Eggs

Maintenance of a *Xenopus* colony is described in (Wu and Gerhart 1991). *Xenopus* females are induced to lay eggs by injections of pregnant mare serum gonadotropin (PMSG), which accelerates the growth of oocytes, followed by human chorionic gonadotropin (HCG) to induce oocyte maturation and egg laying. Routinely, we inject sufficient frogs with PMSG on Fridays for the following week. There should be at least 3 days before HCG injections, which can be made in the evening to allow the frogs to lay overnight. Frogs are then kept for at least 3 months before injecting them again.

The quality of extracts depends upon the eggs from which they are prepared. Each frog can lay several thousand eggs. Good eggs show two well distinguishable poles, a dark animal pole and a clear vegetal one, separated by a thin white zone. They should not activate spontaneously or lyse and become white upon incubation. Batches of eggs containing more than a few percent of poor eggs should be discarded. Low numbers of poor eggs can be removed using a Pasteur pipette. Different batches of eggs can be quite variable, so avoid mixing them. Many laboratories experience a seasonal variation in egg quality, the summer being worse.

Materials and instrumentation

For hormonal stimulation of frogs

- Pregnant mare serum gonadotropin (PMSG, Intervet).
- Human chorionic gonadotropin (HCG, Sigma).
- 1 ml syringes, 27-gauge needles.
- 0.1 M NaCl (4 M stock).

PMSG injection

1. Dilute the PMSG to 200 u/ml in sterile water. It can be frozen and stored at –20 °C.
2. Inject 0.5 ml (100 u) per frog. The frogs are injected subcutaneously in the back near the border of the leg using a new needle.
3. Keep the frogs for 3 to 10 days. Do not feed them or the water in which they will lay their eggs will be fouled.

HCG injection

1. Dilute the HCG to 2000 u/ml in sterile water. It can be stored at 4 °C for up to 1 week.
2. Inject 0.5 ml (1000 u) per frog as above.
3. Keep the frogs in 0.1 M NaCl overnight. It is preferable to maintain the temperature at 16–20 °C. At the lower temperature, egg laying is slower, but the eggs may be of better quality. Use the eggs within a few hours of laying.

2.2 Preparation of Egg Extracts

Xenopus egg extracts can be prepared as 10 000 g supernatants ("low speed extracts"), which still contain membranous fractions required for nuclear envelope formation (see Sect. 2.5 below), or 100 000 g supernatants which are clarified of heavy membranes ("high speed extracts"). The mechanism controlling the timing of Cdc2/Cyclin activation is present in both types of extract, but high speed extracts freeze better. Interphasic egg extracts are prepared by activating the eggs with an electric pulse to release the metaphase arrest and degrade mitotic Cyclins, followed an incubation in cycloheximide to inhibit their re-synthesis. Extracts containing little or no mitotic Cyclins can also be prepared as high speed extracts without the use of cycloheximide, since the high speed centrifugation removes the polysomes and Cyclins are not synthesized in such extracts. However, the extracts need to be prepared rapidly after activation to ensure that Cyclin synthesis does not occur to any great extent. High-speed extracts containing increasing amounts of Cyclin B1 can be prepared by incubating activated eggs at 20 °C without cycloheximide before crushing. Up to 50 min of incubation results in extracts with levels of Cyclin below the threshold required for Cdc2 activation; extracts prepared at 60 min will undergo a single round of Cdc2 activation and inactivation upon incubation. Metaphase arrested extracts can be prepared by omitting activation by electric pulse (step B, below), and ensuring that EGTA is present in the acetate buffer.

Materials and instrumentation

- MMR/4: 25 mM NaCl, 0.4 mM KCl, 0.25 mM $MgSO_4$, 0.5 mM $CaCl_2$, 1.25 mM Hepes, 25 μM EDTA, pH 7.2 (500 ml). A 10x MMR stock is kept at room temperature and diluted 40 times.
- L-cysteine (Sigma): 200 ml of 2.5% cysteine in water, pH 7.8 (neutralized with 6N NaOH). It should be prepared less than one hour before use.
- 50-ml plastic tubes.
- Activation chamber. We use a home-made activation chamber, similar to that described by Murray (1991). The bottom of the chamber and the lid form the two electrodes. They are made of stainless steel, the chamber of Plexiglas. For standard extract preparation, a 10 × 10 cm square chamber, 6 cm high, is convenient. The bottom of the chamber is covered with 1 cm of 1% agarose in MMR/4. The agarose is prepared freshly and the chamber cleaned before each use. The eggs rest on the agarose. The lid lies above the eggs on 0.5 cm high plastic feet fixed at the four corners. There should be enough MMR/4 buffer to cover the eggs and the lid (about 200 ml).
- A power supply that deliver a 12 V A.C. pulse to the chamber.
- Cycloheximide (Sigma): 200 μg/ml in MMR/4 (200 ml).
- Plastic petri dishes (10–15 cm diameter) with a thin layer of 1% agarose in MMR/4 covering the bottom.
- Plastic Pasteur pipette or glass Pasteur pipette with a cut tip (about 3 mm diameter) that has been fire-polished.
- Acetate buffer: 100 mM potassium acetate, 60 mM EGTA, 2.5 mM magnesium acetate, 1 mM DTT, 50 μg/ml cytochalasin D, pH 7.2. The EGTA is important for preparing extracts with active Cdc2, otherwise it may be omitted. The buffer without DTT and cytochalasin D can be stored at 4 °C for some days or in 50 ml aliquots at –20 °C. The cytochalasin D stock is made up at 1 mg/ml in DMSO; the DTT stock is 100 mM in water. They are stored in aliquots at –20 °C and added to the acetate buffer immediately before use. Fill the centrifugation tubes used in step B below with ice-cold acetate buffer before starting to handle the eggs. Keep some more acetate buffer on ice for washing.
- Centrifugation tubes and centrifuges. We use routinely SW 50.1 (5 ml tubes) or SW 60 (4 ml tubes) rotors of Beckman ultracentrifuges. Any swinging-out rotor corresponding to the desired volume would do, for in-

stance a TLS-55 for a table-top ultra-centrifuge (Beckman) or a HB-4 rotor with appropriate adaptors for a Sorvall centrifuge (for the low-speed centrifugation only). Pre-cool the rotors.

- 5-ml syringes and needles that can pierce the centrifugation tubes.
- Stock solutions for the ATP regenerating system: adenosine triphosphate (ATP) 50 mM, pH 7; creatine phosphate (Boehringer) 500 mM in water; creatine kinase (Boehringer) 4 mg/ml in 50% glycerol in water. Store aliquots at –20 °C. 1 µl of each will be added to 50 µl of extract.
- Liquid nitrogen.

Dejellying the eggs

1. Wash the eggs twice in MMR/4 in a 50-ml plastic tube. Allow the eggs to settle to the bottom of the tube between each wash. There should not be more than 30–35 ml of eggs per tube, otherwise the action of cysteine will be very slow. The yield is about 8 to 10 ml of packed eggs after the jelly is removed.

2. Rinse twice with cysteine. Fill up the tube completely the second time and close it. Avoid air bubbles inside the tube. Contact of the eggs with air causes activation.

3. Incubate in cysteine for a few minutes. Keep the tube horizontal in your hands and turn it continuously (about one turn every 2 s). It takes about 2 to 3 min to remove the jelly. Stop the incubation when you see that the eggs tend to sink rapidly.

4. Pour off the cysteine and wash the eggs three times with MMR/4. The eggs now pack rapidly and tightly at the bottom of the tube. Use immediately for activation.

Egg activation by electric pulse

1. Gently pour the eggs from the 50-ml tube into the activation chamber coated with agarose and filled with MMR/4 + cycloheximide. Close the lid of the chamber. The eggs should stay below the lid. Make sure that the buffer covers the lid.

2. Activate the eggs with two 12 V electric pulses, each 1 s long and separated by 5 s.

3. Remove the lid and observe the eggs for 4–6 min. If activation is successful, the pigment at the animal pole should contract so that the black part covers less than half of the egg surface. The extent of contraction is variable from batch to batch and may be easier to see under a dissecting microscope. The eggs rotate under gravity to turn their less dense dark animal pole upwards. Remove eggs which have turned white or have a marbled appearance. If there are too many (more than 5%), it is best to discard the whole batch.

4. Incubate the eggs for 60 to 90 min at 20 °C in MMR/4 + cycloheximide in the petri dishes with a covering of agarose. Discard eggs which have turned bad during this time. The eggs are then ready for extract preparation.

Egg crushing and centrifugation

1. Wash the eggs with one volume of ice-cold acetate buffer.
2. Transfer the eggs to a centrifuge tube filled with ice-cold acetate buffer using a plastic Pasteur pipette (or a glass Pasteur pipette with a cut tip). Allow the eggs to settle and remove as much buffer as possible with a pipette. Alternatively, the eggs may be packed by centrifugation at 100–200 *g* before removing the buffer. Keep everything at 4 °C from now on.
3. Centrifuge at 10 000 *g* for 10 min at 4 °C (9000 rpm in the SW 50.1 rotor, 8000 rpm in the SW 60 rotor). Use maximal acceleration rate to crush the eggs rapidly. The yolk sediments to the bottom of the tube (about 40% of the total volume). The dark pigments are found within and immediately above the yolk. If all pigment is sedimented, the supernatant is yellowish, otherwise it is gray. Residual pigments can be eliminated by a second spin at the same speed (this is not necessary if the extracts are to be subsequently centrifuged at higher speed). Finally, a bright yellow layer of lipids is found at the top (Félix et al. 1993).
4. Pierce the tube with a large needle above the pigment and yolk layers. Remove the viscous and turbid supernatant (without the yellow lipids at the top). Avoid making bubbles in the extract. If the needle is clogged after it has been used to pierce the tube, change it rapidly before taking out the extract.
5. Add the ATP regenerating system: 1 mM ATP, 10 mM creatine phosphate, 80 µg/ml creatine kinase (1/50th volume of each stock solution for one volume of extract).
6. If desired, centrifuge at 100 000 *g* for 60 min at 4 °C (30 000 rpm in a SW 50.1 rotor, 28 000 rpm in a SW 60 rotor). A translucent pellet of polyribosomes and glycogen is found at the bottom of the tube. Heavy membranous components sediment above. The supernatant is still straw coloured and contains membranes and other particulate material like monosomes. Pierce the tube with a needle above the heavy membranes and take the supernatant above this fraction, but not any yellow lipids at the top.
7. Extracts can be kept for a few hours on ice before use. Otherwise, freeze 50 or 100 µl aliquots in 0.5 ml tubes in liquid nitrogen. Freezing of low-speed extracts is more problematic than freezing of 100 000 *g* supernatants, probably because of their higher vesicle content, including lysosomes. Before freezing low-speed extracts, add 5% (by vol) glycerol and protease inhibitors. Hutchison (1993) recommends aprotinin at 10 µg/ml. Extracts are stable for many months stored in liquid nitrogen.

2.3 Activation of Cdc2 in Interphasic Extracts by the Addition of Recombinant Cyclins

All the components other than mitotic Cyclins that are required for the cell cycle are apparently present in *Xenopus* eggs. The activation of Cdc2 can therefore be studied using cycloheximide-treated interphasic extracts by adding back recombinant Cyclins, produced in bacteria, baculovirus-infected insect cells or translated in vitro in reticulocyte lysates (Solomon et al. 1990). The concentration of Cyclin can be controlled, and the post-translational regulation of kinase activation studied in the absence of any effects on Cyclin synthesis. The kinetics of Cdc2 activation for each batch of extract are highly reproducible. High-speed extracts do not seem to differ from low speed extracts in their characteristics of kinase activation and can often be used (Clarke et al. 1992). Experiments can be performed using extracts that have been stored frozen, avoiding the variability between different batches. Such experiments have shown that there is a threshold amount of Cyclin B that needs to be exceeded to trigger Cdc2 activation, below which the Cdc2/Cyclin B complex is kept phosphorylated on all three sites (Solomon et al. 1990; 1992). Even with high concentrations of Cyclin, there is a lag period before kinase activation. Cyclin A, on the other hand, activates Cdc2 rapidly and roughly in proportion to the amount added because it is not subject to phosphorylation at the two inhibitory sites (Clarke et al. 1992; Devault et al. 1992). Other additions may be made to manipulate the kinetics of activation of Cdc2, for instance purified protein phosphatase or their inhibitors. Inhibition of a type-2A protein phosphatase abolishes the lag period and the threshold requirement of Cdc2/Cyclin B activation (Solomon et al. 1990; Clarke et al. 1993).

! Pipette egg extracts gently, and when adding Cyclins or other reagents, mix carefully by flicking the tube. The extract should not be diluted by the addition of more than the minimum necessary – addition of more than 50% of the initial volume may prevent the activation of Cdc2.

Materials and instumentation

- Cycloheximide-treated interphasic egg extract, one 100 µl aliquot stored in liquid nitrogen.
- Cyclin, diluted in Tris buffer (50 mM Tris/HCl, pH 8.0, 100 mM KCl, 5 mM $MgCl_2$, 5 mM dithiothreitol).
- Extraction buffer (EB; 80 mM ß-glycerophosphate, pH 7.3, 20 mM EGTA, 15 mM $MgCl_2$, store in 50-ml aliquots at –20 °C. Just prior to use, add 1 mM Na_3VO_4, 1 mM dithiothreitol and protease inhibitors: 25 µg/ml aprotinin, 25 µg/ml leupeptin, 1 mM benzamidine, 0.5 mM PMSF).
- 96 micro-well plates (Nunclon).
- 0.5-ml microtubes (Eppendorf or similar).
- 0–20-µl pipette and tips.

Activation of Cdc 2

1. Aliquot 19 µl of EB into each well on a microwell plate. Label each row with a letter corresponding to an incubation condition (A, B, C, etc.) and each column with the time of the sample to be taken from each incubation (0, 10, 20, etc.). Keep the plate on ice.
2. Aliquot 10 µl of egg extract into microtubes on ice labeled A, B, C, etc.
3. Place tubes in rack at 20 °C or at room temperature and remove 1 µl into corresponding wells (A0, B0, C0, etc.) on the microwell plate on ice. Mix well by pipetting up and down.
4. Add 1 µl of Cyclin dilution (0–10 µM) to start reaction. Stagger addition to each incubation tube by 15 s. Control incubations without Cyclin should also be used.
5. Remove 1 µl samples from each tube every 10 min into corresponding wells (A10, B10, C10, etc.). Stagger sampling from each incubation tube by 15 s.
6. When completed, either transfer 5 µl from each well to a well on a fresh plate, and carry out kinase assays (Sect. 2.4), or freeze plate at –20 °C until use.

2.4 Assaying of the H1 Kinase Activity of Cdc2

Materials and instrumentation

- $\gamma[^{32}P]$ ATP (0.5–1 × 10^6 cpm/nmol). Take appropriate precautions for use of radioactive material.
- 4 mM ATP solution in water (diluted from a 100 mM stock, pH adjusted to 7.0 using NaOH. Store at –20 °C).
- 20 mg/ml histone H1 solution in water (store at –20 °C).
- Extraction buffer (EB; as Sect. 2.3).
- 96 microwell plate (Nunclon).
- Phosphocellulose P81 paper (Whatman).
- 0–20 µl pipette and tips.
- Tweezers and scissors.
- Beaker with wash solution (150 mM H_3PO_4).
- Technical grade ethanol.
- Scintillation counter and vials.

Kinase reaction

1. Prepare the assay mix of $\gamma[^{32}P]$ATP-histones at 4 °C (sufficient for 60 assays): 3 µl of $\gamma[^{32}P]$ ATP (0.5–1 × 10^6 cpm/nmol), 45 µl of 4 mM ATP, 222 µl of EB, 30 µl of 20 mg/ml histone-H1.
2. Pipette 5 µl of each sample to be assayed into labeled microwell plate, including two blanks of E.B at 4 °C.

3. Warm the micro-well plate to 20 °C. Start each reaction by adding 5 µl of assay mix to each sample, mix well by pipetting up and down several times. Start and stop each reaction at 15 s. intervals. Let each reaction continue for 15 min at 20 °C.
4. Stop the reaction by transferring 6 µl of the reaction mixture onto P81 paper precut into 1.5 × 1.5 cm squares and labeled with a pencil. If a time course is being performed, successive samples from an incubation can be spotted onto a strip of squares.
5. Allow mixture to adsorb for a few seconds, then submerge in a beaker containing about 300 ml of 150 mM H_3PO_4.
6. Remove free ATP by washing the filters for 3 × 5 min in 150 mM H_3PO_4 by swirling the beaker. The histone remains stuck to the P81 paper. The background can be checked by measuring the blanks with a Geiger counter.
7. Rinse briefly in ethanol and air-dry on paper towels.
8. Cut strips and transfer the squares into vials, add scintillation fluid and measure radioactivity in a scintillation counter. Count two 5 µl samples of the assay mix adsorbed onto P81 paper to determine the ATP specific radioactivity.

Calculations

1. Calculate the mean cpm values for the blanks to get the background. Subtract this from the cpm for each sample.
2. Calculate the mean cpm value for 5 µl of assay mix.
3. ATP concentration in the assay is that due to 5 µl of assay mix (3000 pmol)+ that due to 5 µl of diluted extract (250 pmol, assuming 1 mM ATP in undiluted extract).
4. Kinase activity in pmol/min/µl of original extract

$$= \text{cpm} \times \frac{\text{pmol ATP in assay}}{\text{cpm in 5 µl mix}} \times \frac{\text{sample vol.}}{\text{counted vol.}} \times \frac{1}{\text{vol. extract in assay}} \times \frac{1}{\text{assay time}}$$

$$= \text{cpm} \times \frac{3250}{\text{cpm in 5 µl mix}} \times \frac{10\,\text{µl}}{6\,\text{µl.}} \times \frac{1}{0.25\,\text{µl}} \times \frac{1}{15\,\text{min}}$$

$$= \text{cpm} \times \frac{1444}{\text{cpm in 5 µl mix}}.$$

2.5 Formation of Nuclei in Interphasic Egg Extracts

Xenopus egg extracts have been widely used as a model system to study the formation of the nucleus and DNA replication. Low speed extracts are used to provide the membranes required for nuclear envelope formation. Demembranated *Xenopus* sperm heads added to low speed extracts decondense rapidly and attract vesicles which form a complete nuclear envelope after about 60 min, when DNA replication is initiated (Blow and Laskey 1986; Hutchison et al. 1987). Cyclins A and B are apparently not required for this process, since nuclei that replicate their DNA can be formed in interphasic extracts that do not synthesize proteins (Blow and Sleeman 1990; Clarke 1994 unpubl.). Nuclei can also be formed around naked double-stranded DNA (eg., λ phage DNA) added to extracts (Newport 1987). In contrast, synthesis of a complementary strand on single-stranded DNA templates (eg., M13 phage) does not require a nuclear envelope, and occurs rapidly after addition to extracts (Méchali and Harland 1982).

To monitor the cell cycle progression of an extract, demembranated sperm heads can be added at about 100/µl. This concentration is insufficient to affect the activation of Cdc2/Cyclin B. Sperm morphology can be monitored by removing samples for fluorescence microscopy using a DNA stain such as Hoechst 33342 (Sigma B2261) or DAPI (Sigma D1388) (Murray 1991). Membrane formation can be examined under phase contrast microscopy or using a fluorescent lipophilic dye such as DHCC (Sigma D3652) (Newmeyer and Wilson 1991; Sullivan et al. 1993). DNA replication can be monitored by the incorporation of radioactivity from [α-^{32}P]dCTP or [α-^{32}P]dATP into acid-precipitable DNA (Blow and Laskey 1986; Hutchison et al. 1987), or by detecting incorporation of BrdUTP (Blow and Laskey 1986; Hutchison et al. 1987; Leno and Laskey 1991).

Dasso and Newport (1990) have shown that if higher numbers of nuclei are added to "cycling" extracts, the activation of Cdc2/Cyclin B is inhibited. In somatic cells there is a checkpoint mechanism that prevents activation of Cdc2/Cyclin B and entry into mitosis if DNA replication is not successfully completed. This mechanism does not normally operate during the early embryonic cell cycles, but apparently all the components are present, only the concentration of replicating DNA in the early embryo is too low to activate the checkpoint. The concentration of nuclei required to couple Cdc2/Cyclin B activation to DNA replication needs to be titrated for each extract; typically 500–1000/µl inhibit Cdc2 activation, but lower concentrations of nuclei will inhibit in the presence of aphidicolin to arrest DNA replication. This coupling can be performed in cycloheximide-arrested interphasic extracts, forming Cdc2/Cyclin B by the addition of recombinant Cyclin B (Clarke et al. 1993; Clarke unpubl.).

Preparation of Demembranated Sperm Heads. Variations of the method of (Gurdon 1976) have been described by several groups. This protocol is based on that described in Blow and Laskey (1986); Murray (1991); Philpott et al. (1991).

Materials and instumentation

- Buffer XN: 50 mM HEPES-KOH (pH 7.0), 250 mM sucrose, 75 mM NaCl, 0.5 mM spermidine, 0.15 mM spermine. Immediately before use, add 1 mM dithiothreitol, 10 µg/ml leupeptin, 0.1 mM PMSF.
- Lysolecithin (L-α-Lysophosphatidylcholine, Sigma L4129), 2 mg/ml in buffer XN, prepared freshly.
- 3% (mass/vol) bovine serum albumin (BSA) in buffer XN.

Preparation

1. Anaesthetize two male *Xenopus* by immersion in 2% (mass/vol in H_20) 3-amino benzoic acid ethyl ester (Sigma) for about 30 min. Once they are insensible, remove from the anaesthetic and kill by cutting the spinal cord at the back of the neck with a pair of sharp scissors.
2. Incise the skin on both sides of the lower abdomen and remove the testes onto a petri dish containing ice–cold MMR (Sect. 2.2). The testes are white and oval, about 1 cm long. Keep everything at 4 °C or on ice from now on. Dissect away fat and rinse in MMR.
3. Wash twice with XN and remove as much buffer as possible with a pipette. Macerate the testes with scissors.
4. Add 4 ml of XN and homogenize by pipetting up and down with a pipette tip cut to give an opening of about 3 mM.
5. Remove debris by centrifugation for 1 min at 1000 rpm in a Hereaus bench top megafuge (or similar).
6. Centrifuge the supernatant at 4500 rpm for 5 min. The majority of the sperm should sediment.
7. Resuspend the pellet avoiding the red erythrocytes, and wash three times with XN by centrifugation.
8. Resuspend in 1 ml XN and add 50 µl lysolecithin solution. Keep on ice for 10 min. Then add 2 ml of 3% BSA and recover the sperm heads by centrifugation. Wash three times with XN + 0.3% BSA.
9. Finally resuspend in 0.5 ml of XN + 0.3% BSA + 30% glycerol. Count the density in a haemocytometer and adjust to 10^7 sperm heads/µl. Freeze 10 µl aliquots in liquid nitrogen and store at –70 °C.

Acknowledgements. Many of these protocols were first developed by Marie-Anne Félix in Eric Karsenti's laboratory at the European Molecular Biology Laboratory in Heidelberg. I thank Marie-Anne and the other numerous members of the lab who helped and provided thoughtful discussions; in particular Eric for his advise and guidance. My work was funded by fellowships from The Wellcome Trust, EMBL and a grant from the HFSP (to E.K.).

References

Belmont LD, Hyman AA, Sawin KE, Mitchison TJ (1990) Real-time visualization of cell cycle-dependent changes in microtubule dynamics in cytoplasmic extracts. Cell 62: 579–589

Blow JJ, Laskey RA (1986) Initiation of DNA replication in nuclei and purified DNA by a cell-free extract of *Xenopus* eggs. Cell 47:577–587

Blow JJ, Sleeman AM (1990) Replication of purified DNA in *Xenopus* egg extract is dependent on nuclear assembly. J Cell Sci 95:383–391

Clarke PR (1994) Protein phosphatases and the cell cycle: the control of entry into mitosis. Adv Protein Phosphatases 8:163–185

Clarke PR Leiss D, Pagano M, Karsenti E (1992) Cyclin A- and Cyclin B-dependent protein kinases are regulated by different mechanisms in *Xenopus* egg extracts. EMBO J 11:1751–1761

Clarke PR, Hoffmann I, Draetta G, Karsenti E (1993) Dephosphorylation of Cdc25-C by a type-2A protein phosphatase: specific regulation during the cell cycle in *Xenopus* egg extracts. Mol Bio. Cell 4:397–411

Dasso M, Newport JW (1990) Completion of DNA replication is monitored by a feedback system that controls the initiation of mitosis in vitro: studies in *Xenopus*. Cell 61: 811–823

Devault A, Fesquet D, Cavadore J-C, Garrigues AM, Labbé JC, Lorca T, Picard A, Philippe M, Dorée M (1992) Cyclin A potentiates maturation-promoting factor activation in the early *Xenopus* embryo via inhibition of the tyrosine kinase that phosphorylates Cdc2. J Cell Biol 118:1109–1120

Dunphy WG, Brizuela L, Beach D, Newport J (1988) The *Xenopus* Cdc2 protein is a component of MPF, a cytoplasmic regulator of mitosis. Cell 54:423–431

Félix M-A, Clarke PR, Coleman J, Verde F, Karsenti E (1993) *Xenopus* egg extracts as a system to study mitosis. In: Fantes P, Brooks R (eds) The cell cycle: a practical approach. IRL Press, Oxford, pp 253–283

Félix M-A, Pines J, Hunt T, Karsenti E (1989) A post-ribosomal supernatant from activated *Xenopus* eggs that displays post-translationally regulated oscillation of its Cdc2+ mitotic kinase activity. EMBO J 8:3059–3069

Félix MA, Labbe JC, Doree M, Hunt T, Karsenti E (1990) Triggering of Cyclin degradation in interphase extracts of amphibian eggs by Cdc2 kinase. Nature 346:379–382

Gautier J, Norbury C, Lohka M, Nurse P, Maller J (1988) Purified maturation-promoting factor contains the product of a *Xenopus* homolog of the fission yeast cell cycle control gene Cdc2+. Cell 54:433–439

Gautier J, Minshull J, Lohka M, Glotzer M, Hunt T, Maller JL (1990) Cyclin is a component of maturation-promoting factor from *Xenopus*. Cell 60:487–494

Gurdon JB (1976) Injected nuclei in frog oocytes: fate, enlargement, and chromatin dispersal. J Embryol Exp Morphol 36:523–540

Hara K, Tydeman P, Kirschner M (1980) A cytoplasmic clock with the same period as the division cycle in *Xenopus* eggs. Proc Natl Acad Sci USA 77:462–466

Hutchison CJ (1993) The use of cell-free extracts of *Xenopus* eggs for studying DNA replication in vitro. In: Fantes P, Brooks R (eds) The cell cycle: a practical approach. IRL Press, Oxford, pp177–195

Hutchison CJ, Cox R, Drepaul RS, Gomperts M, Ford CC (1987) Periodic DNA synthesis in cell-free extracts of *Xenopus* eggs. EMBO J 6:2003–2010

Leno GH, Laskey RA (1991) DNA replication in cell-free extracts from *Xenopus laevis*. In: Kay BJ, Peng HB (eds) *Xenopus laevis:* practical uses in cell and molecular biology, vol 36. Academic Press, San Diego, pp561–579

Lohka MJ, Masui Y (1983) Formation in vitro of sperm pronuclei and mitotic chromosomes induced by amphibian ooplasmic components. Science 220:719–721

Méchali M, Harland RM (1982) DNA synthesis in a cell-free system from *Xenopus* eggs: priming and elongation on single-stranded DNA in vitro. Cell 30:93–101

Minshull J, Blow JJ, Hunt T (1989) Translation of Cyclin mRNA is necessary for extracts of activated *Xenopus* eggs to enter mitosis. Cell 56:947–956

Minshull J, Golsteyn R, Hill CS, Hunt T (1990) The A- and B-type Cyclin associated Cdc2 kinases in *Xenopus* turn on and off at different times in the cell cycle. EMBO J 9: 2865–75

Murray AW (1991) Cell cycle extracts. In: Kay BJ, Peng HB (eds) *Xenopus laevis:* practical uses in cell and molecular biology, vol. 36. Academic Press, San Diego, pp581–605

Murray AW, Kirschner MW (1989) Cyclin synthesis drives the early embryonic cell cycle. Nature 339:275–80

Newmeyer DD, Wilson KL (1991) Egg extracts for nuclear import and nuclear assembly reactions. In: Kay BK, Peng HB (eds) *Xenopus laevis*: practical uses in cell and molecular biology, vol. 36. Academic Press, San Diego. pp 607–634

Newport J (1987) Nuclear reconstitution in vitro: stages of assembly around protein-free DNA. Cell 48:205–217

Nurse P (1990) Universal control mechanism regulating onset of M-phase. Nature 344: 503–8

Philpott A, Leno GH, Laskey RA (1991) Sperm decondensation in *Xenopus* egg cytoplasm is mediated by nucleoplasmin. Cell 65:569–578

Solomon MJ, Glotzer M, Lee TH, Philippe M, Kirschner MW (1990) Cyclin activation of $p34^{Cdc2}$. Cell 63:1013–1024

Solomon MJ, Lee T, Kirschner MW (1992) Role of phosphorylation in $p34^{Cdc2}$ activation: identification of an activating kinase. Mol Biol Cell 3:13–27

Sullivan KMC, Busa WB, Wilson KL (1993) Calcium mobilization is required for nuclear vesicle fusion in vitro: implications for membrane traffic and IP_3 receptor function. Cell 73:1411–1422

Tuomikoski T, Felix MA, Doree, M Gruenberg J (1989) Inhibition of endocytic vesicle fusion in vitro by the cell-cycle control protein kinase Cdc2. Nature 342:942–945

Verde F, Labbe JC, Doree M, Karsenti E (1990) Regulation of microtubule dynamics by Cdc2 protein kinase in cell-free extracts of *Xenopus* eggs. Nature 343:233–238

Wu M, Gerhart J (1991) Raising Xenopus in the laboratory. In: Kay BL, Peng HB (eds) *Xenopus laevis*: practical uses in cell and molecular biology, vol. 36. Academic Press, San Diego, pp3–18

Part Three

Techniques for Study the Cell Cycle

11 Cloning and Expression of Cell Cycle Genes in *Schizosaccharomyces pombe*

GUILLAUME COTTAREL

1 Background

It is undoubtedly admitted that the study of the cell cycle and yeast genetics cannot be dissociated from each other. Both the fission yeast *Schizosaccharomyces pombe* and the budding yeast *Saccharomyces cerevisiae* are powerful genetic systems to investigate the life cycle in broad terms. In both organisms, a large collection of cell division cycle (cdc) mutants that arrest the cell cycle at a specific point have been isolated. Their characterization has contributed to the elucidation of the sequential events of the cell cycle and to the identification of genes whose activity is required for the progression through the cell cycle. The general principle is that the succession of each cell cycle step is dependent on the completion of the preceding step.

In fission yeast, the cell division cycle temperature sensitive (cdc^{ts}) mutations that prevented entry into mitosis are revealing. The phenotypes of different mutations of $cdc2^{+}$ suggest that this gene encodes a mitotic inducer. The study of the interactions of different mutants affecting entry into mitosis suggests that the activity of the $cdc2^{+}$ gene product is regulated by other proteins. The $cdc25^{ts}$ mutation, which arrests cells at the same point in the cell cycle as $cdc2^{ts}$ mutation, can be rescued by the $wee1^{ts}$ mutation. The growth and division of the $cdc25^{ts}$ $wee1^{ts}$ double mutant is normal at the restrictive temperature, suggesting that *cdc25* is an antagonist of the mitotic inhibitor *wee1* (Nurse 1975). In a similar observation, a specific cold-sensitive allele of *cdc2* is rescued by an allele of *cdc13* (Booher and Beach 1987). The Cdc2-Cdc13 complex is an essential component of a biochemically characterized mitotic activator complex known as the Maturation Promoting Factor (MPF) (Lohka et al. 1988). An active form of MPF induces all mitotic events such as nuclear membrane breakdown (not in yeast, however), chromosome condensation, and spindle assembly. The advances in molecular biology subsequently allowed the isolation of genes from species different than yeast by their ability to complement a cell cycle defect. A few examples include the cloning of human *cdc2* (Lee and Nurse 1987) and of human *cyclin D1* (Xiong et al. 1991). Ultimately, history has showed that the enzymes governing the major events of cell proliferation and division are highly conserved in function and in structure from yeast to human.

2 Principle and Applications

Cloning by complementation of fission yeast mutations is a common method of isolating genes.

Consider a temperature sensitive mutant (tsm) isolated as a cell cycle mutant which is both recessive and temperature-sensitive for growth. The mutant can grow at 30 °C but is unable to make a colony at 36 °C. The tsm mutant is arrested at a specific cell cycle stage where its normal equivalent product begins to function. An *S. pombe* genomic DNA clone or a heterologously (e.g., human cDNA) expressed clone that complements this mutation can be isolated by transforming the tsm strain with a genomic library or an expression library, and subsequently screening for temperature-resistant colonies. Further analysis is needed to prove that the observed complementation is plasmid-specific and is not due to reversion of the tsm locus or to chromosomic suppression. In the case of a plasmid-conferred phenotype, the tsm^+ character cosegregates with the plasmid selection marker mkr^+ (tsm^+, mkr^+ versus tsm^-, mkr^-). In some cases, the gene of interest may integrate into the genome, then revealing only the tsm^+, mkr^+ segregation. In the case of a reversion of the tsm locus, tsm^+ colonies grow independently of the presence of the plasmid marker (tsm^+, mkr^+ and tsm^+, mkr^-).

There are several ways to make yeast cells competent to take up DNA. The alkali method is based on the observation that alkali cations make yeast competent to take up DNA, the subsequent addition of polyethylene glycol and a heat shock trigger DNA uptake. The enzymatic method involves the degradation of the cell wall. The cells freed of cell wall (spheroplasts) are kept intact in a high osmotic pressure solution and treated with polyethylene glycol to promote DNA uptake. After removal of the polyethylene glycol spheroplasts and DNA are resuspended in osmotically stable regeneration agar.

Most of the vectors used for transformation carry as markers the *S. cerevisiae LEU2* and *URA3* or the *S. pombe ura4* genes which complement the *leu1* and *ura4* mutant alleles respectively. The vector origin of replication is either the *S. cerevisiae* 2 μ origin of replication or the *S. pombe* autonomous replicative sequences (ars), with the former yielding a lower copy number than the latter.

The described protocols for fission yeast transformation and gene expression are commonly in use. However, some knowledge of fission yeast genetics and cell biology is preferable. It is obvious that a strain should be completely characterized. The type of medium, the permissive/restrictive temperatures, the amount of drug should be analyzed. As an example, the *mcs4–13* mutant displays a thiabendazole (an inhibitor of microtubule function) sensitivity and a slight cell elongation on minimal medium. The same mutant exhibits a temperature sensitive phenotype when associated *with wee1–50 cdc25–22* mutations on minimal medium, whereas on rich medium is viable at the same restrictive temperature (Molz et al. 1989). Similarly the *cdc13-17* allele is osmotically remedial and was therefore always analyzed on solid medium that lacked sorbitol (Booher and Beach 1988).

3 Methods

Reagents, buffers, media

Lithium acetate 0.1 M
10.20 g in 90 ml of deionized water, adjust to pH 4.9 with acetic acid, bring up to 100 ml with water.

50% Polyethylene Glycol 3350
50 g of PEG in 100 ml of 0.1 M lithium acetate pH 4.9.

MB Medium (1 l)

KH_2PO_4	0.5 g
$MgSO_4.7H_2O$	0.5 g
NaCl	0.1 g
$CaCl_2.2H_2O$	0.1 g
$(NH_4)_2SO_4$	5 g
Glucose	5 g
Potassium acetate	0.36 g
Mineral stock	0.1 ml
Vitamins stock	1 ml
Amino acid as required	150 mg

TE

10 mM Tris-HCl pH 7.4
1 mM EDTA

SP1

1.2 M sorbitol
50 mM sodium-citrate ($Na_3C_6H_5O_7.2H_2O$)
50 mM sodium phosphate ($Na_2HPO_4.7H_2O$)
40 mM EDTA

pH to 5.6, autoclave

SP2

1.2 M sorbitol
50 mM sodium-citrate ($Na_3C_6H_5O_7.2H_2O$)
50 mM sodium phosphate ($Na_2HPO_4.7H_2O$)
pH to 5.6, autoclave

SP3

1.2 M sorbitol
10 mM Tris-HCl
pH to 7.6, autoclave

SP4

1.2 M sorbitol
10 mM Tris-HCl
10 mM calcium chloride
pH to 7.6, autoclave

SP5

20% polyethylene glycol (PEG 3350)
10 mM Tris-HCl
10 mM calcium chloride
pH to 7.6, filter sterilize

Breaking buffer

Triton X-100	2% (v/v)
Sodium duodecyl sulfate	1% (v/v)
NaCl	100 mM
Tris-HCl, pH 8.0	10 mM
EDTA, pH 8.0	1 mM

Minimal Medium (MM) (1 liter), MM medium can also be referred to as EMM or PM (Alfa et al. 1993). Can be obtained from Bio 101 (La Jolla, California).

KH phthalate	3 g
Na_2HPO_4	2.2 g
NH_4CL[1]	5 g
Glucose	20 g
Salts stock (50x)	20 ml
Mineral stocks (10 000x)	0.1 ml
Vitamin stock (1000x)	1 ml

[1] Glutamate (monosodium salt, 5 g/l) can replace the ammonium chloride to aid the growth of some auxotrophs and to increase sensitivity to specific inhibitors.

For solid medium add 20 g/l of agar (Difco Bacto). Adjust to 1 l, autoclave.

For MM + sorbitol plates add 218 g/l of sorbitol, autoclave.

Salts Stock (50x) (1 l)

$MgCl_2.6H_2O$	53.3 g
$CaCl_2.2H_2O$	0.735 g
KCl	50 g
Na_2SO_4	2 g

Filter sterilize, aliquot in small amount, store at 4 °C.

Mineral Salts (10 000x) (1 l)

H_3BO_3	5 g
$MnSO_4$	4 g
$ZnSO_4.7H_2O$	4 g
$FeCl_3.6H_2O$	2 g
$MoO_4.H_2O$	1.6 g
$CuSO_4.5H_2O$	0.4 g
KI	1 g
Citric acid	10 g

Filter sterilize, store at 4 °C.

Vitamins Stock (1000x) (1 l)

Nicotinic acid	10 g
Inositol	10 g
Biotin	10 mg
Pantothenic acid	1 g

Filter sterilize, store at 4 °C.

Yeast Extract Medium (YE) (1 l)

Difco yeast extract	5 g
Glucose	30 g

Amino acids when required at 75–150 mg/l

Phloxin B (Magdala red)

5 mg/ml in distilled water, filter sterilize. Add to medium at a final concentration of 20 mg/l.

3.1 *Schizosaccharomyces pombe* Transformation

Lithium acetate procedure (Okazaki et al. 1990)

1. At permissive temperature, grow the strain in 10 ml YE medium to saturation.
2. Inoculate 100 ml of MB medium and let grow to a density of 0.5–1×10^7 cells/ml.
3. Pellet the culture at 3000 rpm for 5 min.
4. Wash the cells in 40 ml of water, pellet the cells.
5. Resuspend the cells in 0.1 M lithium acetate (pH 4.9) to reach 1×10^9 cells/ml.
6. Transfer 0.1 ml aliquots to Eppendorf tubes, leave at permissive temperature for 60 min.
7. Add 1 μg of transformant DNA in 15 μl TE, vortex gently the cells and incubate at permissive temperature for 60 min.
8. Add 290 μl of PEG 3350/4000 to each tube and vortex gently. Incubate at permissive temperature for 60 min.
9. Heat-shock at 43 °C for 15 min. Leave the tubes at room temperature for 10 min.
10. Pellet the cells at 3000 rpm for 5 s, remove the supernatant.
11. Resuspend the cells in 1.0 ml of 1/2 YE (0.5 ml YE + 0.5 ml water). For adenine and/or uracil prototroph strains it is recommended to add adenine and/or uracil in the YE, since unsupplemented YE contains limited amount of adenine and uracil. Transfer the cell suspension in a flask containing 9 ml of 1/2 YE medium (+/– adenine, +/– uracil) and incubate for 60 min with shaking at permissive temperature.
12. Plate aliquots of 100–300 μl on MM selection plates.

Protoplast procedure (Beach and Nurse 1981)

1. Grow the strain at permissive temperature in 10 ml YE medium until high density.
2. Inoculate 100 ml of MM + 0.5% glucose and supplements, and let grow until cell density reaches 5×10^6–2×10^7 cells/ml.
3. Spin the culture at 3000 rpm for 5 min, decant the supernatant, resuspend the pellet in 10 ml of SP1. Add 20 μl of β-mercaptoethanol, incubate 10 min at room temperature.
4. Pellet the cells at 3000 rpm for 5 min and resuspend in 10 ml of SP2 containing 25 mg of NovoZym 234/lysing enzyme from *Trichoderma harzianum* (Sigma). Incubate at 37 °C until approximately 50% of the cells are

spheroplasts (0.5–1 h): the cells become rounded and are inspected under the microscope.

5. Add 30 ml of SP3, mix by shaking, pellet at 3000 rpm for 5 min. Wash the pellet twice with 30 ml of SP3.
6. Resuspend the pellet in 1 ml of SP4.
7. Use 100 µl of protoplasts with 1–10 µg of transforming DNA, leave at room temperature for 15 min. At this stage, to increase transformation efficiency, 100 µl of SP4 containing lipofectin at 66 g/ml, can be added (Allshire 1990). Incubate for 15 additional min.
8. Add 1 ml of SP5, mix by shaking the tube and incubate at room temperature for 15 min.
9. Pellet the cells at 2000 rpm for 5 min. Carefully remove the supernatant; the pellet will appear as a film at the bottom of the tube.
10. Resuspend the cells in 200–500 µl of SP4 and spread various volumes if necessary on dried MM + sorbitol and selection agar plates.

Transformants should appear within 2–5 days at around 30 °C.

Modifications

Others valuable *S. pombe* transformation protocols have been published and are here simply quoted: lithium chloride procedure (Bröker 1987), rapid lithium acetate (Elble 1992), improved lithium acetate (Kanter-Smoler et al. 1994), cryopreserved competent cells (Bröker 1993), electroporation procedures (Hood and Stachow 1990; Prentice 1992).

3.2 *Schizosaccharomyces pombe* Selection of the cdc⁺ Genes

As mention earlier, each mutant or each approach to select new cell cycle control genes is strain dependent. Nevertheless, some principles to clone genes in a temperature-sensitive mutant strain can be enunciated. After transformation with the library and the control vectors, the cells are plated on plasmid selection medium and allowed to grow at the permissive temperature. Once the colonies are visible they are replica-plated on Magdala red (phloxin B) containing medium and incubated at restrictive temperature. As soon as the colonies are visible (e.g., 24 h at 36 °C), select the pink growing colonies among the red colonies. The drug acts as a "metabolic marker". Gene rescued cells, actively dividing, are able to exclude the Magdala red and appear pink. The nonrescued cells which accumulate in the compound cannot or barely grow and appear red. Another possibility is to plate the cells after transformation on plasmid selection medium and let grow at permissive temperature for 24–36 h. Then, the plates are transferred at restrictive temperature. Inspection of the plates is done every 24 h to identify the few growing colonies.

3.3 Analysis of Plasmid-Conferred Phenotype

This step is to determine if your positive clones are plasmid-dependent for survival. Integration of the plasmid could occur and must be identified.

Plasmid instability test

1. Streak your positive clones on selective medium and let grow at permissive temperature.
2. Streak or patch the further growing candidates on nonselective medium and let grow at permissive conditions (permissive temperature, no drug, no plasmid selection). Repeat this step at least three times. These steps relax plasmid selection and allow plasmid loss.
3. Streak for numerous single colonies on YE plates.
4. When single colonies are visible, replica-plate onto selective and nonselective plates and incubate at permissive and restrictive temperatures. For example: incubate at restrictive temperature the YE plates to select for temperature resistance (ts^+), at permissive temperature the MM plates to select for the plasmid (mkr^+). When colonies are visible score for the correlation between the recovery of the phenotype and the presence of the marker (ts^+ mkr^+ versus ts^- mkr^-).

Integrated Plasmid Identification

In a few cases, plasmid instability has not been observed among positive candidates (Hoffman, pers. comm.) or the gene of interest could not be cloned when present on a multicopy plasmid (Matsumoto and Beach 1991). In both cases, transformants carrying the gene of interest possessed an integrated rather than autonomous copy of the plasmid, in such cases integration occurs by homologous recombination. Integration of the plasmid must be confirmed by Southern blot. Indeed, reversion of the tsm and of the auxotrophic marker loci can occur.

Rapid isolation of yeast chromosomal DNA is done according to the "smash and grab" protocol (Hoffman 1993).

Plasmid identification

1. Grow a 10-ml culture of your transformants in YE until saturation.
2. Pellet the cells at 3000 rpm for 5 min. Remove the supernatant and resuspend in 0.5 ml water.
3. Transfer the resuspended cells to a microcentrifuge tube and spin 5 s. Pour off supernatant and disrupt pellet by vortexing briefly.

4. Resuspend cells in 200 µl breaking buffer. Add about 200 µl of glass beads (0.45 µm diameter) and 200 µl phenol/chloroform/isoamyl alcohol and vortex at high speed for 3 min.
5. Add 200 µl of TE, vortex briefly.
6. Microcentrifuge 5 min at high speed. Transfer the aqueous layer to a clean microcentrifuge tube. Add 1 ml of 100% ethanol, mix by inversion.
7. Microcentrifuge 3 min at high speed. Remove the supernatant and resuspend the pellet in 0.4 ml TE.
8. Add 30 µl of 1 mg/ml RNAseA, mix, incubate 5 min at 37 °C.
9. Add 10 µl of 4 M ammonium acetate and 1 ml 100% ethanol. Mix by inversion.
10. Microcentrifuge 3 min at high speed. Discard the supernatant and dry the pellet. Resuspend DNA in 100 µl TE.
11. Digest 5 µl of the DNA with restriction enzymes which linearize the vector in a final volume of 20 µl.
12. Southern blots are done according to established protocols by using the vector without insert as a probe. Depending on the restriction enzyme used, bands of various sizes are detected. This is use to characterize the vector plus flanking genomic DNA.

3.4 Plasmid Recovery from Yeast

Unstable plasmid recovery (Hoffman 1993)

1. Grow your selected transformants in 2 ml of MM medium lacking the supplement for proper plasmid selection (e.g., without leucine if you plasmid vector carries the *LEU2*$^+$ gene) until the cell density is higher than 1×10^7 cells/ml.
2. Transfer the culture to a microcentrifuge and pellet the cells at 3000 rpm for 5 s, remove the supernatant. Disrupt the pellet by vortexing briefly.
3. Resuspend the pellet in 200 µl of breaking buffer, add about 200 µl of glass beads and 200 µl of phenol/chloroform/isoamyl alcohol. Vortex 2 min at high speed.
4. Microcentrifuge at high speed for 5 min.
5. Transform competent *Escherichia coli* with 1–2 µl of the aqueous layer. The use of electrocompetent *E. coli* cells is recommended. Plate on the appropriate selection plates.

Integrated plasmid recovery

1. From the restriction analysis analyzed in step 12 (above), select the enzyme which liberates the *E. coli* selectable marker and origin of replication joined to a genomic DNA fragment of appropriate size. It should be emphasized that, due to the recombination process, the integrated plasmid is flanked on one side by the wild-type genomic DNA originated from the plasmid and on the other by the mutated locus. Further analysis is later required to confirm the identity of the wild-type fragment.
2. From step 12 ending the "Plasmid identification" procedure take 5 µl and digest with the selected restriction enzyme.
3. Inactivate the restriction enzyme and ligate 1–3 µl of the digestion reaction with T4 DNA ligase in a final volume of 20 µl.
4. Transform competent *E. coli* with 1–2 µl of the ligation reaction. The use of electrocompetent *E. coli* cells is recommended. Plate on the appropriate selection plates.

3.5 Gene Expression

Genomic DNA Expression

As mentioned previously, most of the fission yeast vectors used for transformation carry the *S. cerevisiae LEU2* and *URA3* or the *S. pombe ura4* genes as markers which complement the leu1 and ura4 mutant alleles respectively (for an extensive list see Moreno et al. 1991). New *ura4*, *LEU2*, and *URA3*-based vectors for easy manipulation of DNA have recently appeared (Weilguny et al. 1991; Barbet et al. 1992; Cottarel et al. 1993). In addition, recent reports mentioning new and useful markers have increased the plasmid collection. The selection marker (from *S. pombe* or *S. cerevisiae*) and the corresponding mutant are listed:

Gene	Mutation	Vector	Reference
his2	his2-		Weilguny et al. (1991)
his7	his7-	pEA500	Apolinaro et al. (1993)
CAN1	can1-	pIRT2-CAN1	Ekwall and Ruusala (1991)
HIS3	his5-	pSP3	G. Cottarel (1995)
LYS2	lys1-	pSP4	G. Cottarel (1995)

Controlled Gene Expression

Overexpression and/or heterologous gene expression can be done in *S. pombe*. Constitutive, inducible and repressible promoters are available.

Constitutive Promoters

Several version of plasmids are now available and some are refereed.

Promoter	Level of expression	Reference
SV40 early	Moderate	Jones et al. (1988)
adh	High	Russell and Nurse (1986); McLeod et al. (1987); Bröker and Bäuml (1989)
CaMV	Moderate	Faryar and Gatz (1992); Forsburg (1993)

Inducible Promoters

Several versions of the vectors are available, the *nmt1* promoter exists with three levels of expression (Basi et al. 1993).

Promoter	Induced/repressed condition	Reference
fbp	Off in MM + 8% glucose On in MM + 0.1% glucose + 3% glycerol	Hoffman and Winston (1989)
Multiple Glucocorticoid Response elements	On with 2 μM dexamethasone	Picard et al. (1990)
CaMV-tet	On in 10 mg/ml tetracycline and + tet repressor	Faryar and Gatz (1992); Forsburg (1993)
nmt1	Off in MM + thiamine (5 μg/ml) On in MM without thiamine.	Maundrell (1990); Maundrell (1993); Forsburg (1993)

Acknowledgment. G. C. thanks Charles Hoffman (Boston College) for pertinent comments.

References

Alfa C, Fantes P, Hyams J, McLeod M, Warbrick E (1993) Experiments with fission yeast: a laboratory course manual. Laboratory Press, Cold Spring Harbor

Allshire RC (1990) Introduction of large linear minichromosomes into *Schizosaccharomyces pombe* by an improved transformation procedure. Proc Natl Acad Sci USA, 87: 4043–4047

Apolinaro E, Nocero M, Jin M, Hoffman CS (1993) Cloning and manipulation of the *Schizosaccharomyces pombe his7⁺* gene as a new selectable marker for molecular genetic studies. Curr Genet 24:491–495

Barbet N, Muriel WJ, Carr AM (1992) Versatile shuttle vectors and genomic libraries for use with *Schizosaccharomyces pombe*. Gene 114:59–66

Basi, G, Schmid E, Maundrell K (1993) TATA box mutations in the *Schizosaccharomyces pombe nmtl* promoter affect transcription efficiency but not the transcription start point or thiamine repressibility. Gene 123:131–136

Beach D, Nurse P (1981) High-frequency transformation of the fission yeast *Schizosaccharomyces pombe*. Nature (London) 290:140–142

Booher R, Beach D (1987) Interaction between *cdc13⁺* and *cdc2⁺* in the control of mitosis in fission yeast; dissociation of the G_1 and G_2 roles of the *cdc2⁺* protein kinase. EMBO J 6:3441–3447

Booher R, Beach D (1988) Involvement of *cdc13⁺* in mitotic control in *Schizosaccharomyces pombe*: possible interaction of the gene product with microtubules. EMBO J 7: 2321–2327

Bröker M (1987) Transformation of intact *schizosaccharomyces pombe* cells with plasmid DNA. BioTechniques 5:516–518

Bröker M (1993) Rapid transformation of cryopreserved competent *Schizosaccharomyces pombe* cells. BioFeedback 15:599–600

Bröker M, Bäuml O (1989) New expression vectors for the fission yeast *Schizosaccharomyces pombe*. FEBS Letter 248:105–110

Cottarel G, Beach D, Deuschle U (1993) Two new multi-purpose multicopy *Schizosaccharomyces pombe* shuttle vectors, pSP1 and pSP2. Curr Genet 23:547–548

Cottarel G (1995) The *Saccharomyces cerevisiae HIS3* and *LYS2* genes complement the *Schizosacchaformyces* pombe *his5-303* and *lys1-131* mutations, respectively: new selectable markers and multi-purpose multicopy shuttle vectors pSP3 and pSP4. Curr Genet (in press)

Ekwall K, Ruusala T (1991) Budding yeast CAN1 gene as a selection marker in fission yeast. Nucleic Acid Res 19:1150

Elble R (1992) A simple and efficient procedure for transformation of yeasts. BioTechniques 13:18–20

Faryar K, Gatz C (1992) Construction of a tetracycline-inducible promoter in *Schizosaccharomyces pombe*. Curr Genet 21: 345–349

Forsburg SL (1993) Comparison of *Schizosaccharomyces pombe* expression system. Nucleic Acid Res 21:2955–2956

Hoffman CS (1993) Preparations of yeast DNA, RNA and proteins. In: Current protocols in molecular biology. Edited by Ausubel, Brent, Kingston, Moore, Seidman, Smith and Struhl.

Hoffman CS, Winston F (1989) A transcriptionally regulated expression vector for the fission yeast *Schizosaccharomyces pombe*. Gene 84:473–479

Hood MT, Stachow C (1990) Transformation of *Schizosaccharomyces pombe* by electroporation. Nucleic Acid Res 18:688

Jones RH, Moreno S, Nurse P, Jones NC (1988) Expression of the SV40 promoter in fission yeast: identification and characterization of an AP-1 like factor. Cell 53:659–667

Kanter-Smoler G, Dahlkvist A, Sunnerhagen P (1994) Improved method for rapid transformation of intact *Schizosaccharomyces pombe* cells. BioTechniques 16:798–799

Lee MG, Nurse P (1987) Complementation used to clone a human homologue of the fission yeast cell cycle control gene *cdc2*. Nature London 327:31–35

Lohka MJ, Hayes MK, Maller JL (1988) Purification of maturation-promoting factor, an essential regulator of early mitotic event. Proc Natl Acad Sci USA 85:3009–3013

Matsumoto T, Beach D (1991) Premature initiation of mitosis in yeast lacking RCC1 or an interacting GTPase. Cell 66:347–360

Maundrell K (1990) nmt1 of fission yeast a highly transcribed gene completely repressed by thiamine. J Biol Chem 265:10 857–10 864

Maundrell K(1993) Thiamine-repressible expression vectors pREP and pRIP for fission yeast. Gene 123:127–130

McLeod M, Stein M Beach, D (1987) The product of the *mei3*$^{+}$ gene, expressed under the control of the mating-type locus, induces meiosis and sporulation in fission yeast. EMBO J 6:729–736

Molz L, Booher R, Young P, Beach D (1989) *cdc2* and the regulation of mitosis: six interacting *mcs* genes. Genetics 122: 773–782

Moreno S, Klar A, Nurse P (1991) A guide to yeast genetics and molecular biology. Methods Enzymol 194:795–823

Nurse P (1975) Genetic control of cell size at cell division in yeast. Nature (London), 256:547–551

Okazaki K, Okazaki N, Kume K, Jinno S, Tanaka K, Okayama H (1990) High-frequency transformation method and library transducing vectors for cloning mammalian cDNAs by *trans*-complementation of *Schizosaccharomyces pombe*. Nucleic Acid Res 18: 6485–6489

Picard D, Schena M, Yamamoto KR (1990) An inducible expression vector for both fission and budding yeast. Gene 86:257–261

Prentice HL (1992) High efficiency transformation of *Schizosaccharomyces pombe* by electroporation. Nucleic Acid Res 20:621

Rusell P, Nurse P (1986) *cdc25+* functions as an inducer in the mitotic control of fission yeast. Cell 45:145–153

Weilguny D, Prætorius M, Carr A, Egel R, Nielsen O (1991) New vectors in fission yeast: application for cloning the *his2* gene. Gene 99:47–54

Xiong YT, Connolly B, Futcher B, Beach D (1991) Human D-type cyclin. Cell 65:691–699

12 Preparation of Synchronous Cultures of the Fission yeast, *Schizosaccharomyces pombe*

JAMES CREANOR

1 Background

Many of the advances that have been made in the understanding of cell division and mitotic controls have been achieved by cell cycle studies of yeast and an essential tool in many of these has been the production of synchronous cultures. These cultures have been made by a variety of techniques, some of which can be universally applied and others that have been developed to suit a particular cell type. The fission yeast, *Schizosaccharomyces pombe,* has become a much favored organism for many of these studies, and the purpose of this chapter is to describe several of the methods that are currently available for the production of synchronous cultures in this yeast.

2 Principle and Applications

S. pombe grows in length only during the cell cycle, so a measurement of length is sufficient to position the cell fairly accurately in the cycle. The cells are cylindrical in shape and cell length varies from about 7 µm at birth to about 14 µm at the time of cell division. During the first 75% of the cycle the cells increase in length in an approximately exponential manner. For the next 25% of the cycle there is no increase in length or volume and the nucleus divides at the beginning of this stage. Shortly afterwards, at about 85% of the way through the cycle, the septum appears transversely across the middle of the cell. Finally, the cell splits at the septum and forms two daughter cells. Cell synchrony can be maintained in *S. pombe* for three or four cycles, unlike the situation in *S. cerevisiae,* where the decay of synchrony is usually very rapid.

There are two ways of producing a synchronous culture:

- A subpopulation containing cells of the same size, and therefore at the same stage of the cell cycle, can be separated by physical means from the rest of the culture (selection methods).
- The entire population can be subjected to a treatment that induces all of the cells to divide at the same time (induction methods).

These techniques have their own particular advantages and difficulties and the choice of method has to be determined to a great extent by the event or process that is being studied and by the available apparatus. The details that will be given are described using *S. pombe* but many of them can be applied with slight modification to *Saccharomyces cerevisiae* and to other yeasts.

The cell cycle can also be studied by separating an entire culture into discrete populations of cells, within each of which the cells are at a similar cycle stage; for a discussion on this method see Mitchison (1988). It has been used successfully with budding yeast, but it does not seem to work so well with *S. pombe,* possibly because cells remain stuck together at the end of the cycle.

3 Selection Methods

Cells are collected from an asynchronous culture, and one age group, usually the smallest cells at the beginning of the cell cycle, is separated from the rest of the population. This fraction of cells is then grown as a synchronous culture. It is possible to produce only small cultures using this technique since the optimum degree of synchrony is obtained by selecting 5–10% of the original asynchronous culture. However, the subsequent growth of the cells after selection is likely to represent a normal cell cycle and asynchronous control cultures can be made that can monitor, to some extent, the effect of the synchronization technique.

3.1 Velocity Separation in Tubes

This was the first technique to be described for the production of synchronous cultures (Mitchison and Vincent 1965) and is relatively quick and easy to carry out. The requirements are:

Materials

- A 4-l culture of *S. pombe* strain 972h$^-$ growing in Edinburgh Minimal medium, EMM 3 (see Appendix).
- 10–40% sucrose gradient (w/w) or a 7.5–30% lactose gradient (w/w) made up in EMM 3.
- A filter apparatus to hold a filter paper of 15 cm.
- Whatman No. 50 filter paper (Whatman Scientific Ltd., St Leonard's Road, 20/20 Maidstone, Kent. ME16 0LS).
- A centrifuge with a swing-out head capable of holding a 100-ml gradient tube.
- 5-ml syringe with a long 19G needle.
- 100–150 ml fresh or conditioned medium.

Standard procedure

1. Harvest about 10^{10} cells by filtration or centrifugation from a 4-l exponential phase culture. A Whatman No. 50 filter paper in a membrane holder is both quick and cheap to use. This step normally takes about 15 min.
2. Scrape off the cells using a small spatula and resuspend in about 4 ml of medium.
3. Layer the cell suspension on top of an 80-ml linear gradient of 10–40% sucrose or 7.5–30% lactose (made up in medium) prepared in a 100-ml centrifuge tube.
4. Centrifuge the cells in a swing-out head until the cell layer has moved about half way down the tube which will take about 5 min at 900 *g* at 25 °C.
5. Remove the top layer of cells (in 5–10 ml) using a 5-ml syringe with a long 19G needle and add these to warm medium which has been retained from the filtration of the original culture.

The yield from this technique is about 150 ml of synchronous culture containing approximately 1.5×10^6 cells/ml. If a larger culture than this is required then obviously more than one gradient can be used and the starting culture can be increased in size. Synchrony can be monitored either by measuring cell number increase over time using a Coulter Counter or by counting the percentage of septated cells under low power, dark ground illumination.

Control

A culture used as a control for most of the artifacts that might have been produced while making the synchronous culture, can be made by carrying out the above procedures until the gradient has been centrifuged (step 4). Then, instead of removing the top layer containing the small cells, the gradient is mixed and a culture is made by inoculating a sample from the mixed up gradient into conditioned medium at the same cell density as was used in preparing the synchronous culture.

The following points related to this method should be noted:

- This is a simple method that requires readily available apparatus.
- Collection can be done more quickly using a continuous flow centrifuge but such a machine is expensive and does not improve the technique significantly.
- Lactose is probably the best gradient material since it is not metabolized by *S. pombe*. Other materials can also be used, e.g., sorbitol, glycerol, Ficoll, Percoll, etc.

- The method can easily be scaled down by using smaller tubes (e.g., 15 ml) and in this case a 5–15% sucrose gradient will suffice.
- If cells clump, then it is impossible to obtain separation of single cells at the start of the cycle.
- This method can easily be applied to other yeast with minor modifications. In the case of *S. cerevisiae,* it may be necessary to discard the top fraction of the gradient since this often contains small unbudded cells that are either slow growing or do not divide.

The main disadvantage of this method, apart from the low yield, is that metabolic perturbations can be seen in control cultures (Creanor and Mitchison 1979). These can persist for several hours and often occur with a frequency similar to the duration of the cell cycle. One possible explanation for this is that the cells are starved during the lengthy collection procedure and while spinning on the gradient and, for this reason, they may approach entry into stationary phase.

3.2 Preparation of Synchronous Cultures by Elutriation

When it became apparent that synchronous cultures prepared by separation of cells in tube gradients exhibited medium or long term perturbations, a method was sought that would overcome the effects of cell concentration and starvation, thought to be the main causes of these perturbations. The most successful technique was with the Beckman JE–5.0 Elutriator Rotor which works by a counter-current method in which yeast cells are kept suspended in the rotor cell as a result of the opposing forces of medium flow and centrifuge speed. By adjusting the flow rate and the speed of the centrifuge, a fraction of small cells at the beginning of the cycle can be pumped out from the top of the elutriation chamber.

The following apparatus is used:

Material

- Beckman J–6M/E centrifuge with bubble trap and pressure gauge built into the inflow line (Beckman Instruments Ltd., Progress Road, High Wycombe, Bucks. HP12 4JL).
- A pulse-free peristaltic pump, e.g. Cole Parmer Masterflex (C.P. Instrument Co. Ltd., P.O. Box 22, Bishop's Stortford, Herts. CM23 3DX).
- Beckman JE–5.0 rotor with 40-ml elutriation chamber; a smaller chamber size can be used (4 ml) but this will decrease the size of the synchronous culture produced.
- 4 l of exponential phase cells growing in EMM 3.
- Several liters of pre-warmed medium.

Procedure

1. Pump the system full of medium and remove all air bubbles.
2. Increase the rotor speed to 4000 rpm.
3. Pump the cells into the rotor at about 150 ml/min from the exponential culture that is kept in a warm water bath adjacent to the centrifuge. It will take about 20 min to load 2×10^{10} cells. The effluent medium can be retained for pumping out fractions at the end and for diluting the final culture.
4. Using the stroboscope, adjust the pump speed to ensure that the cells are retained in the chamber.
5. Check that the effluent medium is free of cells.
6. When all of the cells are collected in the rotor, reduce the centrifuge speed to 3000 rpm. and the pump speed accordingly to prevent cells being removed from the chamber. Alteration of the centrifuge speed will cause considerable mixing in the chamber, so leave for a few minutes to equilibrate at the new setting.
7. Gradually increase the pump speed until cells appear in the effluent medium and collect fractions of about 100 ml. Examine these under the microscope to check for a uniform population of small cells.
8. Pool the best fractions and dilute if necessary to give a synchronous culture of $1–2 \times 10^6$/cells ml. A good degree of synchrony can be obtained by selecting about 5% of the original culture, but obviously the smaller the fraction selected, the better the synchrony.

There are several advantages with this method. The temperature in the chamber is largely determined by the temperature of the medium being pumped through it and so there is minimal risk of temperature shock during collection of the cells. There is also little risk of starvation since the medium in the chamber is replaced about every 10 s and it has been shown that a fraction containing 3×10^9 *S. pombe* cells will continue to grow in the 4-ml chamber with a normal doubling time of 150 min for as long as medium is circulated through the system (Creanor and Mitchison 1979).

After each use, the rotor chamber has to be cleaned out and disinfected because of the complex nature of the circulation system. Alcide Exspor (Life Science Laboratories) is suitable for this and it is usually left in the rotor overnight and washed out with 1 l of sterile distilled water just before collecting the cells. EMM 3 is a fairly acid medium and there is a tendency for the stainless steel parts of the rotor to show signs of corrosion after prolonged use, so it is important to clean and dry the machine after each use.

Although this method is likely to produce balanced synchronous cultures with the minimum of perturbation, nevertheless the effect of the procedure on subsequent cell growth should be monitored on asynchronous control cultures. This is accomplished by using the same technique as described above, but instead of pumping out the small cells, the entire contents of the rotor chamber are pumped into medium (most easily accomplished by switching off the centrifuge) and diluted to the same cell density as was used in the synchronous culture.

3.3 Preparation of Synchronous Cultures Using a Zonal Rotor

A zonal rotor Type A (MSE) has been used for producing large synchronous cultures. This technique has been more or less superseded by the elutriation system but it will be described here since the apparatus is widely available.

Material

- Zonal Rotor Type A (MSE Scientific Instruments, Manor Royal, Crawley. RH10 2QQ, Sussex).
- MSE 6-l centrifuge.
- Large gradient maker capable of making a 1500-ml gradient.
- Peristaltic pump.
- 20 l of an exponential culture growing in EMM 3.
- 2.5 l of fresh or conditioned medium.
- 2.5 l of 40% sucrose (w/w) made up in EMM 3.
- 1 l of 15% sucrose (w/w) made up in EMM 3.

Procedure

1. Prepare a 15–40% sucrose gradient of 1500 ml in the rotor with a peristaltic pump while the rotor is running at 600 rpm.
2. Harvest about 10^{11} cells using a Whatman filter paper as described above and resuspend in 40 ml medium. The filtered medium can be retained for subsequent dilution of the synchronous culture.
3. Load the cells into the center of the rotor using a 50-ml syringe and overlay with about 30 ml of medium which will displace the cell suspension into the top of the gradient.
4. Accelerate the rotor to 2000 rpm and the cells will begin to move through the gradient.
5. When the bottom of the cell suspension has moved about 75% of the way to the edge of the rotor, reduce the centrifuge speed to 600 rpm.
6. Remove the cells by pumping 40% sucrose into the outer edge of the rotor and collect fractions of 50 ml displaced from the center of the rotor.
7. Examine the cells microscopically and pool those fractions that contain small cells from the beginning of the cycle, diluting if necessary with fresh or conditioned medium to give $1–2 \times 10^6$ cells/ml.

It takes a considerable time to prepare a synchronous culture by this method and the cells are highly concentrated during collection and centrifugation. It is important therefore to prepare control cultures to test the effect of the procedure on the subsequent growth of the culture. Material other than sucrose can be used to prepare the gradient, and there is evidence that using a nonmetabolized sugar can reduce perturbations in growth.

3.4 Conclusions

The simplest method of producing a synchronous culture is by velocity sedimentation in a tube and in most cases this does not produce lasting perturbations in growth. However, the method of choice is undoubtedly the elutriation system, although this is an expensive piece of equipment and requires skill to operate.

4 Induction Methods of Synchronization

In these methods, the entire culture is induced by some means or other to divide at the same time. Thus large numbers of cells can easily be synchronized but subsequent cell cycles may be distorted in some way by the procedure and control cultures can be difficult to make.

4.1 Synchronous Cultures Prepared Using Chemical Inhibitors

Chemical inhibitors have been used to prepare synchronous cultures of *S. pombe* (Mitchison and Creanor 1971) and the principle is to apply a block for sufficient time to cause all of the cells to pile up at the block point (usually DNA synthesis). The chemical is then removed by filtration and washing and the cells go on to progress synchronously through the cell cycle. Hydroxyurea and deoxyadenosine inhibit DNA synthesis in *S. pombe* and both have been used successfully to prepare synchronous cultures, although significant side effects can make the interpretation of the results difficult. However, the production of synchronous cultures using hydroxyurea will be described.

Standard procedure

1. Grow up a mid-log phase culture of cells in EMM 3 at a temperature of 30–35 °C.
2. Add 11 mM hydroxyurea and continue to grow the cells for a generation time.

3. Filter the culture using any suitable membrane filter and wash with warm, fresh medium to remove the inhibitor.
4. Resuspend the cells in warm, fresh medium and continue growing the culture, measuring the degree of synchrony either by counting cell number increase over time or scoring the percentage of dividing cells.

The degree of synchrony obtained by this very simple method is good and compares well with the best synchronous cultures prepared by selection. However, it must be repeated that the cell cycles following the release of the block are not normal with respect to many of the parameters that have been measured (e.g., the synthesis of many enzymes). Care must therefore be taken in interpreting results obtained in synchronous cultures made by this method.

4.2 Induction Synchrony Using *cdc* Mutants

Any temperature sensitive cell division cycle (*cdc*) mutant of *S. pombe* can, in theory, be used to produce a synchronous culture simply by holding the culture at the restrictive temperature for sufficient time to let all of the cells pile up at the block point. In practice, however, not all *cdc* mutations are readily reversible, and many of them will not resume cell division synchronously when returned to the permissive temperature. The best mutants for preparing synchronous cultures of this nature are *cdc 2–33* and *cdc 25–22*, both of which cause cells to become blocked at the end of G_2.

The procedure that can be applied to these and to other suitable mutants is as follows:

Procedure

1. Grow up a mid-log phase culture of the mutant in EMM 3 in a water bath at the permissive temperature (normally 25–28 °C).
2. Shift the culture to a temperature of 35 °C and leave the cells growing for at least a generation time; in practice the best synchrony is obtained by leaving at the restrictive temperature for 210–240 min.
3. Return the culture to the permissive temperature and the cells will go on to divide with a high degree of synchrony.

Using this method, synchrony decays rapidly after the first synchronous division probably because the cells are at least twice as large as normal when restored to the permissive temperature, but are as variable in size as in a normal asynchronous culture. The subsequent cycles of individual cells are shorter than normal and vary according to their size as the cells attempt to return to normal length as quickly as possible.

4.3 Induction of Synchrony by Heat Shocks

A brief mention should be made of one of the earliest induction methods of making synchronous cultures, namely by applying repetitive heat shocks. This was developed for *Tetrahymena* (Zeuthen 1964) but it can be applied successfully to *S. pombe*. The technique is very simple and is most conveniently done by having an automatic device to carry out the temperature shocks.

Procedure

1. Grow a culture to mid-log phase at a temperature of 32 °C in a shaking water bath.
2. Transfer to a bath at 41 °C for 30 min.
3. Replace the culture in the bath at 32 °C and leave for 130 min.
4. Repeat the heat shock procedure five or six times.
5. At the end of these procedures the culture will divide with a reasonable degree of synchrony.

Obviously very large synchronous cultures can be made using this technique, but it is likely that there will be long-term perturbations as a result of the periods spent at the high temperature. These are difficult to measure with control cultures, since the nature of the treatment is always to produce a culture that will divide synchronously.

5 Concluding Remarks

Synchronous cultures can be prepared by selection or induction and a description has been given of the more commonly used methods in the fission yeast *S. pombe*. Typical cultures prepared by some of the techniques described are shown in Fig. 1. There are advantages and disadvantages with either method and care should be taken to monitor the effect of the synchronization procedure when this is possible.

The method of choice when looking at cell cycle patterns is the elutriation system, since this seems to have least effect on the subsequent growth of the culture. However, the apparatus is expensive and requires skill in its use. Large numbers of dividing cells are most easily obtained using any of the induction methods but the cell cycles after the induction treatment are likely to be abnormal ones. The best advice is to use different methods for preparing synchronous cultures if they are available, and where possible, to prepare control cultures to monitor the effects of the synchronization technique.

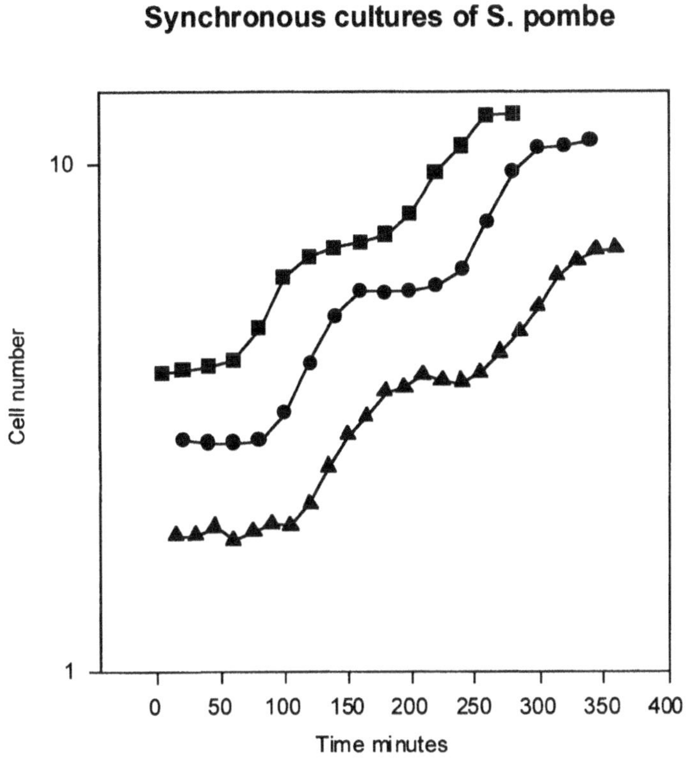

Fig. 1. Three synchronous cultures prepared by different methods. *Solid triangles* Culture prepared by selection using a sucrose gradient; *solid circles* culture prepared by selection using the elutriation rotor; *solid squares* culture prepared by induction of *cdc2–33* after block and release

A novel method for eliminating the effect of imperfect synchrony has recently been described (Creanor and Mitchison 1994) and can be applied to most data obtained from studies of synchronous cultures.

References

Creanor J, Mitchison JM (1979) Reduction of perturbations in leucine incorporation in synchronous cultures of *Schizosaccharomyces pombe* made by elutriation. J Gen Microbiol 112:385–388

Creanor J, Mitchison JM (1994) The kinetics of H1 histone kinase activation during the cell cycle of wild-type and *wee* mutants of the fission yeast *Schizosaccharomyces pombe*. J Cell Sci 107:1197–1204

Mitchison JM (1988) Synchronous cultures and age fractionation. In: Campbell J, Duffus JH (eds) Yeast: a practical approach. IRL Press, Oxford pp 51–63

Mitchison JM, Vincent WS. (1965) Preparation of synchronous cultures by sedimentation. Nature 205:987–989

Mitchison JM, Creanor J (1971) Induction synchrony in the fission yeast *Schizosaccharomyces pombe*. Exp Cell Res 67:368–374

Zeuthen E (1964) The temperature-induced division synchrony in *Tetrahymena*. In: Zeuthen E (ed) Synchrony in cell division and gowth. Interscience Publishers, New York, pp 99–158

Appendix

Edinburgh Minimal Medium (EMM 3) for growing *S. pombe*.

	Distilled water	1 Liter
	Glucose	20 g
	Ammonium chloride	5 g
	Potassium hydrogen phthalate	3 g
	Disodium hydrogen phosphate	1 g
	Sodium sulfate	0.1 g
	Calcium chloride. $2H_2O$	0.015 g
Vitamins		
	Inositol	0.01 g
	Nicotinic acid	0.01 g
	Calcium pantothenate	0.001 g
	Biotin	0.01 mg
Trace elements		
	Boric acid	0.05 mg
	Manganese sulfate	0.4 mg
	Zinc sulfate	0.4 mg
	Ferric chloride	0.2 mg
	Molybdic acid	0.16 mg
	Potassium iodide	0.1 mg
	Copper sulfate	0.04 mg
	Citric acid	1 mg

Vitamins and trace elements can be made up as stock solutions at 1000 *x* concentrates: 1 ml is then used to make 1 l of EMM 3.

Part Four

Techniques for Studying the Cell Cycle of Any Species

13 Identification of Novel Cell Cycle Targets Using Small Molecule Ligands

VIVIAN BERLIN AND M. ISABEL CHIU

1 Background

Complementation of yeast *cdc* and *cln* mutants as well as biochemical approaches have been used to identify and characterize the major components of the mammalian cell cycle machinery, the cyclins and cyclin-dependent kinases (cdks). More recently the two-hybrid system (or interaction trap) has been used to identify novel regulators of the cell cycle in which a known component of the cell cycle is used as "bait" to search for protein interactors (see chapter by Fiore et al., Part Four). For example, the latter approach was used to identify $p21^{Cip/Waf1/Sdi1/Pic1}$, a small molecular weight inhibitor of cdks (Harper et al. 1993), Cdi1, a dual-specificity phosphatase that interacts with both Cdc2 and Cdk2 (Gyuris et al. 1993; Hannon et al. 1994), and $p16^{Ink4/Mts1}$, a Cdk4 inhibitor (Serrano et al. 1993).

We have recently extended the utility of the two-hybrid system to identify novel cell cycle targets using small molecule ligands (Chiu et al. 1994). Rapamycin is a model compound since it specifically inhibits the G1-to-S transition in lymphocytes (Bierer et al. 1990), certain nonlymphoid cells (Albers et al. 1993), and the yeasts, *Saccharomyces cerevisiae* (Heitman et al. 1991) and *Candida albicans* (Morris 1992). The similar effects of rapamycin in animal cells and yeast suggest that the biochemical steps affected by rapamycin are conserved.

Rapamycin and FK506, structurally related macrolides, are potent immunusuppressants that block distinct steps in T cell activation (Bierer et al. 1990). Both exert their action when bound to the cytosolic protein FKBP12 (Harding et al. 1989; Koltin et al. 1991). FK506, complexed with FKBP12, binds to and inhibits the calcium-dependent phosphatase, calcineurin (Liu et al. 1991), which is required for lymphokine gene expression stimulated by the binding of antigen to the T cell receptor (Tocci et al. 1989). By contrast, rapamycin interacts with distinct targets required for subsequent lymphokine-induced cell division (Dumont et al. 1990). In this chapter, we describe the approach we took to use the two-hybrid system to identify the targets of the FKBP12/rapamycin complex which can serve as a guide to identify the targets of other small molecule ligands, e.g., peptides, hormones and synthetic organic compounds.

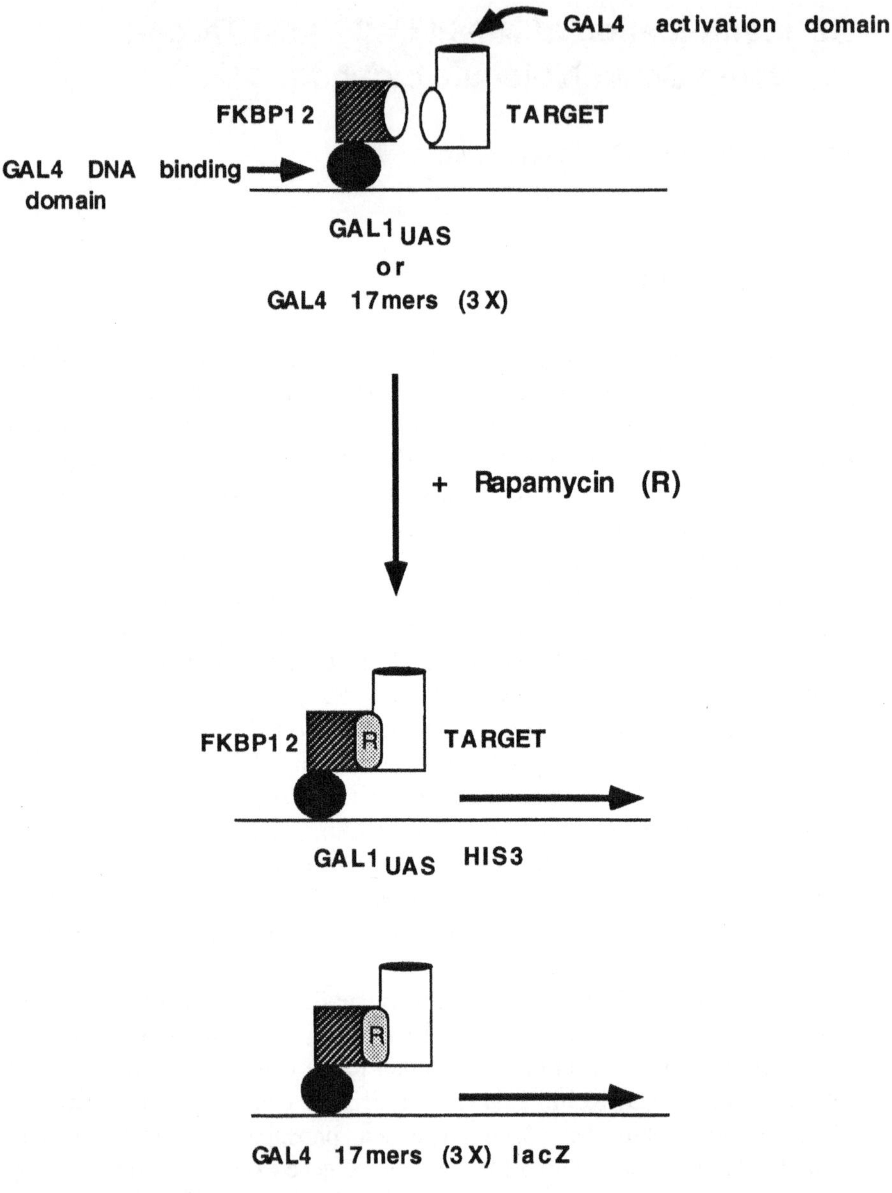

Fig. 1. Ligand-dependent transactivation in the two hybrid system. Rapamycin mediates interaction of FKBP12 fused to the GAL4 DNA binding domain, and the target fused to the GAL4 activation domain. The tripartite complex activates transcription of the reporter genes, *HIS3* and *lacZ*, under the control of transcriptional activation sequences containing Gal4 binding sites (*GAL1* UAS and Gal4 17mers)

2 Principle and Applications

The two-hybrid system is designed to identify proteins that interact with one another. Protein-protein interactions are detected by the expression in yeast of two reporter genes, such as yeast *HIS3* and bacterial *lacZ*. Transactivation of the reporters is dependent upon the formation of a complex between proteins fused to a DNA binding domain and to an activation domain. If interaction of the two hybrids occurs, the yeast strain will be prototrophic for histidine and contain detectable β-galactosidase activity (Bartel et al. 1993). Ligand-dependent transactivation is an additional criterion for identifying such targets (Fig. 1), i.e., targets are identified that mediate *HIS3* and *lacZ* expression in the presence, but not in the absence, of a particular drug. The ligand in question may complex directly with one or both hybrid proteins, mediating a tripartite interaction. Another application of this approach would be to use the two-hybrid system to design and test small molecule ligands that modulate specific protein-protein interactions that have potential therapeutic value.

3 Methods

3.1 System Validation

The use of the two-hybrid system to detect protein-protein interactions mediated by a particular class of small molecule ligands may be feasible only in certain situations. An issue of concern is whether fusion of the ligand receptor to the GAL4 DNA binding domain or LexA affects the ability of the receptor to bind ligand and exert its biological effect. Experiments can be designed to address this concern. For example, a variety of different "baits" can be constructed in which the ligand receptor is fused to the LexA or the GAL4 DNA binding domain, containing or lacking a polyglycine linker at the fusion junction. Mutant ligand receptors that alter the affinity of the receptor ligand complex for the downstream target, if available, can be included in preliminary studies. Interaction of wild type and mutant baits with a known target fused to an activation domain can then be assayed. For example, we tested the ability of wild type and mutant FKBP12 baits to bind to FK506 and interact with calcineurin fused to either the VP16 or the GAL4 activation domain. Interaction was tested at a variety of FK506 concentrations (Table 1).

A second test of function can be performed if yeast contains a homologue of the ligand receptor of interest which, when deleted in yeast, produces an assayable phenotype, such as drug resistance. In a yeast strain containing a deletion for the endogenous gene encoding the ligand receptor, one can assess

Table 1. FK506 mediates interaction of human FKBP12 with human calcineurin in the two hybrid system

Plasmids		Growth on no-histidine			Color on X-GAL	
Binding domain fusion	Activation domain fusion	FK506 (μg/ml)			No FK506	FK506, 0.2 μg/ml
		0	0.02	0.2		
pGAL4-hFKBP12	pGAL4-hCNA	–	+	++	White	Blue
pGAL4-hFKBP12*	pGAL4-hCNA	–	++	+++	White	Blue
pGAL4-hFKBP12	pGAL4-FKI1	+	–	–	Blue	White
pGAL4-SNF1	pGAL4-hCNA	–	–	–	White	White
pGAL4-FKBP12	pGAL4	–	–	–	White	White

The yeast strain Hf7C (gift of G. Hannon),containing the *HIS3* and *lacZ* reporter genes shown in Fig. 1, was transformed with the indicated pairs of plasmids. The binding domain fusions contain the GAL4 DNA binding domain fused to the human FKBP12 cDNA, wild type or mutant (*), or to the yeast *SNF1* gene. The FKBP12 mutant increases the sensitivity of yeast to rapamycin (V. Berlin, unpubl. results). The *GAL4* activation domain is unfused or fused to human calcineurin (hCNA) or to FKI1 (FKBP12 interactor) isolated from a two hybrid screen for drug-independent FKBP12 interactors (V. Berlin, unpubl. results). Growth on media lacking histidine and detectable β-galactosidase activity (blue color on media containing X-GAL) is evidence of interaction.
– no growth; +, ++, +++, poor, moderate, and good growth, respectively, assessed by the rate of appearance and size of colonies.

Table 2. FKBP12 fusions mediate rapamycin toxicity in yeast

Binding domain fusion plasmid	Growth on media containing rapamycin (ng/ml)					
	15.6	31.2	62.5	125	250	500
pLexA-hFKBP12	+	+	+	+	+	+/–
plexA-G_6-hFKBP12	+	+	+	+/–	+/–	–
pLexA-da	+	+	+	+	+	+
pGAL4-hFKBP12	+	+	+	+	+	+
pGAL4-hFKBP12*	+	+	+	+/–	+/–	+/–
pGAL4-G6-hFKBP12	+	+	+	+	+	+
pGAL4-G6-hFKBP12*	+	+	+	+/–	+/–	+/–
pGAL4	+	+	+	+	+	+
hFKBP12	+/–	–	–	–	–	–

Yeast strains containing a deletion of *FKB1*, the gene for FKBP12, were transformed with the indicated human FKBP12 fusion constructs, a LexA fusion containing an unrelated insert (LexA-da) or the GAL4 DNA binding domain alone. The human FKBP12 sequences were wild type or mutant (*), as described in the legend to Table 1. Certain constructs contain a spacer of six glycine residues (G6) at the fusion junction. The human FKBP12 cDNA, not fused to a DNA binding domain, but expressed from the yeast *GAL1* promoter, was included in the analysis. Transformants were plated on media containing various concentrations of rapamycin.
– no growth; + growth; +/– slow growth and reduced colony size.

the ability of the receptor baits to confer drug sensitivity. For example, we tested the ability of wild type and mutant FKBP12 baits to complement a deletion of *FKB1*, the yeast gene for FKBP12. Since a deletion of *FKB1* confers resistance to rapamycin, we analyzed the effect of the FKBP12 baits on the ability of yeast to grow on media containing different concentrations of rapamycin (Table 2).

3.2 Components of the Two-Hybrid System

The original two-hybrid system (Fields and Song, 1989) has undergone modification by a variety of groups (e.g., Vojtek et al. 1993). The current systems use either the LexA or the GAL4 DNA binding domain and a variety of activation domains such as VP16 (Vojtek et al. 1993), B42 (Gyuris et al. 1993), or the GAL4 DNA binding domain (Chevray and Nathans 1992; Bartel et al. 1993). Each of the components of the two-hybrid system is described below.

Yeast Strains

The most commonly used two-hybrid yeast strains contain the *HIS3* and *LacZ* reporter genes integrated in the chromosome, under the control of the *GAL4* or *LexA* transcriptional activation sequences. Variations on the theme include the use of the *LEU2* gene as reporter, maintenance of the *lacZ* reporter gene on a plasmid, and control of reporter expression levels by manipulating the number of DNA binding sites upstream of the reporter genes (see chapter by Fiore et al., Part Four). These strains can be further modified using standard gene-disruption techniques (Rothstein 1991). For example, it may be desirable to conduct the two-hybrid screen using a yeast strain deleted for the endogenous drug receptor, such as we did for *FKB1* (Chiu et al. 1994). Briefly the coding sequences of the gene to be disrupted contained on a linear fragment are replaced with a selectable marker. Since many of the two-hybrid yeast strains are ade2$^-$, *ADE2* can be used as the selectable marker for making disruptions. The DNA fragment containing the disruption flanked by noncoding sequences is used to transform a two-hybrid yeast strain to adenine prototrophy. Southern analysis can be performed and mutant phenotypes (e.g., drug resistance) can be assayed to verify that yeast transformants contain the gene disruption. The yeast strain used for the two-hybrid screen must also be auxotrophic for the two selectable markers encoded by the bait and library plasmids. For example, strain L40 (Vojtek et al. 1993) carries the mutant alleles *trp1-901* and *leu2-3,112*, allowing complementation by the *TRP1* (bait plasmid) and *LEU2* (library plasmid) markers.

Bait Plasmids

Depending on the two-hybrid system used, coding sequences of the ligand-binding protein are fused in frame to either the LexA or to the GAL4 DNA binding domain. Fusions can be constructed that contain or lack a polyglycine linker at the fusion junction. Introduction of a polyglycine linker to separate the DNA binding domain from the ligand receptor may facilitate proper folding of each domain. The bait plasmids should be sequenced to verify that the coding sequences are fused in frame with the DNA binding domain and that no mutations have arisen during cloning. If possible, the functionality of the baits should be tested, as described above for the FKBP12 baits. Before proceeding with the library screens, each bait should be tested against the vector plasmid encoding the activation domain alone (fish vector) in the two-hybrid yeast strain to control for background activity. For example, in the LexA system, our mutant FKBP12 bait paired with pVP16 alone showed significant background transactivation of the *HIS3* and *lacZ* reporters in the absence of library inserts, and was not used for the screen, whereas the wildtype FKBP12 bait behaved appropriately (not shown). If background is a problem, one could try including only a portion of the receptor coding sequence in the bait fusion, and test each bait empirically. In addition to the DNA binding domain, bait plasmids carry a marker allowing for their selection upon transformation into yeast. For example, both the LexA (Vojtek et al. 1993) and Gal4 (Bartel et al. 1993) bait plasmids carry the *TRP1* gene.

Library Vectors

The library plasmid carries a polylinker downstream of the activation domain into which cDNAs are cloned. cDNAs in-frame with the activation domain coding sequence are expressed as fusion proteins in yeast. The fraction of such cDNAs in the library can be increased with directionally cloned libraries (Hannon et al. 1993), but certain two-hybrid cDNA libraries are nondirectional, i.e., both ends of the cDNA fragments contain the same restriction site for cloning. Library plasmids also carry a selectable marker, different from that of bait plasmids, to allow for their selection upon transformation into yeast. For example, both the VP16 (Vojtek et al. 1993) and Gal4 (Hannon et al. 1993) activation domain plasmids carry the *LEU2* gene.

cDNA Libraries

cDNA libraries, cloned as activation-domain fusions, can be generated from random- or oligo(dT)-primed polyA$^+$ RNA, which in turn can be size-selected to contain long or short inserts. The advantage of short, random-primed libraries is that they allow for rapid identification of peptide domains sufficient for interaction without bias towards carboxy-terminal sequences. With oligo(dT)-primed libraries, individual cDNAs may not be full length, specifically missing 5' ends. In either case, it may be necessary to use the clone obtained in the two-hybrid screen to probe a cDNA library to obtain a full length clone. In instances where the supply of RNA is limited, PCR amplification of cDNAs can dramatically increase the representation of such clones in the library, particularly if the cDNAs are pre-selected to be in the range best-suited for efficient PCR amplification (less than 1 kb). The disadvantages are that large domains may be missed, and that the rate of false positives may be higher because short clones may have interacting properties absent in the parent native protein (Vojtek et al. 1993; S. Hollenberg, pers. comm.).

3.3 Materials

For standard DNA manipulations and bacterial transformations the reader is referred to Sambrook et al. (1989), and to Sherman (1991) for general background on yeast protocols. Media pertinent to the methods outlined in the previous section are defined below:

The following are recipes for solid media; to make the liquid versions of each simply omit the 2% agar. Unless noted otherwise, "%" means g/100 ml.

Media

YPD. 1% yeast extract, 2% peptone, 2% dextrose, 2% agar.

Drop-Out Medium (Minus X). 0.17% yeast nitrogen base (without amino acids or ammonium sulfate), 0.5% $(NH_4)_2SO_4$, 2% dextrose, 2% agar, plus all amino acids (concentrations as in Sherman, 1991) except those specified. For example, Minus Trp and Leu media comprises the above ingredients but without tryptophan and leucine.

Bacterial Minimal Medium. 1.05% K_2HPO_4, 0.45% KH_2PO_4, 0.1% $(NH_4)_2SO_4$, 0.05% sodium citrate, 1 mM $MgSO_4$, 0.2% dextrose, 2% agar, and appropriate supplements. For example, to select for Leu^+ bacterial colonies harboring the library plasmid in strain MC1066, we supplement the above medium with 100 µg/ml ampicillin, 0.4 mM uracil and 0.4 mM tryptophan (and no leucine). The genotype of strain MC1066 is *ΔlacI-A x 74 galU galK strepA*^{r} *trpC9830 leuB6 pyrF::tn5*(kan^r) Rec^+ Ara^+. The *trpC*, *leuB*, and *pyrF* mutations are complemented by the *Saccharomyces cerevisiae TRP1*, *LEU2*, and *URA3* genes, respectively.

Yeast Plates for X-Gal. same as for drop-out medium above, plus the following after autoclaving and cooling to 60 °C: for 1 l, add 43 ml of 1 M K_2HPO_4, 27 ml 1 M KH_2PO_4, and 1 ml 100 mg/ml X-Gal in DMF.

3.4 Library Screens for Ligand-Target Genes

Transformation

Yeast Transformation. The yeast strain harboring the ligand receptor fused to a DNA binding domain is transformed with a cDNA library expressed as fusions to an activation domain. Transformants are plated on media selecting for the bait and library plasmids, which for both the LexA/VP16 system (Vojtek et al. 1993) and the Gal4 system (Bartel et al. 1993) is media lacking tryptophan and leucine (Minus Trp and Leu). Yeast transformation is performed using a standard lithium acetate procedure (Becker and Lundblad 1994). Although this protocol calls for a 15-min heat shock treatment, the optimal length of time for this step is strain dependent. For the yeast strain YPB2 (Bartel et al. 1993) and YGH1 (G. Hannon, pers. comm.) 30 min is optimal. Approximately 1–5 × 10^4 transformants/μg of DNA are obtained using this procedure. To assure complete representation of the cDNA library, the total number of yeast transformants screened should be several times the number of individual recombinants in the library. Yeast transformants plated on Minus Trp and Leu media are grown for 3 days at 30 °C pooled and stored as frozen stocks in 50% glycerol.

Selection for Ligand-Dependent Protein-Protein Interactors. The following protocol is used to screen library transformants for ligand-dependent targets.

Screening transformants

1. An aliquot of the pooled yeast transformants is outgrown in YPD (Sherman 1991) at 30 °C, for one doubling. The initial inoculum should not be higher than 0.4 A_{600} OD units (1 OD unit is approximately 2 × 10^7 cells/ml).

2. Cells are plated onto selective media, aiming at one million per 15 cm plate. The selective media is synthetic complete media (Sherman 1991) lacking leucine, tryptophan and histidine and containing the ligand of interest. In some cases, the inclusion of 3-aminotriazole in the selective media may be necessary (see Sect. 4, Discussion). To monitor the number of transformants plated, a series of dilutions is plated onto Minus Trp and Leu plates. Cells are incubated at 30 °C until His^+ colonies appear, for 3–10 days depending on the selective conditions used for the screen.

3. His^+ transformants are screened for expression of *lacZ*. This can be accomplished in one of two ways. X-GAL can be added directly to phosphate-buffered (pH 7.0) selective plates (see Sect. 3.3, Materials) so that the His^+ and $lacZ^+$ phenotypes are monitored simultaneously. This method allowed detection of *lacZ* expression in the LexA system with our baits. A more sensitive method of detection is to lift His^+ transformants onto nitrocellulose filters (Schleicher and Schuell, Keene, NH), dip the filters for 30 s in liquid nitrogen to permeabilize cells, and place the filters on a petri dish lid lined with Whatman paper (# 1) and prewet with Z buffer (60 mM

$Na_2HPO_4.7H_20$, 60 mM $NaH_2PO_4.H_20$, 10 mM KCl, 1 mM $MgSO_4.7H_20$) containing 1 mg/ml X-GAL (5-bromo-4-chloro-3-indolyl-β-D-galactopyranoside). Blue color can be detected in a few minutes in some cases, or only after overnight incubation in other cases. Viable cells can be recovered from the filters using this protocol.

4. His^+ $lacZ^+$ transformants are retested for growth on media lacking histidine (Minus His) and containing or lacking ligand. Candidates which grow in the presence, but not in the absence, of ligand are chosen for the next step.

5. To eliminate false positives, each candidate is grown in non-selective medium (YPD) to permit loss of the bait (carrying the Trp^+ marker) and the library (Leu^+) plasmids. Derivatives of each candidate which had lost either the bait or the library plasmid or both (Trp^-Leu^+; Trp^+Leu^-; Trp^-Leu^-, respectively) or neither (Trp^+Leu^+) are tested again for growth on Minus His plates containing ligand. Candidates for which only the Trp^+Leu^+ derivative grows, while the other three derivatives do not, are characterized further.

6. DNA is prepared from candidates showing a plasmid-linked, ligand-dependent His^+ phenotype (Hoffman and Winston 1987). The DNA is used to transform *E. coli* strain MC1066, a leucine auxotroph, to leucine prototrophy to recover the library plasmids encoding the putative ligand target (see Sect. 3.3, Materials).

7. Plasmid DNA recovered in *E. coli* is used to transform yeast strains containing the original bait or an unrelated bait. Nonspecific interactors (those conferring a His^+LacZ^+ phenotype when paired with the unrelated bait fusion) are eliminated from further analysis, while clones specific for the original bait are retained.

8. Candidate clones which fulfill all the above criteria are analyzed by DNA sequencing on both strands, starting with oligonucleotides from vector sequences as primers.

4 Discussion

Using the methods described here, we have successfully adapted the two-hybrid approach to identify cellular protein targets of small molecule ligands. In reconstruction experiments we showed that this system can detect interaction between FKBP12 and calcineurin, in the presence of FK506. From screening cDNA libraries using FKBP12 fusions as bait in the presence of rapamycin, we identified several cellular proteins which specifically interact with the FKBP12/rapamycin complex, one of which is the mammalian homologue of yeast TOR, which we named RAPT1 (Chiu et al. 1994). The latter finding is supported by biochemical approaches in which glutathione-affinity chromatography was used to recover GST-FKBP12 complexes from cell lysates in the presence of rapamycin (Brown et al. 1994; Sabatini et al. 1994; Sabers et al. 1995). In addition, the two-hybrid approach allowed for the rapid identification of a small region in RAPT1 that is sufficient for interaction with the FKBP12/rapamycin complex.

In our screens, we have chosen to pool library transformants prior to plating onto selective media rather than subjecting them to selection directly. The advantage of this approach is that a single library transformation allows for an unlimited number of screens and as many variations on selection conditions. For example, one may need to conduct the screen with varying concentrations of 3-aminotriazole, an inhibitor of His3 activity whose presence increases the levels of *HIS3* expression required for histidine prototrophy. Alternatively, one may choose to conduct the library screen using different concentrations of ligand, as we have done with rapamycin. As a separate experiment, the same library transformants can be screened to identify cellular proteins that interact with the ligand receptor in the absence of ligand.

Once positive interactors have been identified, the cloned sequences can be compared to other sequences in the database. When interpreting the results of such comparisons, one must guard against false positives resulting from spurious fusions to the wrong strand of the cDNA which have led to the creation of protein fusions that bear no biological significance, especially when using a bidirectionally cloned cDNA library. A useful tool uniquely available to our approach is the use of ligand titrations to characterize the interactions with positive clones; for example, a weak interactor would require higher concentrations of ligand to confer His^+ and $lacZ^+$ than a stronger interactor.

We have discussed points of particular interest when applying the two-hybrid method to the identification of protein interactions mediated by a small molecule ligand. For general details about the two-hybrid method, the reader is referred to the chapter by Fiore et al. (Part Four) and to the literature cited herein.

References

Albers MW, Williams RT, Brown EJ, Tanaka A, Hall FL, Schreiber SL (1993) FKBP-rapamycin inhibits a cyclin-dependent kinase activity and Cyclin D1-Cdk association in early G1 of an osteosarcoma cell line. J Biol Chem 268:22825–22829

Bartel PL, Chien C-T, Sternglanz R, Fields S (1993) Using the two hybrid system to detect protein-protein interactions. In Hartley DA (ed) Cellular interactions in development: a practical approach. Oxford University Press, Oxford pp 153–179

Becker DM, Lundblad V (1994) In: Ausebel FM, Brent R, Kingston RE, Moore DD, Seidman JG, Smith JA, Struhl K (eds) Current protocols in molecular biology, vol 2, pp 13.7.1–13.7.2 John Wiley & Sons

Bierer BE, Mattila PS, Standaert RF, Herzenberg LA, Burakoff SJ, Crabtree G, Schreiber SL (1990) Two distinct signal transmission pathways in T lymphocytes are inhibited by complexes formed between and immunophilin and either FK506 or rapamycin. Proc Natl Acad Sci USA 87:9231–9235

Brown EJ, Albers MW, Shin TB, Ichikawa K, Keith CT, Lane WS, Schreiber SL (1994) A mammalian protein targeted by G1-arresting rapamycin-receptor complex. Nature 369:756–758

Chevray PM, Nathans D (1992) Protein interaction cloning in yeast: identification of mammalian proteins that react with the leucine zipper of Jun. Proc Natl Acad Sci USA 89:5789–5793

Chiu MI, Katz H, Berlin V (1994) RAPT1, a mammalian homolog of yeast TOR, interacts directly with the FKBP12/rapamycin complex Proc Natl Acad Sci USA 91:12 574–12 578

Dumont FJ, Melino MR, Staruch MJ, Koprak SL, Fischer PA, Sigal NH (1990) The immunosuppressive macrolides FK-506 and rapamycin act as reciprocal antagonists in murine T cells. J Immunol 144:1418–1424

Fields S, Song, O-K (1989) A novel genetic system to detect protein-protein interactions. Nature 340:245–246

Gyuris J, Golemis E, Chertkov H, Brent R (1993) Cdi1, a human G1 and S phase protein phosphatase that associates with Cdk2. Cell 75:791–803

Hannon GJ, Demetrick D, Beach D (1993) Isolation of the Rb-related p130 through its interaction with CDK2 and cyclins. Genes Dev 7: 2378–2391

Hannon GJ, Casso D, Beach D (1994) KAP: A dual specificity phosphatase that interacts with cyclin-dependent kinases. Proc Natl Acad.Sci USA 91:1731–1735

Harding MW Galat A, Uehling DE, Schreiber SL (1989) A receptor for the immunosuppressant FK506 is a cis-trans peptidyl-prolyl isomerase. Nature 341:758–760

Harper JW, Adami GR, Wei N, Keyomarsi K, Elledge SJ (1993) The p21 Cdk-interacting protein Cip1 is a potent inhibitor of G1 cyclin-dependent kinases. Cell 75:805–816

Heitman J, Movva NR, Hall MN (1991) Targets for cell cycle arrest by the immunosuppressant rapamycin in yeast. Science 253:905–909

Hoffman CS, Winston F (1987) A ten-minute DNA preparation from yeast efficiently releases autonomous plasmids for transformation of *Escherichia coli*. Gene 57:267–272

Koltin Y, Faucette L, Bergsma DJ, Levy MA, Cafferkey R, Koser PL, Johnson RK, Livi GP (1991) Rapamycin sensitivity in *Saccaromyces cerevisae* is mediated by a peptidyl-prolyl *cis-trans* isomerase related to human FK506-binding protein. Mol Cell Biol 11: 1718–1723

Liu J, Farmer JD, Lane WS, Friedman J, Weissman I, Schreiber SL (1991) Calcineurin is a common target of cyclophilin-cyclosporin A and FKBP-FK506 complexes. Cell 66: 807–815

Morris RE (1992) Rapamycins: antifungal, antitumor, antiproliferative, and immunosuppressive macrolides. Transplantation Rev. 6:39–87

Rothstein R (1991) Targeting, disruption, replacement and allele rescue: integrative transformation in yeast. Methods Enzymol 194:281–301

Sabatini DM, Erdjument-Bromage H, Lui M, Tempst P, Snyder SH (1994) RAFT1: a mammalian protein that binds to FKBP12 in a rapamycin-dependent fashion and is homologous to yeast TORs. Cell 78:35–43

Sabers CJ, Martin MM, Brunn GJ, Williams JM, Dumont FJ, Wiederrecht G, Abraham R (1995) Isolation of a protein target of the FKBP12-rapamycin complex in mammalian cells. J Biol Chem 270:815–822

Sambrook J, Fritsch EF, Maniatis T (1989) Molecular Cloning. A Laboratory Manual. 2nd edn. Cold Spring Harbor Laboratory Press, Cold Spring Harbor

Serrano M, Hannon GJ, Beach D (1993) A new regulatory motif in cell-cycle control causing specific inhibition of cyclin D/Cdk4. Nature 366:704–706

Sherman F (1991) Getting started with yeast. In : Guthrie C, Fink GR (eds) Guide to yeast genetics and molecular biology. Methods Enzymol 194:3–20

Tocci MJ, Matcovich DA, Collier KA, Kwok P, Dumont F, Lin CS, DeGudicibus S, Siekierka JJ, Chin J, Hutchinson NI (1989) The immunosuppressant FK-506 selectively inhibits expression of early T cell activation genes. J Immunol 143: 718

Vojtek AB, Hollenberg SM, Cooper JA (1993) Mammalian ras interacts directly with the serine/threonine kinase raf. Cell 74:205–214

14 Bacterial Expression and in Vitro Assay of Cell Cycle Kinases

LISA CONNELL-CROWLEY AND WADE HARPER

1 Background

There is currently widespread interest in the mechanisms that regulate the activation of Cyclin-dependent kinases (Cdks). Monomeric Cdks have no detectable activity but are fully activated by association with a Cyclin subunit and phosphorylation of a conserved threonine residue. Cak (Cdk-activating kinase) is the enzyme responsible for phosphorylating Cdks on the activating threonine (Fesquet at al. 1993; Poon at al. 1993; Solomon at al. 1993). Negative regulation of Cdc2-Cyclin B complexes is achieved by phosphorylation of tyrosine 15 in Cdc2 by tyrosine kinases of the Wee1 family. While most studies investigating Cdk activation have employed some kind of cell lysate system containing endogenous Cak activity, recent developments now make it possible to produce active Cdk-Cyclin complexes in vitro using highly purified reagents generated in bacteria. The primary advantage of using bacteria is that they contain far fewer protein kinases than eukaryotic cells, making it possible to prepare Cdk and Cyclin components in their unphosphorylated state. This facilitates the study of both positive and negative regulation via phosphorylation. Here we describe methods for purification of Cdk2, Cdc2, Cyclins, and other cell cycle proteins expressed in bacteria, as well as assays for histone H1 kinase activity of Cdk2 and Cdc2. We hope that these protocols will be a guide to specific individual needs.

2 Principal and Applications

Two general strategies have been used for bacterial production of activatable Cdc2 and Cdk2 protein kinases.

- Expression as a fusion with affinity tags such as glutathione-S-transferase (GST) (Connell-Crowley at al. 1993; Poon at al. 1993;), or
- Expression as authentic or epitope tagged proteins followed by conventional purification (Connell-Crowley at al. 1993).

The major advantage of using fusion proteins is the ease with which the recombinant protein can be purified. While only a fraction of the soluble fusion protein may be folded properly for Cyclin association and catalytic activity, the presence of the GST component does not appear to affect subunit interactions in the fraction containing a properly folded kinase domain. Importantly, the solubility of individual Cdks or the same Cdk from different species expressed in bacteria can vary widely. For example, human Cdk2 is partially soluble in bacteria (when expression is induced at 25 °C) and can be purified from this fraction in an activatable form while human Cdc2 is almost completely insoluble (Connell-Crowley at al. 1993).

A-, B-, and E-type Cyclins have been produced in bacteria in a form capable of activating Cdks, typically as GST or protein A fusion-proteins (Solomon at al. 1990; Koff at al. 1992; Poon at al. 1993). GST-Cyclin B and GST-Cyclin E are poorly soluble but sufficient quantities of protein can be produced for in vitro studies. Human and bovine Cyclin A are quite soluble and have been purified as protein A and histidine-tagged fusion proteins (Poon at al. 1993), or as the authentic 54 kDa protein (Connell-Crowley at al. 1993).

In all cases examined thus far, maximal activation of Cdks requires phosphorylation of a conserved threonine residue (T160 in Cdk2 and T161 in Cdc2) by Cak. Thus, the use of this enzyme is critical for the production of highly active complexes in vitro. While the catalytic subunit of Cak has been identified as $p40^{MO15}$ (Fesquet at al. 1993; Poon at al. 1993; Solomon at al. 1993), this protein is catalytically inactive when produced as a GST fusion in bacteria (Poon at al. 1993). This presumably reflects the requirement that MO15 associate with its specific activator and/or be phosphorylated (Poon at al. 1993). However, it is possible to activate GST-MO15 using *Xenopus* egg extracts (Poon at al. 1993), facilitating activation of Cdks in vitro. Using purified proteins, Cdk2 can be phosphorylated by Cak in the absence of Cyclin but is still not active unless Cyclin is bound. Phosphorylation in the absence of Cyclin may be unique to this purified system as Cyclin binding is a prerequisite for phosphorylation of Cdks by Cak in crude cell extracts (Solomon at al. 1990). One difference between Cdk2 and Cdc2 is that the former can be partially activated by Cyclin A in the absence of T160 phosphorylation (Connell-Crowley at al. 1993). Thus, the bacterial system can be used to make active Cdk2-Cyclin A complexes without the necessity of producing Cak.

3 Methods

Buffers

L1

20 mM Tris-HCl (pH 7.5), 10% sucrose, 1 mM PMSF, 5 µg/ml leupeptin, and 5 µg/ml antipain.

L2
50 mM Tris-HCl (pH 7.4), 0.25 mM NaCl, 50 mM NaF, 10 mM NaPPi, 0.1% NP40, 2 mM PMSF, 0.1% aprotinin, 2 mM sodium orthovanadate, and 20 μM leupeptin.

L3
50 mM Tris-HCl (pH 7.5), 2 mM EDTA, 1 mM DTT, 0.25 mM PMSF, 1 μg/ml leupeptin, 2 μg/ml aprotinin, 15 μg/ml benzamidine.

Buffer A
20 mM Tris-HCl (pH 7.5), 30 mM NaCl, 1 mM EDTA, 1 mM DTT, 1 mM PMSF, 1 μg/ml leupeptin, and 1 μg/ml of antipain.

Buffer B
1 mM sodium phosphate (pH 6.5 or 7.0), 10 mM $MgCl_2$, 1 mM EDTA, 1 mM DTT, and 1 mM PMSF.

Buffer C
10 mM Tris-HCl (pH 7.5), 30 mM NaCl, 1 mM PMSF, and 1 μg/ml leupeptin.

Buffer D
25 mM HEPES (pH 7.4), 1 mM EDTA, 2 mM DTT, and 10% glycerol.

Buffer E
PBS (170 mM NaCl, 3 mM KCl, 10 mM Na_2HPO_4, and 2 mM KH_2PO_4) supplemented with 250 mM KCl, 0.1% Tween-20, 1 mM DTT, 0.25 mM PMSF, 1 μg/ml leupeptin, 2 μg/ml aprotinin, and 15 μg/ml benzamidine.

EB
20 mM potassium EGTA, 15 mM $MgCl_2$, 10 mM DTT, 80 mM potassium β-glycerophosphate (pH 7.3).

NETN
20 mM Tris-HCl (pH 8.0), 100 mM NaCl, 1 mM EDTA, 0.5% NP-40, 1 mM PMSF, 5 μg/ml leupeptin, and 5 μg/ml antipain.

3.1 Purification of HA-Cdk2

This protocol outlines the purification of human Cdk2 with a 16-amino acid hemagglutinin A (HA) epitope tag fused at the N-terminus (Connell-Crowley at al. 1993). The T7 RNA polymerase system is used for expression (Studier at al. 1990). While ~10–20% of the total cell protein is Cdk2, <50% of this is soluble even when induced at 25 °C. In addition, there is sometimes variability in expression levels among individual transformants in the BL21(DE3) strain used and reduced expression levels are often experienced after cryopreservation of strains. It is therefore prudent to use freshly transformed strains and to check several colonies for expression levels in small scale experiments, including cells lacking a plasmid as a negative control, prior to preparing large scale cultures. HA-Cdk2 migrates at approximately 35.5 kDa and should be a prominent band on SDS-PAGE of total cell proteins if the induction is good.

Purification protocol A

1. Inoculate a 1-l culture with a 10-ml overnight culture of pET HA-Cdk2 in BL21(DE3) cells and incubate with shaking at 37 °C. Induce with 0.4 mM IPTG at OD_{600} ~0.8–1.0. Incubate the cells overnight with shaking at 25 °C.
2. Pellet the cells by centrifugation at 5000 rpm for 5 min. Resuspend the cell pellet in 70 ml of ice cold buffer L1 and add 0.1 mg/ml lysozyme. Incubate 45 min on ice. All subsequent steps should be performed on ice or at 4 °C.
3. Sonicate the lysate 5 × 15 s on ice while adding protease inhibitors between the sonications to reduce proteolysis. The lysate should become less viscous. Remove the insoluble material by centrifugation at 14000 rpm for 20 min. At this point, the extract can be stored in aliquots at –70 °C if desired.
4. Dialyze the soluble extract 3 h at 4 °C against buffer A (1 liter for the equivalent of 0.5 l of culture).
5. Pour a 15-ml Trisacryl SP column (IBF Biotechnics, Columbia, MD) and a 20-ml Toyopearl 650C DEAE column (Supelco, Bellefonte, PA). Equilibrate both at 4 °C with several column volumes of buffer A.
6. Load the dialyzed extract (cleared if necessary) onto the Trisacryl SP column. Cdk2 does not bind this column, so only the material that flows through the column and a wash step with one column volume of buffer A are collected.
7. Immediately load the flow through and the wash onto the DEAE column. Cdk2 also does not bind this column so, again, only the flow through and a wash step are collected. At this point, 50–70% of the total protein should be Cdk2, depending on the level of induction. This can be stored at –70 °C if desired.
8. The last step of purification uses an ATP affinity column which removes the remaining bacterial contaminants and Cdk2 that is not properly folded. Pour a 1–2 ml ATP agarose column (type IV, Sigma, St. Louis, MO) and equilibrate with buffer B.

9. Add 10 mM $MgCl_2$ to the extract to facilitate Cdk2 binding to the column. Load the extract (we load the equivalent of 150 ml of culture) and wash with four column volumes of buffer B. Elute the Cdk2 with buffer B containing 0.2 M NaCl. The Cdk2 should begin to come off the column after the first 2–4 ml. Analyze 15 µl aliquots from 0.5 ml fractions on 12% SDS-PAGE.
10. Pool fractions containing Cdk2 and dialyze in buffer D. This preparation is stable for more than a year at –70 °C.

Typical yields are 300 to 800 µg of >95% homogeneous Cdk2 per 150 ml of culture. Cdk2 purified in this manner has exceedingly low histone H1 kinase activity ($<10^{-17}$ mol/min/µg Cdk2) (see below). While this system works well with wild-type Cdk2 and mutants at T160, some active site mutants including D127A, D127E, R126A, and R150A are not bound efficiently to the ATP agarose, presumably due to alterations in the ATP binding site.

3.2 Purification of GST-Cdk2

Activatable human Cdk2 can be easily purified as a GST-fusion (Connell-Crowley at al. 1993). GST-Cdk2 that is not properly folded, as well as bacterial contaminants, can be removed by purification on an ATP affinity column.

Purification protocol B

1. Inoculate and induce a 1-l culture of the expression strain, as outlined in the previous protocol, using 0.1 mM IPTG for induction.
2. Pellet the cells by centrifugation at 5000 rpm for 5 min and resuspend in 70 ml of ice cold NETN. All subsequent steps should be performed on ice.
3. Sonicate the extract 5 × 15 s on ice while adding protease inhibitors between sonications to reduce proteolysis. Centrifuge at 14000 rpm for 20 min to remove the insoluble material.
4. Wash 2 ml of glutathione-S-sepharose (Pharmacia) several times with NETN + 0.5% powdered milk. Add 1.4 ml glutathione-S-sepharose (1:1 in NETN + 0.5% powdered milk) to the lysate. Rotate for 30 min at 4 °C.
5. Centrifuge the glutathione beads briefly to pellet and wash three times with NETN to remove the unbound proteins. It is often convenient to place the glutathione beads in 4 × 1.5-ml centrifuge tubes during the final washing steps. Analyze 10 µl of beads by SDS-PAGE. GST-Cdk2 migrates at approximately 58 kDa. As a control, include a sample of GST protein purified in a similar manner. Preparations at this stage will contain

substantial quantities of GST breakdown products (20–40% of the quantity of Cdk2) but these can be removed by subsequent chromatography.

6. Elute the GST-Cdk2 from the beads by incubating with 20 mM glutathione in 50 mM Tris-HCl (pH 9.6) and 120 mM NaCl (use 1 ml per 0.2 ml of beads) for 10 min at 4 °C. A typical yield is 1 mg GST-Cdk2 per liter of culture.
7. Immediately dialyze the eluted GST-Cdk2 against 1 l of buffer D containing 10 mM NaCl. Store in small aliquots at –70 °C.
8. After elution, dialyze against 1 l of buffer B and follow steps 8–10 of the HA-Cdk2 purification protocol. Note: in this case, GST-Cdk2 is eluted from the ATP column with 0.4 M NaCl. Typically, half of the GST-Cdk2 eluted from the glutathione beads is recovered at a purity of >90%.

3.3 Purification of Human Cyclin A

Purification protocol C

1. Transform BL21(DE3) with pET-Cyclin A. Induction, cell lysis, and clearing are performed as described for HA-Cdk2. Soluble proteins can be stored at –70 °C or processed immediately.
2. Precipitate the Cyclin A by slowly adding finely ground ammonium sulfate to the soluble extract (80% wt/vol) at 4 °C with stirring. The solution should become cloudy. Recover the precipitate by centrifugation at 12 000 rpm for 20 min.
3. Dissolve the precipitate in 20 ml of ice cold buffer A and dialyze against 3 × 1 l of the same buffer.
4. Apply the dialyzate to a 20-ml Toyopearl 650C DEAE column (Supelco, Bellefonte, PA) that has been equilibrated in buffer C. Wash with 75 ml of buffer C containing 100 mM NaCl. Elute the Cyclin A with 75 ml of buffer C containing 250 mM NaCl, collecting 5-ml fractions. Identify the fractions that contain Cyclin A by immunoblotting or by SDS-PAGE. Human Cyclin A migrates at approximately 54 kDa.
5. Load the fractions containing Cyclin A onto a 4-ml Fast S Sepharose column (Pharmacia, Piscataway, NJ) equilibrated with buffer C. Elute the bound proteins with a 100-ml NaCl gradient (0.1–0.6 M NaCl) at 0.8 ml/min. Identify the fractions containing Cyclin A by immunoblotting or by SDS-PAGE. Cyclin A should be a major protein at this point.
6. Pool the Cyclin A fractions and dilute twofold into 10 mM Tris-HCl (pH 7.5), 1 mM EDTA, and 1 mM DTT. Remove any particulate by filtration through a cellulose-acetate filter or by centrifugation.

7. Apply fractions to an anion exchange column (e.g., Pharmacia Mono-Q) equilibrated with buffer C. Elute the bound proteins with a 0–500 mM NaCl gradient for 40 min at 0.8 ml/min. Cyclin A elutes at ~250 mM NaCl.

8. Dialyze peak fractions against buffer C containing 10% glycerol and store in small aliquots at −70 °C. The typical yield is 5 mg/l of culture with a purity of >95%. This preparation lacks detectable contaminating histone H1 kinase activity and is stable at −70 °C for more than 1 year, although multiple freeze thaw cycles are not advised.

3.4 Purification of GST-Cyclins

Cyclins A, B, D, and E are readily purified as GST-fusion-proteins in *E. coli* using a number of available protocols (Solomon at al. 1990; Koff at al. 1992; Poon at al. 1993; Hannon at al. 1994). In general, a large percentage of the induced protein is insoluble but sufficient material for analytical experiments are readily obtained. In most of the experiments reported, the GST fusion protein is used in excess of Cdks and it is not clear what proportion of the total protein is folded properly.

3.5 Purification of GST-p40^{MO15} and GST-Cdc2

While human Cdc2 is highly insoluble when expressed in *E. coli*, *Xenopus* Cdc2 and *Xenopus* MO15 can be easily purified as soluble GST-fusion proteins (Poon at al. 1993). To enrich for properly folded Cdc2 or to activate MO15, these GST-fusion proteins can be incubated with a *Xenopus* egg extract.

Purification protocol D

1. Inoculate a 400 ml culture with a 2–5-ml overnight culture of *Xenopus* GST-Cdc2 or GST-MO15 in BL21(DE3) cells and incubate at 30 °C with shaking. When the OD_{600} reaches 0.8–1.0, induce the culture with 100 µM IPTG. Incubate overnight at 23 °C.

2. Harvest the cells by centrifugation at 5000 rpm for 5 min. Lyse the cells in 25 ml of buffer L3 with 2 mg/ml lysozyme for 15 min at 4 °C. All subsequent steps should be carried out on ice or at 4 °C.

3. Sonicate the lysate 5 × 15 min on ice while adding protease inhibitors to reduce proteolysis. Centrifuge at 18 000 *g* for 30 min to remove insoluble material.

4. Filter the supernatant through a 0.45-µm cellulose-acetate filter and load onto a 0.5-ml glutathione-S-sepharose column (Pharmacia, Piscataway, NJ) equilibrated with buffer E.

5. Wash the column with 5 ml of buffer E followed by 1.5 ml of PBS with 1 mM DTT. Analyze 10 µl of the glutathione beads by SDS-PAGE (GST-MO15 migrates at 66 kDa, GST-Cdc2 migrates at 60 kDa). A typical yield is 2–5 µg/10 ml beads.

Enrichment

Enrichment for Properly Folded GST-Cdc2. *Xenopus* Cdc2 purified from *E. coli* is not readily activated by Cyclin and MO15. However, Poon et al. (1993) have shown that incubation of purified GST-Cdc2 in *Xenopus* egg extract will produce activatable Cdc2, presumably by providing an environment in which it can fold properly.

Mix the purified GST-Cdc2 coupled to glutathione beads with 100 µg of *Xenopus* egg extract and incubate for 30 min at 23 °C. Recover the beads by brief centrifugation and wash three times with buffer E.

Activation

Activation of GST-MO15. Incubate 1–10 µg of GST-MO15 with 100 µg *Xenopus* egg extract for 30 min at 23 °C in the presence of 1 mM ATP. Under these conditions, a small percentage of the M015 will associate with the endogenous activator protein and become activated. Recover the GST-MO15 beads by brief centrifugation and wash three times with buffer E and two times with 10 mM Tris-HCl (pH 7.5) containing 10 mM $MgCl_2$. These beads are used for activation of soluble Cdk-Cyclin complexes (see below).

Alternatively, GST-MO15 can be eluted from glutathione beads prior to incubation with *Xenopus* extracts and the activated protein subsequently recovered using glutathione beads (Poon at al. 1993).

3.6 Purification of Human Wee1

An active 49 kDa form of human Weel can be produced in bacteria as a GST-fusion (Parker and Piwnica-Worms 1992).

Purification protocol E

1. Grow a 100-ml culture of JM109/pGST-Weel overnight. Dilute 1:10 into 1 l of LB and shake at 37 °C for 1 h.

2. Induce expression with 1 mM IPTG and incubate for 3 h at 30 °C.

3. Collect cells by centrifugation and resuspend in 50 ml of buffer L2. Sonicate 5 ×15 s and centrifuge at 10 00 g for 20 min.

4. Adsorb GST-Weel onto glutathione beads as described for GST-Cdk2. Analyze 10 µl of washed beads by SDS-PAGE. The GST-Weel fusion protein migrates at approximately 76 kDa and should be present at about

2–5 μg/10 μl beads. As a control, also include a sample of GST protein purified in a similar manner.

3.7 Assay for Direct Activation of Cdk2 by Cyclin A

Cdk2 can be partially activated by Cyclin A (but not Cyclin B) in the absence of T160 phosphorylation (Connell-Crowley at al. 1993). Monomeric Cdk2 has a specific activity of $<10^{-17}$ mol/min/μg. In the presence of stoichiometric amounts of pure Cyclin A, Cdk2 is activated $>10^{5}$-fold (16×10^{-12} mol/min/μg). This corresponds to ~1.8% activity of the fully activated, T160-phosphorylated form. While this low specific activity requires higher levels of Cdk2/Cyclin complexes for routine assay (~100 nM vs. 1 nM), the relative ease with which large amounts of these nonphosphorylated complexes can be generated makes this a viable approach for many kinds of analytical experiments.

Cdk2 activation protocol A

1. Prepare 1 μM stocks of Cdk2 and Cyclin A in 1X kinase buffer: 20 M Tris-HCl (pH 7.4), 10 mM $MgCl_2$. EB buffer (see below) can also be used.
2. Add 2.5 μl each of diluted Cdk2 and Cyclin A to 18 μl of kinase buffer (or EB) and incubate on ice for ~10 min. Add 1 μl of histone H1 (1 mg/ml) and initiate the reaction by addition of 1 μl of an ATP premix [4 μl (γ-^{32}P) ATP at 3000 Ci/mmol, 1 μl 10 mM ATP, and 7 μl kinase buffer]. The final concentration of (γ-^{32}P) ATP is 30 μM at 0.3 Ci/pmol.
3. After 60 min, add 25 μl 2× SDS-PAGE buffer and analyze samples by SDS-PAGE (12%) and autoradiography. Incorporated radioactivity can be measured by excising and scintillation counting H1 bands or by gel scanning. Typically, activity is high enough to have strong H1 bands in a 10–16-h exposure.

It is prudent to do controls in which the Cyclin or the Cdk2 is omitted to ensure that there is no contaminating kinase present. In addition, it is important to perform preliminary titration experiments using a constant amount of Cdk2 and increasing amounts of Cyclin A in order to determine the optimal Cyclin A levels for activation, as the percentage of properly folded Cyclin A may be sensitive to the exact method of purification and/or storage. In our hands, Cdk2 is fully activated by stoichiometric quantities of Cyclin A. The histone H1 concentration used is slightly below the K_m value measured for the unphosphorylated complex (K_m ~1.4 μM) and can be adjusted if necessary. In some situations it is useful to rapidly check Cdk2/Cyclin complexes for kinase activity. This can be done using the filter binding kinase assay described later in this chapter which avoids the time necessary for running and exposing SDS gels.

3.8 Assay for Activation of Cdk-Cyclin Complexes by Cak

The following protocol, adapted from Solomon at al. 1993, could be used for both purified Cak and GST-MO15 activated by incubation in *Xenopus* extracts.

Activation protocol B

1. Prepare a reaction mix containing 20 nM Cdk2 or Cdc2, 20–50 nM Cyclin A or B, 2 mM ATP, and 1 mg/ml ovalbumin in EB.
2. Mix 5 µl of the Cdk-Cyclin premix with 5 µl of Cak (at ~1 nM) and incubate for 30 min at room temperature. Cak stocks should be diluted into EB prior to use. This quantity of Cak should be sufficient to activate >90% of the Cdk complex in this time period.
3. Dilute reaction mixtures five-fold into EB. These can then be assayed for histone H1 kinase activity immediately or, alternatively, activatable complexes can be stored at –70 °C. It is prudent to include controls lacking either Cak or Cyclin. Phosphorylation of Cdk2 on T160 leads to increased mobility on SDS-PAGE which can be detected on immunoblots of Cak-treated Cdk2.
4. Assay the activated complex (10 µl of the diluted Cak-reaction mixture) using the conditions indicated above.

3.9 Rapid Kinase Assays Using Filter Binding

Filter binding assays have been used extensively for both threonine/serine and tyrosine kinases. This assay takes advantage of the fact that many substrates used to detect kinase activity have a net positive charge and can therefore bind to phosphocellulose paper after labeling with ^{32}P. For routine Cdk2 assays, we use a peptide derived from histone H1 (AKAKKTPKKAK) at 50 µM. In some cases, protein substrates (e.g., histone H1) also bind to phosphocellulose. The primary advantage of the filter binding assay is that it is much faster than methods requiring electrophoresis, and can also be highly sensitive. The most common problem encountered with filter binding assays is that high background levels of radioactivity remain bound to filters even in the absence of added kinase. This is sometimes due to the presence of basic ^{32}P-labeled impurities in commercial ^{32}P-ATP preparations that can bind to phosphocellulose. The use of high quality, purified ATP is suggested.

Peptide kinase assay

1. After incubation at 37 °C, spot 10 μl of each reaction on 1" by 1" squares of phosphocellulose paper (grade P 81, Whatman).
2. Allow the phosphocellulose squares to dry and wash with phosphoric acid wash buffer (10 ml of concentrated H_3PO_4 in 1 l of water) until the negative controls have <800 cpm of ^{32}P (as detected using a hand-held Geiger counter). Typically, eight washes of 150 ml is sufficient for ten phosphocellulose squares. During in vitro assays, the Cyclin subunit is often phosphorylated. However, in control experiments, we have found that phosphorylated Cyclin A does not bind to phosphocellulose and does not contribute to background in this assay.
3. Allow the phosphocellulose squares to dry and measure the radioactivity by scintillation counting.

3.10 Phosphorylation of Cdc2 Complexes by Wee1

Cdc2 phosphorylation assay

1. Mix Cdc2/Cyclin B complexes (0.2–1 μg) in 20 μl of 50 mM Tris-HCl (pH 7.4), 50 mM NaCl, 1 mM DTT, 10 μM ATP, 40 μCi [γ-^{32}P] ATP, and 10 mM $MgCl_2$ with 20 ng of GST-Wee1. Incubate 15–30 min at 30 °C.
2. To assay for Cdc2 phosphorylation, remove 5 μl of the assay mixture and analyze by 12.5% SDS-PAGE followed by autoradiography.
3. To assay for inhibition of Cdc2/Cyclin B, perform histone H1 assays as describe above for Cdk2/Cyclin A.

References

Connell-Crowley L, Solomon MJ, Wei N, Harper, JW (1993) Phosphorylation independent activation of human cyclin-dependent kinase 2 by cyclin A in vitro. Mol Biol Cell 4: 79–92

Fesquet D, Labbe J-C, Derancourt J, Capony J-P, Galas S, Girard F, Lorca T, Shuttleworth J, Doree M, Cavadore J-C (1993) The MO15 gene encodes the catalytic subunit of a protein kinase that activates cdc2 and other cyclin-dependent kinases (Cdks) through phosphorylation of Thr161 and its homologues. EMBO J 12:3111–3121

Hannon GJ, Demetrick D, Beach D (1993). Isolation of the Rb-related p130 through its interaction with CDK2 and cyclins. Genes Dev 7:2378–2391

Koff A, Giordano A, Desai D, Yamashita K, Harper JW, Elledge SJ, Nishimoto T, Morgan DO, Franza R, Roberts JM (1992) Formation and activation of a cyclin E-cdk2 complex during the G1 phase of the human cell cycle. Science 257:1689–1694

Parker LL, Piwnica-Worms H (1992) Inactivation of the p34cdc2-cyclin B complex by the human wee1 tyrosine kinase. Science 257:1955–1957

Poon RYC, Yamashita K, Adamczewski JP, Hunt T, Shuttleworth J (1993) The cdc2-related protein kinase $p40^{MO15}$ is the catalytic subunit of a protein kinase that can activate $p33^{cdk2}$ and $p34^{cdc2}$. EMBO J 12:3123–3132

Solomon MJ, Glotzer M, Lee TH, Philippe M, Kirschner MW (1990) Cyclin activation of $p34^{cdc2}$. Cell 63:1013–1024

Solomon MJ, Lee T, Kirschner MW (1992) Role of phosphorylation in $p34^{cdc2}$ activation: identification of an activating kinase. Mol Biol Cell 3:13–27

Solomon MJ, Harper JW, Shuttleworth J (1993) CAK, the $p34^{cdc2}$ activating kinase contains a protein identical or closely related to $p40^{MO15}$. EMBO J 12:3133–3142

Studier WF, Rosenberg AH, Dunn JJ, Dubendorff JW (1990) Use of T7 RNA polymerase to direct expression of cloned genes. Methods Enzymol 185:60–89

15 Heterologous Expression and Secretion of Cell Cycle Proteins Using the Yeast *Picchia pastoris*

GUILLAUME COTTAREL

1 Background

Bacterial and baculoviral expressions are the two current systems of choice for protein expression (see Hannon and Connel-Crowley and Harper, this Vol.). However, these methods cannot combine the same advantages (see discussion in Hannon, this Vol.) Here, the use of an emerging protein expression technology which conciliates the major benefits of bacterial and baculoviral protein expression is described. Proteins can be produced in high amounts either intra- or extracellularly, allowing an extremely effective protein purification. Moreover, as it uses an eukaryotic host protein processing, folding and post-translational modifications can occur and, as the host is a yeast cell, maintenance and growth even in high volume (fermentation) are fast, easy and economical.

Pichia is with 56 species the largest ascoporogenous genus (Kreger-van Rij 1989). Its natural habitat is mainly in association with plants. The species used in this protocol, *Pichia pastoris*, was first isolated in 1919 from the exudate of a chestnut tree in France and appears to be common in tree fluxes in European forests. *P. pastoris* was also found in fluxes of elms and oaks in forests in Canada and the United States (Phaff and Starmer 1989). *P. pastoris* is similar to the budding yeast *Saccharomyces cerevisiae* and is amenable to molecular genetic manipulations wich were developed for *S. cerevisiae*. *P. pastoris* is a methylotrophic yeast due to its ability to utilize methanol as a carbon and energy source (Rose 1989). As explained below, this characteristic is essential.

In this chapter, I mainly refer to the *P. pastoris* expression kit provided by Invitrogen (San Diego, California) and to several reference articles (Cregg et al. 1993; Buckholz and Gleeson 1991) which report that heterologous protein expression can yield as much as 12 g/l intracellularly (tetanus toxin fragment C) or 4 g/l extracellularly (human serum albumin).

It should be mentioned that heterologous cell cycle protein expression might interfere with the *P. pastoris* cell cycle control machinery as overexpression of cell cycle regulators was shown to delay or activate events such as mitosis (unpubl. results).

2 Principles and Applications

The first enzyme in the methanol pathway, the alcohol oxidase, coded by the *AOX1* gene, oxidizes methanol to formaldehyde. AOX is undetectable in cells cultured on carbon sources such as glucose, glycerol or ethanol, but constitutes approximately 30% of total soluble protein in methanol-grown cells (Couderc and Baratti 1980). Levels of *AOX1* are controlled at the level of mRNA expression: in methanol-grown cells about 5% of the polyA+ RNA is transcribed from the *AOX1* gene. The *AOX1* gene has been isolated (Ellis et al. 1985) and its promoter is part of the expression vectors provided by the Invitrogen kit. Heterologous protein expression in *P. pastoris* can be either intracellular or secreted. Secretion requires the presence of a signal sequence to target the protein into the secretory pathway. In the provided kit, two different vectors with different secretion signals are included. One carries the native *P. pastoris* signal from the acid phosphatase gene *PHO1*. The other carries the secretion signal from the *S. cerevisiae* α-mating factor (Clare et al. 1991). The vectors carry the *P. pastoris HIS4* gene as selectable marker for transformation. Transformation of a *his4*$^-$ *P. pastoris* strain is done by standard protocols as explained in this book (see Cottarel, this Vol.). In order to stabilize foreign protein-producing strains, vectors are integrated into the genome through a recombination event between the *AOX1* sequence in the *P. pastoris* vector and its genomic locus. This results in the replacement of the *AOX1* gene with the gene of interest. The strains deleted at the *AOX1* locus must rely on the weakly expressed *AOX2* gene to metabolize methanol (Cregg et al. 1989). *AOX1* and *AOX2* genes and deduced proteins share high homology at the nucleotide and amino acid levels (Koutz et al. 1989). Selection of the HIS+ transformants is done by plating the cells on medium lacking histidine. The screen for transformants bearing the integration at the correct locus is simply done by streaking colonies from the transformation plate on a –his, +glycerol and a –his, +methanol plate. Colonies which grow poorly on the –his, +methanol plate no longer have an intact *AOX1* gene and have a methanol utilization-deficient (mut–) phenotype. Pilot expression analysis to identify the best producer ist performed by analzying protein expression on SDS-PAGE after the cells are grown for several days in methanol-containing media. To maximize protein expression multiple copies of the construct can be naturally integrated into the genome, however, the natural event occurs at a low frequency. Southern analysis of the isolated genomic DNA should identify such cases. Another approach is to engineer multiple expression cassette copies into a single vector (review in Cregg et al. 1993).

This chapter focuses only on the protocol which provides protein expression and secretion. It is emphasized that normally *P. pastoris* secretes an extremely low level of proteins. As a consequence, the majority of protein in the media is produced due to the presence of the engineered construct.

Structure and nucleotide sequences of the vectors as well protocols for proper gene cloning are provided by the Invitrogen kit.

3 Materials

Buffer and stock solutions

- TE: 10 mM Tris-HCl pH 7.4, 1 mM EDTA pH 8.
- LiCl/TE: 0.1 M LiCl in TE.
- PEG: 40% polyethylene glycol 3350 in LiCl/TE buffer.
- 5× YNB: 134 g yeast nitrogen without amino acids with ammonium sulfate in 1 l water.
- 500× B: 20 mg biotin in 100 ml water. Store at 4 °C.
- 10× D: 200 g D-glucose in 1 l water, autoclave.
- 10× M: 5 ml methanol in 95 ml water. Filter sterilize, store at 4 °C.
- 10× GY: 100 ml glycerol in 900 ml water.
- 1 M PP: 132 ml of 1 M K_2HPO_4 and 868 ml of 1 M KH_2PO_4 and verify that the pH is 6.
- 15% Glycerol: 150 ml glycerol in 850 ml water. Sterilize by autoclaving, 1.5 ml aliquot into cryotubes.

Media

- BMGY: 10 g yeast extract, 20 g peptone in 600 ml water, autoclave. Add 100 ml 1 M PP, 200 ml 5× YNB, 2 ml 500× B, 100 ml 10× GY. Store at 4 °C.
- BMMY: 10 g yeast extract, 20 g peptone in 600 ml water, autoclave. Add 100 ml 1 M PP, 200 ml 5× YNB, 2 ml 500× B, 100 ml 10× M. Store at 4 °C.
- MD plates: 15 g agar in 700 ml water, autoclave. Add 200 ml 5× YNB, 2 ml 500× B, 10 ml 10× D. Homogenize with a steering bar, pour onto the plates. Store at 4 °C.
- MM plates: 15 g agar in 700 ml water, autoclave. Add 200 ml 5× YNB, 2 ml 500× B, 10 ml 10× M. Homogezine with a steering bar, pour onto the plates. Store at 4 °C.
- YPD: 10 g Bacto-yeast extract, 20 g Bacto-peptone, 20 g dextrose in 1 l water, autoclave.

4 Methods

Lithium transformation

This protocol is derived from the procedure described by Ito et al. (1983) for *S. cerevisiae*. This method does not favor multicopy integration of the DNA. If such an event is desired, the spheroplast method should be used (Invitrogen kit).

1. Inoculate 50 ml YPD with 1 ml of an overnight culture of GS115 provided by the Invitrogen kit. Shake at 30 °C until a density of about 5×10^7 cells/ml is achieved.
2. Pellet the cells at 1500 rpm for 10 min.
3. Wash the cells with 10 ml sterile water.
4. Pellet the cells at 1500 rpm for 10 min.
5. Wash the cells with 10 ml LiCl/TE buffer.
6. Pellet the cells at 1500 rpm for 10 min.
7. Resuspend in 2 ml LiCl/TE buffer and incubate for 1 h at 30 °C with agitation.
8. To 0.1 ml GS115-component cells add 1–20 μg of transforming DNA previously digested with an appropriate restriction enzyme to favor integration at the *AOX1* locus. Carrier DNA, such as salmon sperm DNA, can be added, the final volume of DNA should not exceed 20 μl.
9. Incubate at 30 °C for 30 min with agitation.
10. Add 0.7 ml 40% PEG and vortex briefly.
11. Incubate at 30 °C for 30 min with agitation.
12. Heat shock at 37 °C for 5 min.
13. Pellet the cells.
14. Resuspend the cells gently in 0.1 ml water and spread on MD plates.
15. Colonies will appear after 2–3 days at 30 °C.

Screening for His+/Mut– transformants

This step is for the identification of the *AOX1*-disrupted transformants, an event which occurs at a frequency of 5–35% for this His+ transformants.

1. Pick transformants with sterile toothpicks and streak lightly on MD and MM plates. As control, streak the albumin-secreting protein-expressing GS115/His+ Mut– strains and intracellular GS115/HIs+ Mut+ β-gal strains provided by the Invitrogen kit.
2. Incubate the plates at 30 °C for 2–3 days.
3. Score the plates and identify the candidates which grow normally on MD plates but show little or no growth on MM plates as does the albumin-secreting GS115/His+ Mut– strains.

Identification of efficient secretion transformants

1. Inoculate 10 ml BMGY with your selected clones and incubate with vigorous shaking at 30 °C until saturation (48 h). One tube is inoculated with GS115/His+ Mut– as control.

2. Pellet the cells at 1500 rpm for 10 min.
3. Discard the media and resuspend the pellet with 2 ml BMMY.
4. Incubate with vigorous shaking at 30 °C for 2 days.
5. Transfer 0.2 ml of cells into an Eppendorf tube. Return the remainder of the culture into the shaker.
6. Pellet the 0.2 ml aliquot in a microcentrifuge for 5 s at maximum speed.
7. Analyze a few μl of the supernatant by SDS-PAGE, followed by Coomassie blue staining.
8. A time-course expression study can be performed. Repeat the last three steps for a total of 6 days. Fresh media (BMMY) can be added after cell centrifugation for renewed protein secretion.
9. The properly expressed colne(s) hence identified are stored in glycerol (0.5 ml saturated BMGY culture is mixed with 1.5 ml 15% glycerol) and kept at –70 °C. The strain(s) can be reactivated at any time by streaking some frozen cells on YPD plates.

Pilot expression of secreted protein

1. Inoculate 10 ml BMGY with your strain(s) and incubate overnight at 30 °C.
2. Inoculate 1 l BMGY with the 10 ml preculture. Use a large flask and vigorous agitation for proper aeration for 2 days.
3. Verify that the OD_{600} is between 10–20. Dilution of the culture should be done for proper OD_{600} measurement.
4. Pellet the cells at 1500 rpm for 10 min.
5. Resuspend the cells with 100 ml BMMY and induce protein expression and secretion in a large flask and with vigorous agitation for proper aeration for 2 days.
6. Analyze protein expression by SDS-PAGE, followed by Coomassie blue staining. BMMY media can be replaced to renew protein expression.
7. Once the proper level of secretion is reached, pellet the whole culture and collect the supernatant. Analyze protein expression by SDS-PAGE, followed by Coomassie blue staining.
8. Protein concentration can be done with various standard methods (e.g., ammonium sulfate precipitation, centrifuge concentrator).

References

Buckholz RG, Gleeson AG (1991) Yeast systems for the commercial production of heterologous proteins. Bio/Technology 9:1067–1072

Clare JJ, Romanos MA, Rayment FB, Rowedder JE, Smith MA, Payne MM, Sreekrishna K, Hendwood CA (1991) Production of mouse epidermal growth factor in yeast: high-level secretion using *Pichia pastoris* strains containing multiple genes copies. Gene 105: 205–212

Couderc R, Baratti J (1980) Oxidation of methanol by the yeast *Pichia pastoris*: purification and properties of alcohol oxidase. Agric Biol Chem 44:2279–2289

Cregg JM, Madden KR, Barringer KJ, Thill GP, Stillman CA (1989) Functional characterization of the two alcohol oxidase genes from the yeast *Pichia pastoris*. Mol Cell Biol 9: 1316–1323

Cregg JM, Vedvick TS, Raschke WC (1993) Recent advances in the expression of foreign genes in *Pichia pastoris*. Bio/Technology 11:905–910

Ellis SB, Brust PF, Koutz PJ, Waters AF, Harpold MM, Gingeras TR (1985) Isolation of alcohol oxidase and two other methanol regulatable genes from the yeast *Pichia pastoris*. Mol Cell Biol 5:1111–1121

Ito H, Fukuda Y, Murata K, Kimura A (1983) Transformation of intact yeast cells treated with alkali cations. J Bacteriol 153:163–168

Koutz PJ, Davis GR, Stilman C, Barringer K, Cregg JM, Thill G (1989) Structural comparison of the *Pichia pastoris* alcohol oxidase genes. Yeast 5:167–177

Kreger-van Rij (1989) Classification of yeasts. In: Rose AH, Harrison JS (eds) The yeasts. Biology of yeasts, vol 1. Academic Press, New York

Phaff HJ, Starmer WT (1989) Yeasts associated with plants, insects and soil. In: Rose AH, Harrison JS (eds) The yeasts. Biology of yeasts, vol 1. Academic Press, New York

Rose AH (1989) Responses to the chemical environment. In: Rose AH, Harrison JS (eds) The yeasts. Biology of yeasts, vol 2. Academic Press, New York

16 SV40-Based in Vitro DNA Replication Assay

ANINDYA DUTTA AND ELLEN WINCHESTER

1 Background

In the absence of well defined origins of DNA replication from animal cells, the biochemical approach to study DNA replication as related to the cell cycle, has been confined to two systems. One is the replication of exogenous DNA in *Xenopus* egg extracts (see chapter by P. Clarke, Part Two), and the other, which is the subject of this chapter, is the in vitro replication of DNA based on the replication of the papovavirus SV40.

A specially prepared extract of human HeLa or 293 cells is capable of supporting the replication of a plasmid containing the origin of replication of SV40 (ori), provided that the reaction contains enough deoxyribo- and ribonucleotides (particularly a source of ATP), divalent cations and the SV40 encoded origin binding protein T antigen (reviewed in Kelly 1988; Stillman 1989; Hurwitz et al. 1990). An extract from cells synchronized in the G1 phase of the cell cycle is poorer in its ability to support DNA replication than an extract from cell in S phase. Therefore the mechanism of the G1-S transition can theoretically be dissected using this system. Further, activators of the G1-S transition, the cdk kinases, have been shown to activate DNA replication in G1 extracts, while known inhibitors of the G1-S transition, like the tumor suppressor protein p53, and the Cyclin-Cdk interactor protein p21, have been shown to inhibit DNA replication in extracts from asynchronous cells.

The advantages of the SV40 based system for studying cell cycle regulation are that the components are purified to near homogeneity, and the individual events that take place successively at the origin of replication can be separately studied. Thus, the exact mechanism by which cell cycle regulators influence the replication machinery can be dissected. The disadvantage is that a major component in the SV40 based replication machinery is the virally encoded T antigen which executes several functions (ori recognition, untwisting of the DNA, recruitment of replication factors, helicase activity, ATPase activity) that are carried out in vivo by unknown proteins at cellular origins of replication. These cellular proteins are likely to be targets of regulation at the G1-S transition, but since their role is supplanted by T antigen, this regulation cannot be studied with the SV40 based system. Further, several cell-cycle regulators directly influence T antigen function (e.g., Cdc2 kinase, p53), and any effect observed in the SV40 in vitro DNA replication system could be complicated by a direct action of this regulator on T antigen. Finally, several regulatory factors present *in vivo*, e.g. nuclear membrane, nuclear matrix, chromatin, and inhibitory factors not extracted in the cell lysates, are also likely to be missed in the SV40 based system. Nevertheless, the disadvantages are still outweighed by the ability to replicate DNA with a few purified factors and the ability to dissect the steps of the replication reaction.

2 Principle and Applications

The decreased replication activity of extracts from G1 cells was reported by Roberts and D'Urso, who also showed that the defect was in the ability to support the unwinding of double-stranded DNA at the origin of DNA replication (Roberts and D'Urso 1988). Addition of extracts from S phase cells complemented the activity, and D'Urso et al. demonstrated that the "S phase promoting factor" was Cyclin A and its associated kinase Cdk2 (D'Urso et al. 1990).

The unwinding of the origin of DNA replication requires the SV40 encoded T antigen and only one cellular protein, the single-stranded DNA binding factor RPA. RPA was identified and purified by virtue of its absolute requirement for SV40-based DNA replication. It is a complex of three polypeptides of 70 kDa, 34 kDa and 13 kDa, all of which are required to support SV40 based in vitro DNA replication. The abundance of RPA does not change between extracts of cells in G1 and in S, but the 34-kDa subunit is phosphorylated in extracts of cells in S phase. We have shown that at least one of the cellular kinases that phosphorylate RPA p34 in vitro is a member of the Cyclin-Cdk family of protein kinases, and have reproduced the observation of D'Urso et al. that addition of this kinase stimulates DNA replication activity of extracts from G1 cells (Dutta and Stillman 1992). However, we have not been able to add RPA pre-phosphorylated by Cdc2 to a G1 extract and stimulate DNA replication. Also, the Cdc2 kinase significantly stimulated the origin-unwinding activity of extracts from G1 cells, but this stimulation was minimal if pure RPA and T antigen were used. This suggests the existence of costimulatory factors in the G1 extract which augment the effect of Cdc2 kinase on the replication machinery. While we do not think that the stimulation is mediated by phosphorylation of T antigen by Cdc2 (the baculovirus produced T antigen is fully phosphorylated on the Cdc2 substrate site), it is difficult to rule out this possibility completely. Activation of SV40 DNA replication in G1 extracts has also been obtained by the addition of cellular protein phosphatase 2A, although in this case the effect is mediated through de-phosphorylation of T antigen (Virshup et al. 1989).

Figure 1b shows a technical point which is important for the study of the G1-S transition using this system. The amount of T antigen added to the re-

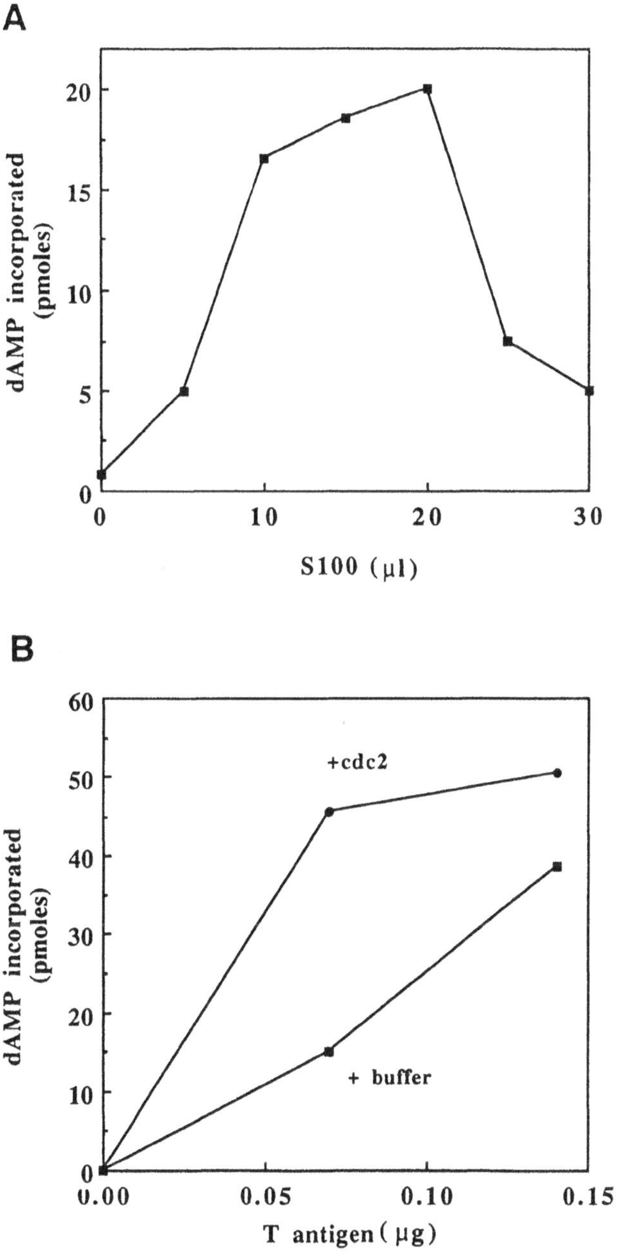

Fig. 1 A, B. Incorporation of dAMP into polynucleotide in the SV40-based in vitro DNA replication system. **A** Effect of increasing quantities of S100 (16.3 μg protein/μl) from an asynchronous culture of 293 cells, keeping the amount of T antigen constant at 1.2 μg per 50 μl reaction. **B** Effect of Cdc2 kinase in an extract from G1 HeLa cells is dependent on the amount of T antigen present in the reaction. At higher concentrations of T antigen the stimulatory effect of Cdc2 was not seen. This reaction was done in a volume of 25 μl using the indicated amounts of T antigen and 55 μg of S100 extract. The pmol incorporated are reported as for a 50-μl reaction. The batch of T antigen is different from the one used in **A**, and probably accounts for the difference in incorporation

action is important for the study of regulatory effects. In vitro biochemical systems are usually optimized for the greatest amount of activity, often resulting in a dispensability of stimulators of the activity. This has been noted for in vitro transcription reactions when attempts were made to reproduce the transcriptional stimulatory activity of enhancers and enhancer binding proteins (Laybourn and Kadonaga 1991). In vitro DNA replication is no exception, and in this case, the requirement of cellular replication stimulatory activities may become evident only when the T antigen concentration in the reaction is titrated down. An additional mechanism by which requirement for stimulatory activities may become evident in the in vitro reactions, is the execution of the transcription or replication assays on DNA templates assembled into chromatin (Laybourn and Kadonaga 1991; Cheng et al. 1992).

In contrast to the study of positive acting factors at the G1-S transition, factors present in G1 extracts which inhibit DNA replication may be more easily studied using the in vitro DNA replication system. A recent success has been the discovery that the growth suppressor protein p21 (which is transcriptionally induced by the tumor suppressor protein p53) directly inhibits the SV40 based DNA replication system via its interaction with and inhibition of the essential DNA replication factor PCNA (Waga et al. 1994). p53 itself also inhibits the SV40-based DNA replication reaction, although the effect is mostly mediated by its interaction with and inhibition of T antigen (Friedman et al. 1990). We have reported that p53 interacts with the RPA holocomplex and that the p53-bound RPA fails to bind single-stranded DNA (Dutta et al. 1993). The significance of this phenomenon in the normal G1-S transition is unclear.

The above examples indicate the wealth of approaches derived from the SV40 based DNA replication reaction used to study cell cycle control. In this chapter we will describe the methods for (1) obtaining a replication competent 293 cell extract, (2) preparing T antigen produced in a baculovirus expression system, and (3) in vitro DNA replication using the cell extract and T antigen. To prepare cell cycle staged extracts, the cells used to make the replication extract are synchronized by centrifugal elutriation (as described the chapter by Marracino and Keng, Part One; also see Dutta and Stillman 1992).

3 Method

3.1 293 Cell Suspension Culture

293 cells are grown in suspension in Joklik's modification of DMEM containing 5% calf serum. The inoculation density is 1×10^5 cells/ ml, and the cells are split or harvested at $5–10 \times 10^5$ cells/ml. For a standard preparation we use 8 l of suspension culture which yield a cell pellet of 10 ml.

Materials

- Hemocytometer for counting the cells
- Trypsin/EDTA is added to a small aliquot of cells to separate cluster of cells before counting.
- PBS is used to wash the cell pellet.
- Six 750-ml IEC centrifuge tubes.
- IEC PR7000 centrifuge (or equivalent).

Protocol

1. Chill 10 to 12 50-ml conical tubes (Falcon, Becton/Dickinson) on ice.
2. Pour out equal volumes of the culture into six 750 ml IEC centrifuge tubes. Spin in a IEC PR7000 rotor (or equivalent) at 1800 rpm for 8 min at 10 °C. Carefully pour the supernatant into waste flasks leaving the cell pellet and a small amount of medium. Resuspend each pellet in the residual medium and transfer to 50-ml conical bottom tubes on ice. Repeat the process until the entire suspension culture is transferred into 2–3, 50 ml conical tubes.
3. Pellet the cells in the 50-ml tubes at 1800 rpm, at 4 °C, for 10 min in a Beckman GS-6KR centrifuge (or equivalent). Remove the supernatant from each tube. Suspend each pellet in 15 ml PBS and combine in a single tube. Repeat centrifugation and remove supernatant. Yield should be an approximately 10 ml cell pellet.

3.2 Preparation of S100 Lysate

Solutions

Hypotonic lysis buffer

Stock	ml/100 ml
1 M HEPES pH 8.0	2
2 M KCl	0.25
1 M $MgCl_2$	0.15
1 M dithiothreitol	0.05

The cell pellet from the 293-suspension culture harvest is lysed in a hypotonic buffer and the S100 lysate containing replication factors separated from nuclei and membranes.

Materials

- The cell pellet from the suspension culture harvest (approximately 10 ml) is kept on ice.
- The Hypotonic Buffer should always be prepared fresh.
- A 40-ml Dounce Homogenizer (with Pestle B Wheaton), chilled on ice.
- A clinical tabletop centrifuge (IEC) in the cold-room at setting #4 is used to pellet the cells.

- 30 ml Corex tubes , Sorvall RC5B centrifuge and SS34 rotor (or equivalent).
- Beckman ultracentrifuge, a SW55.1 rotor (or equivalent).

Protocol

1. Resuspend cell pellet in an equal volume of hypotonic buffer (approximately 10 ml). Spin in a clinical (IEC) centrifuge in the cold room at setting #4 for 10 min. Remove supernatant.
2. Resuspend the pellet again in an equal volume of hypotonic buffer and let sit on ice for 10 min.
3. Homogenize the cells in a chilled 40-ml Dounce Homogenizer with a B Pestle; 30 strokes (down plus up equals one stroke) should rupture the cells. You can check under a microscope to make sure the cells are completely ruptured. Let the homogenate sit on ice for 30 min.
4. Transfer the homogenate to the 30-ml Corex tube (on ice). Spin in a RC5B Sorvall centrifuge in a SS34 rotor at 12 000 rpm for 10 min at 4 °C. Transfer the supernatant to a new tube. The pellet from this spin contains the nuclei. Label with date and "293 Nuclear Pellet" and freeze.
5. Adjust NaCl concentration of supernatant to 0.1 M by adding appropriate volume of 5 M stock NaCl solution.
6. Split the supernatant into three SW55.1 Beckman ultracentrifuge tubes. Centrifuge at 32 000 rpm for 1 h at 4 °C (100 000 *g* for 1 h).
7. Divide supernatant into 0.5 ml aliquots in 1.5-ml Eppendorf tubes. Label with date and "293 S100" and freeze. Measure the concentration of protein in the extract by the Biorad dye binding assay.

3.3 Preparation of T Antigen

Solutions TD (Buffer)

Stock	ml/100 ml
1 M Tris. HCl, pH 7.4	2.5
5 M NaCl	2.72
2 M KCl	0.285
0.2 M Na_2HPO_4	0.35

Buffer B

Stock	ml/200 ml
1 M Tris. HCl, pH 8.0	10
5 M NaCl	6
0.2 M EDTA	1
Glycerol	20
0.1 M PMSF	2

Buffer C

Stock	ml/200 ml
1 M Li Cl	100
0.2 M EDTA	1
Glycerol	20
0.1 M PMSF	2

Buffer D

Stock	ml/100 ml
1 M Pipes pH 7.4	1
5 M NaCl	0.1
0.2 M EDTA	0.5
Glycerol	10
0.1 M PMSF	1

Buffer E. Make just before using!

	per 20 ml
Triethylamine	56.5 µl
Glycerol	2 ml
pH should be approx. 10.8	

Buffer F

Stock	ml/2 l
1 M Pipes pH 7.0	20
1 M DTT	2
0.2 M EDTA	1
5 M NaCl	2
Glycerol	200
0.1 M PMSF	10

Protocol

1. Infect SF9 cells (800 ml, approx. 1.0–1.5 x 10^6 cells/ml) with baculovirus expressing T antigen (see also chapter by G. J. Hannon, Part Four):
   - Spin down cells in *sterile* 0.75 l bottle, 2000 rpm, 10 min (IEC PR7000).
   - Resuspend pellet in 100 ml baculovirus culture supernatant (10^9 plaque-forming unit).
   - Incubate 1 h, 27 °C, gently swirling every 15 min.
   - Return infected cells to culture flask and add 700 ml fresh medium.
   - Allow infection to proceed approx. 48 h–60 h.
2. Harvest cells
   - Spin down cells, keep supernatant as virus stock. Resuspend pellet in 25 ml TD buffer. Transfer to 50-ml screw cap tube. Spin in clinical centrifuge, setting 4, 3 min. Aspirate supernatant.
3. Lyse cells
   - Resuspend in 30 ml B buffer. Add 10% NP40 to final concentration of 0.5%.
   - Put on ice 30 min. Invert several times every 10 min. Spin 10 000 rpm, 10 min in Corex tubes. Keep supernatant and save a 50-µl aliquot to run on gel (crude).
4. Immunoaffinity purify T antigen
   - Load supernatant onto 4B-Cl Sepharose (Pharmacia) column (2 ml) equilibrated in Buffer B. Allow flowthrough to pass onto Protein A Sepharose (Pharmacia) column (2 ml).
   - Run flowthrough over PAS again and allow to pass onto PAS-mAb419 column (2 ml). A standard method should be followed for covalently coupling mouse ascites containing monoclonal antibody to T antigen, mAb419 (Harlow et al. 1981), to protein A Sepharose beads (see chapter by Pagano, Part Four). Pass flowthrough 2× more over PAS-419. Save flowthrough for gel.
   - Wash PAS-419 with 200 ml Buffer C, then 100 ml Buffer D.
   - Elute 0.5 ml fractions with Buffer E. Immediately add 50 µl 0.5 M Pipes pH 7.0 to each fraction and leave on ice.
   - Add 5 µl each fraction to 200 µl Biorad protein assay dye reagent diluted 1:5. Mix and pool peak fractions (visual assessment of blue color).
   - Dialyze peak overnight in 1 l buffer F, 4 °C.
   - Wash both the PAS and PAS-419 columns with Buffer E, then Buffer B and store in coldroom in presence of sodium azide.

Note!

1. Make fresh 4B-Cl column each time.
2. Should also change PAS column every four to five preparations.
   - Run aliquots of the cell lysate, PAS-419 flowthrough, and peak elution fractions on an SDS-polyacrylamide gel and visualize proteins by staining with Coomassie blue dye.

3.4 DNA Replication Reaction

An example of DNA replication using varying amounts of S100 extract is given below (Fig. 1a). The incorporation of radioactive dAMP into polynucleotide is assayed by spotting the reaction products on DE81 paper and washing off unincorporated nucleotides with phosphate buffer. The incorporation usually peaks at 20–25 μl of S100 in a 50 μl reaction. A titration curve for T antigen should be done by keeping the S100 quantity constant and varying the T antigen concentration. For studies on G1-S transition, we have found that the stimulation of DNA replication by Cdc2 kinase is greater at low concentration of T antigen (Fig. 1b).

Material

- DE81 chromatography paper from Whatman.
- pSVO10 or pSVO11 plasmid (Dutta and Stillman 1992), or similar plasmid which contains an origin of DNA replication from SV40.
- S100 extract and T antigen.

Buffer

5 × replication mix	(μl)	Final concentration in reaction (mM)
1 M HEPES pH 7.8	45.0	30.0
1 M DTT	0.75	0.5
1 M $MgCl_2$	10.5	7.0
0.1 M ATP pH 7.0	60.0	4.0
20 mM GTP	15.0	0.2
20 mM UTP	15.0	0.2
20 mM CTP	15.0	0.2
10 mM dGTP	15.0	0.1
10 mM dCTP	15.0	0.1
10 mM TTP	18.0	0.12
10 mM dATP	3.75	0.025
1 M phosphocreatine	60.0	40.0
Water		30.0
Total	303.0	

- Creatine phosphokinase (CPK) 10 mg/ml in 10 mM Tris HCl pH 8.0 and 50% glycerol.
- 0.5 M Phosphate buffer for washing DE81 paper, $Na_2HPO_4$142 G in 2 l water.

!

Precaution. The HEPES solution is stored at 4 °C after filter sterilization. The pH of HEPES should be carefully adjusted with 1M KOH for a successful replication reaction. If good incorporation is not obtained, the pH of the HEPES stock should be checked after dilution to 30 mM and should be 7.6. The nucleotide solutions should also be adjusted to a pH of 7.0 to 7.4.

Protocol

1. On ice prepare a master-mix (volumes in µl)

1 reaction	12 reactions	
5 × replication mix	9.9	118.8
10 mg/ml CPK	0.1	1.2
300 ng/µl pSV010	1.0	12.0
$\alpha^{32}P$ dATP	0.05	0.6
Water	1.95	23.4
Total	13.0	156.0

2. In nine Eppendorf microfuge tubes (1–9), mix the following on ice (volumes in µl):

Tube	Water	Master mix (1–2 µg/µl)	293 S100T antigen	
#1	36	13	0	1
#2	31	13	5	1
#3	26	13	10	1
etc. to 8, with 35 ml of 293 S100				
#9	17	13	20	0

3. Incubate at 37 °C for 60 min.
4. Transfer tubes to ice.
5. Cut a rectangle of DE81 paper. Mark and number a grid of ten squares of 1 cm × 1 cm.
6. Spot 5 µl of each reaction on the corresponding square of DE81 paper. Do not allow spots to dry.
7. Wash DE81 paper successively with 0.5 M Na_2HPO_4 (5 min each with three changes), with water (5 min with one change) and with 100% ethanol (5 min with one change).
8. Dry paper, cut squares, put in vial, add liquid scintillation fluid and count in a liquid scintillation counter. Also, for measurement of specific activity, count one square on which has been spotted 1.3 µl of master mix and dried without washing.
9. Specific activity of ^{32}P in reaction (cpm/pmole) = cpm of 1.3 µl master mix/125.
10. pmol of dAMP incorporated in each reaction = (cpm of DE81 paper × 10)/specific activity.
11. The reaction without T antigen should have no incorporation of dAMP.

Acknowledgements. Our research is supported by grants from the American Cancer Society, and the Massachusetts Division of the American Cancer Society. Special thanks to various members of Bruce Stillman's laboratory (Cold Spring Harbor Laboratory) for several of the protocols in this chapter.

References

Cheng LZ, Workman JL, Kingston RE, Kelly TJ (1992) Regulation of DNA replication in vitro by the transcriptional activation domain of GAL4-VP16. Proc Natl Acad Sci USA 89:589–593

D'Urso, G, . Marraccino RL, Marshak DR, Roberts JM (1990) Cell cycle control of DNA replication by a homologue from human cells of the $p34^{Cdc2}$ protein kinase. Science 250: 786–791

Dutta A, Stillman B (1992) *Cdc2* family kinases phosphorylate a human cell DNA replication factor, RPA, and activate DNA replication. Embo J 11:2189–2199

Dutta A, Ruppert JM, Aster JC, Winchester W(1993) Inhibition of DNA replication factor RPA by p53. Nature 365: 9–82

Friedman PN, Kern SE, Vogelstein B, Prives C(1990) Wild-type, but not mutant, human p53 proteins inhibit the replication activities of simian virus 40 large tumor antigen. Proc Natl Acad Sci USA 87:9275–9279

Harlow E, Crawford LV, Pim DC, Williamson NM (1981) Monoclonal antibodies specific for simian virus 40 tumor antigens. J of Virol 39:861–869

Hurwitz J, Dean FB, Kwong AD, Lee S-H (1990) The in vitro replication of DNA containing the SV40 origin. J Biol Chem 265:18043–18046

Kelly TJ (1988) SV40 DNA replication. J Biol Chem 263:17889–17892

Laybourn PJ, Kadonaga JT (1991) Role of nucleosomal cores and histone H1 in regulation of transcription by RNA polymerase II. Science 254:238–45

Roberts JM, D'Urso G (1988) An origin unwinding activity regulates initiation of DNA replication during mammalian cell cycle. Science 241:1486–1489

Stillman B (1989) Initiation of eukaryotic DNA replication in vitro. Ann Rev Cell Biol 5: 197–245

Virshup, DM, Kauffman MG, Kelly TJ (1989) Activation of SV40 DNA replication in vitro by cellular protein-phosphatase 2A. Embo J 8:3891–3898

Waga S, Hannon G, Beach D, Stillman S (1994) The p21 inhibitor of Cyclin dependent kinases controls DNA replication by interaction with PCNA. Nature 369:574–578

17 From Sequence to Function – Structural Motifs of Cell Cycle Proteins

JENS W. ECKSTEIN

1 Background

The discovery of cell cycle proteins followed an unusual course in that most of them were tracked down by physiological and genetic observations, rather than by following their enzymological footprints. Initially, for quite a few of them, neither catalytic functions nor substrates were known, and until today a lot of these questions remain elusive. A remarkable feature of cell cycle proteins lies in their functional conservation throughout a wide range of organisms and in their structural conservation of catalytic sites and binding motifs. This conservation is so high, that, in some cases, cell cycle proteins of distantly related organisms can rescue deletion mutants of each other. Since the breakthrough discovery of MPF, more and more genes involved in cell cycle control are emerging quickly (Kirschner 1992).

Currently, a total of roughly 100 cell cycle proteins [cyclins, cyclin-dependent kinases (Cdks), enzymatic (Cdc25, Wee1, Cak), and non enzymatic (p21, p16, p9) Cdk regulators] have been isolated from all kinds of organisms and their sequences have been deposited in databases. Accessing this information, aligning sequences, and recognizing sequence motifs are powerful tools for the discovery and characterization of new cell cycle proteins.

2 Principle and Applications

Using the huge amounts of information in electronic databases and combining them with knowledge from protein structure and enzymology requires a certain command in computer science and INTERNET. This chapter can give only a glimpse into this information jungle, and I refer to the scientific literature for a review (Henikoff 1993).

Although all molecular biologists today know the basics of connecting to sequence databases via INTERNET or of sending out alignment requests to FASTA or BLAST, not everybody has taken time to explore the INTERNET in search of new databases and software applications. However, I highly recommend doing so, since INTERNET offers access to state-of-the-art software and databases. Moreover, the face of INTERNET changes so rapidly, that many instructions given in this chapter will no longer be up to date by the time the reader holds it in his hands. Consequently, this chapter's main purpose is to encourage the reader to browse the INTERNET for news.

Table 1 summarizes some important database servers and FTP sites (File Transfer Protocol). An updated list of servers, databases, and software can be accessed at the FTP site "expasy.hcuge.ch" (129.195.254.61) in the /database/info directory. The public domain software tools like Gopher (University of Minnesota) or Fetch (Dartmouth College) are a valuable help to explore databases and to download files through INTERNET.

In addition to databases, FTP sites offer a variety of biocomputing software tools. Section 3.5 describes a practical, hands-on session with the program MacPattern (Fuchs 1991), which can be used to search the PROSITE database (Bairoch 1991). This program runs on Apple Macintosh computers, but similar programs are available for MS-DOS and Unix machines (see list at expasy). In any case, be sure to read the Read Me or Information for Users files before downloading and running software from FTP sites. These files contain important information about restrictions and rules of public domain software.

Table 1. Major FTP servers

Organization	Databases[a]	Address
National Center for Biotechnology Information (NCBI)	GenBank, SWISS-Prot, PIR, PROSITE, PKCDD	ncbi.nlm.nih.gov 130.14.20.1
European Molecular Biology Laboratory (EMBL)	EMBL, SWISS-Prot, PROSITE, PKCDD	ftp.embl-heidelberg.de 192.54.41.33
Basel Biozentrum Biocomputing server (EMBnet SWISS)	EMBL, SWISS-Prot, GenBank, PIR, PROSITE, PKCDD	nic.switch.ch 130.59.1.40
National Institute of Genetics (NIG, Japan)	DDBJ, EMBL, GenBank, SWISS-Prot, PIR, PROSITE	ftp.nig.ac.jp 133.39.16.66
Human Genome Data Base / Johns Hopkins U.	GDB, OMIM	ftp.gdb.org 192.239.77.3
Protein Data Bank (PDB)	PDB	pdb.pdb.bnl.gov 130.199.144.1

[a] GenBank = GenBank nucleic acid db, SWISS-Prot = Swiss-Prot protein sequence db; PIR = Protein information resource (NBRF protein sequence db); Prosite = Dictionary of protein sites and patterns; PKCDD = Protein kinases catalytic domain db; EMBL = European Molecular Biology Laboratory nucleotide sequence db; DDBJ - DNA data bank of Japan, GDB = Human genome db; OMIM = Online Mendelian inheritance in man; PDB = Protein data bank (3d structures).

Currently, two databases can be used for motif searches in proteins, the PRINTS database (Attwood and Beck 1994) and the PROSITE database (Bairoch 1991). The PRINTS database is derived from the OWL composite sequence database. OWL is a non-redundant protein sequence database formed from SWISS-Prot, PIR, GenBank translations, and the Brookhaven databank. For more information, access the directory /pub/database/owl at the FTP site "s-ind2.dl.ac.uk". At the time this chapter was written, the latest PRINTS release was from November 1st, 1993.

PROSITE is derived from the SWISS-Prot database. The current release is version 12.1 from October 1994, the 13.0 release is planned for March 1995. In the following I will concentrate on the PROSITE database, as a representative example. PROSITE consists of two files, PROSITE.DAT and PROSITE.DOC. Both files are ASCII files and can be opened by text editors like MS Word. PROSITE.DAT contains the essential information of a motif and can be read by computer programs like MacPattern (Fuchs 1991). PROSITE.DOC contains textual information that describes the motif and its derivation in greater detail. The version of MacPattern is described in Section 3.2. MacPattern can read a variety of sequence file formats as long as they are text files, and I refer to the manual that can be loaded through FTP at the EMBL server.

3 Motifs and Functions of Cell Cycle Proteins

3.1 Cyclins

To date, the cyclins constitute the group of cell cycle proteins with the most entries. On the order of 50 protein sequences of A, B, C, D, and E type cyclins from different organisms are deposited in SWISS-Prot. Four different motifs can be described for the cyclin family[1]:

1. The "cyclin box" is a 150-amino acid sequence in the middle of the cyclins that contains three distinct patterns[2]. These are: a 32-amino acid pattern,
[MN]-R-x(2)-[LIVM]-x(2)-[FYW]-[LIVM]-x(8)-[LIVMC]-x(4)-[LIVMFY]-x(2)-[STAGC]-LIVMFYQ]-x-[LIVMFY](2)-D-[RK]-[LIVMFYW],

---

[1] C-type cyclins deviate significantly from other cyclins and some of the motifs are not conserved in this member of the cyclin family.

[2] Protein motifs are in the Prosite format. The one letter amino acid code is used, x represents any amino acid. When more than one amino acid is possible, the one letter codes are in brackets, e.g., [AV] stands for alanine or valine. Repeats are indicated by a number in (), i.e., x(2) stands for x-x and [AV](2) for A-V-A-V.

a 21 amino acid pattern approximately 10 amino acids downstream,
L-Q-L-[VLI]-G-x(8)-K-[FMLY]-[EK]-E-x(3)-P,
and a 4 amino acid pattern 21 residues further downstream,
M-E-x(6)-L-x(3)-L.

2. The so-called destruction box, a 10-amino acid pattern at the N-terminus, is important for the proteolytic breakdown of A and B-cyclins:
R-x-[ITVA]-L-[GDEA]-[DENV]-[ILV]-[GQST]-[SN].

Cyclins are regulatory subunits of the cyclin-dependent kinases and have no catalytic site. The function of conserved patterns involves most probably protein-protein interaction and recognition. Cyclins bind to their partner cyclin-dependent kinases and also offer docking sites to proteins that act as inhibitors for several of the cyclin-dependent kinases/cyclin complexes (i.e., p21, p16). In the case of Cyclin A, a minimal region for the binding to Cdc2 and for MPF activity has been mapped (Lees and Harlow 1993). A mutant cyclin A from *X. laevis* with a 173-amino acid N-terminal deletion, a 112-amino acid C-terminal deletion, and a deletion between residue numbers 241 and 275 was still able to trigger maturation of *Xenopus* oocytes. The remaining residues 189 through 241 and 275 through 320 are situated in the cyclin box.

The first motif in the cyclin box is reminiscent of a motif found in protein tyrosine phosphatases about 45 amino acids downstream of their active site. Galaktinov and Beach (1991) have speculated that cyclins could possibly "donate" this domain to the Cdc25 phosphatase, thus becoming a functional part of Cdc25.

A and B-type Cyclins are degraded via the ubiquitin-dependent pathway. Ubiquitin molecules are covalently bound to lysine residues of the target protein by specific conjugating enzymes (Glotzer et al. 1991). These ubiquitin flags are recognized by huge multiprotein complexes called proteasomes and lead to rapid degradation. About 10 amino acids downstream from the destruction box, several lysine residues are present, which are probably ubiquitination sites.

3.2 Cyclin-Dependent Kinases

Cyclin-dependent kinases or Cdks are members of the protein kinase family and share two highly conserved patterns with them – the so-called phosphate-anchor (or G-x-G-x-x-G motif) and a sequence motif surrounding a catalytically important aspartate.

1. The glycine-rich ATP binding motif is in the extreme N-terminal of the catalytic domain in the vicinity of a conserved lysine residue, which is known to hydrogen bond to one of the ATP phosphates:
   [LIV]-***G***-x-***G***-x-[FYM]-[S***G***]-x-V.
2. A 13 residue pattern can be found around a conserved aspartate:
   [LIVMFYC]-x-[HY]-x-***D***-[LIVMFY]-K-x(2)-N-[LIVMFYC](3).
   Additional patterns are typical for the cyclin-dependent kinase family.
3. A conserved lysine residue and a second conserved aspartate, followed by a highly conserved F-G-L motif, are present just downstream of pattern b:
   [LIV]-***K***-[LV]-[AG]-[***D***E]-F-G-L-[AG]-R.
4. Twelve amino acids downstream from motif (3.) is the following pattern surrounding a conserved tryptophane:
   [TS]-x-[EV]-[VI]-[VL]-T-L-***W***-Y-R-[SAP]-P.
5. In the N-terminus of the Cdc2/Cdk2 family is the "PSTAIRE" motif; occasionally those kinases are referred to as PSTAIRE kinases. This region has been used very successfully for raising antibodies against Cdc2/Cdk2 kinases. Some changes occur in the other Cdk kinases. The following motif was derived from human Cdks:
   P-[SIL]-[***T***S]-[***A***T]-[***I***VL]-***R***-***E***.

Cyclin-dependent kinases possess all the typical motifs of ATP-dependent protein kinases. The comparison of the crystal structures of cAMP-dependent protein kinase (Knighton et al. 1991) and Cdk2 (DeBondt et al. 1993) show very similar folding of the proteins and remarkable structural homology. In fact, some key residues involved in binding and catalysis are within the motifs described above and probably play similar roles in all members of the family. I will describe the function of four key residues in the context of Cdc2/Cdk2. Lysine 33[3] forms hydrogen bonds to the α- and γ-phosphates of ATP. Glutamate 51 forms a hydrogen bond to lysine 33, and glutamate 145 forms hydrogen bonds to Mg^{2+} and lysine 33. Threonine 160 is a site which has to be phosphorylated by Cdk-activating kinase (Cak) (Fesquet et al. 1993; Solomon et al. 1993). It is situated in a loop, also called T-loop, which covers the ATP-binding cleft when unphosphorylated. Phosphorylation leads to a substantial conformational change; the T-loop swings out of the ATP-binding cleft and thus leads to activation of the kinase. Two additional regulatory phosphorylation sites are situated in the glycine-rich phosphate binding motif. In Cdc2 and Cdk2, phosphothreonine 14 and phosphotyrosine 15 are dephosphorylated by the Cdc25 protein phosphatase, which leads to the activation of these kinases. In the case of Cdc2/ClnB, these two dephosphorylations are necessary to trigger G2/M progression into mitosis.

---

[3] Numbering refers to Cdk2.

Using charged-to-alanine scanning mutagenesis, Ducommun and colleagues (Ducommun et al. 1991a, 1991b) have mapped single residues and residue clusters in Cdc2 that are important for the interaction with Cyclin A and with the small Cks protein, also known as $p13^{suc1}$ in *S. pombe* (Draetta et al. 1987). Three clusters have been identified for cyclin A binding: K33A-K34A-R36A, E38A-E40A-E41A-E42A and R50A-E51A. Three clusters are very sensitive to $p13^{suc1}$ binding: K20A-R22A-H23A-K24A, K56A-E57A-R59A-H60A and R215A-R218A. All alanine cluster mutants result in more than 95% loss of Cdk2 kinase activity in histone H1 kinase assays, except for the last cluster mutant R215A-R218A, which leads only to a 45% decrease in kinase activity.

Figure 1 shows the approximate positions of some of those clusters. The structural model of Cdc2 was obtained by homology modeling analogous to Marcote et al. (1993). Additional structural information was derived from the crystal structure of Cdk2 (DeBondt et al. 1993)[4] and integrated into the model (J. Eckstein, unpublished).

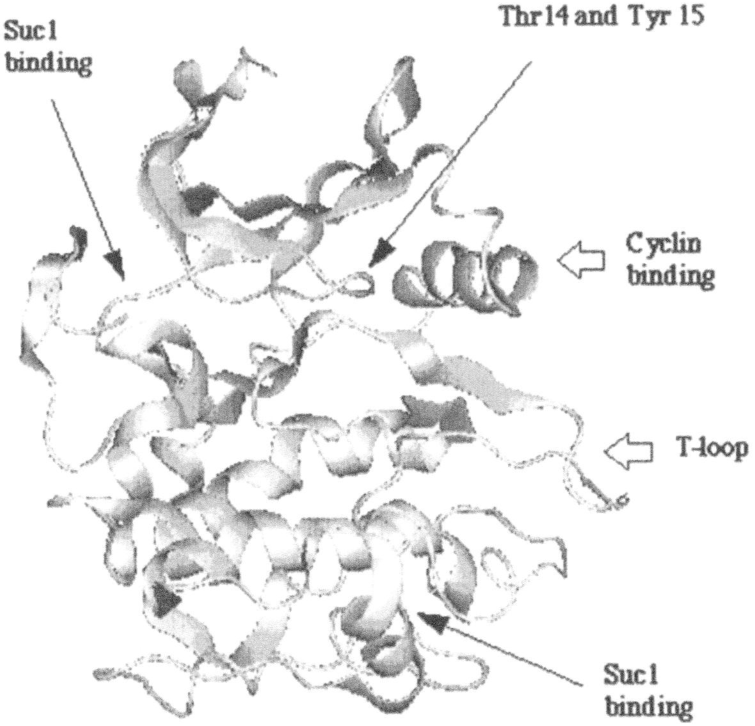

Fig. 1. Ribbon cartoon of the three-dimensional model of Cdc2 as derived from homology modeling

[4] The coordinates of the Cdk2 structure were not available in the Protein Data Bank when this chapter was written

3.3 Cdc25 Phosphatases

Members of the protein phosphatase family, that are homologous to the product of the fission yeast cdc25 gene, dephosphorylate P-Thr 14 and P-Tyr 15 in the Cdc2/cyclin B complex (Millar et al. 1986; Dunphy and Kumagai 1991; Kumagai and Dunphy 1991). This event leads to activation of the Cdc2/cyclin B kinase and plays an important role as a checkpoint at the G2/M transition. Cdc25 homologues have been isolated from more than ten different species. The function of Cdc25 is conserved throughout most eukaryotes despite a relatively low homology in the amino acid sequences.

In some organisms several Cdc25 isoenzymes have been isolated. Currently, three human homologues are known, Cdc25 A, B and C. It has been shown that human Cdc25 C is the active phosphatase at G2/M and activates Cdc2/cyclin B in vivo. Human Cdc25 A is believed to be active at an earlier stage in the cell cycle, and its substrates might be Cdk2/cyclin A and/or Cdk2/cyclin E kinase. The role of Cdc25 B is still unclear.

Three conserved motifs are located in the C-terminus of Cdc25:

1. The active site motif of Cdc25 contains the typical 8 amino acid motif surrounding the catalytic nucleophile of protein phosphatases (H-C-x-x-x-x-x-R).
   [FIS]-[***H***L]-**C**-E-[FYH]-S-x-x-***R***-x-P-x-[ML]-x-[RLK]-x-x-R-x-x-D-R.
2. The second motif is about 40 residues upstream of the active site:
   D-C-R-[FY]-[PE]-Y-E-[FY]-[EDNLT]-G-G-H-I-x-x-[AS]-x-N-[IL].
3. The third motif, situated just 7 amino acids downstream of motif a. is a 33 residue long pattern rich in aromatic amino acids:
   Y-P-[ED]-L-[HYFTDS]-Y-P-[DE]-x-[YAL]-[LVI]-L-x(2)-[NG]-G-Y-[KR]-x-F-[FY]-x(6)-C-[EDYQ]-P-2x-Y.

As mentioned above, Cdc25 contains the active site motif of the protein tyrosine phosphatase family. Mutagenesis experiments (Gautier et al. 1991) and, more recently, a report about the 3-dimensional crystal structure of PTP1B (Barford at al. 1994), have shown that the cysteine and the arginine are essential for substrate binding and catalysis. The two residues are in hydrogen-bonding distance in PTP1B and possibly form an ion pair. This would activate the cysteine and make it a better nucleophile. Work from Jack Dixon's laboratory has shown that the cleavage and hydrolysis of the phosphate ester bond proceeds via the formation of a covalent cysteine phosphate intermediate (Guan and Dixon 1991). The replacement of cysteine by serine leads to a catalytically incompetent protein. Interestingly, as observed by mutagenesis of Cdc25, the neighboring histidine does not play an essential role in catalysis or binding, and mutant proteins were found to be competent as phosphatases. In the crystal structure of PTP1B, histidine is part of a hydrogen bonding network and thus might play a structural role in positioning the active site cysteine. The arginine most likely has two functions, activation of cysteine and phosphate binding. Cdc25 mutants, in which arginine was replaced by alanine, had practically no activity in a p-NPP assay and lost the ability to bind to Cdc2/cyclin B or Cdk2/cyclin A complex (X. Xu and J. Eckstein, unpubl. results).

The residues in the loop between cysteine and arginine show relatively high conservation throughout the Cdc25 family. This loop is probably critical for the positioning of cysteine and arginine as well as for substrate recognition. Replacement of the *Drosophila* Cdc25 E-F-S-S-E motif by the loop sequence S-A-G-V-G from *Drosophila* PTP, however, did not change substrate specificity effectively (Gautier et al. 1991). Just downstream of the active site motif follows a stretch containing two arginine patches. It can be speculated that these positively charged patches are important for the interaction of Cdc25 with Cdks and/or cyclins. At least for some members of the Cdc25 family, it has been shown that phosphorylation is required to gain full phosphatase activity. The mechanism of this phosphorylation is still unclear. In the case of human Cdc25 C, phosphorylation might be accomplished by its substrate Cdc2/cyclin B kinase in a autocatalytic feedback loop (Hoffmann et al. 1993).

3.4 Other Cell Cycle Proteins

To date, there are many more proteins known to be involved in the cell cycle. Phosphorylation and dephosphorylation play an important role as a regulatory mechanism in the cell cycle, as we have seen in the case of the Cdks and Cdc25. At least two more protein kinases should be mentioned in this context, Weel (Igarashi et al. 1991; McGowan and Russell 1993) and Cak (Fesquet et al. 1993; Solomon et al. 1993). Weel counteracts Cdc25 and phosphorylates Tyr 15 on Cdc2. Cak phosphorylates threonine 161 in Cdc2 (160 in Cdk2), as discussed in Section 3.2. Both proteins exhibit the typical protein kinase motif G-x-G-x-x-G.

A protein phosphatase named Cdil has been described recently (Gyuris et al. 1993). Although its substrate is not yet identified, it seems to play a distinctly different role in the cell cycle than does Cdc25. It shares the protein phosphatase active site motif (H-C-Y-G-G-L-G-R), but differs considerable from Cdc25.

Very recently small inhibitory proteins, named p21 and p16, associated with cyclin-dependent kinases have attracted much attention (Harper et al. 1993; Serrano et al. 1993; Xiong et al. 1993). A third protein called p13^{suc1} was first described in yeast (Brizuela et al. 1987), but members of the same family were later found in other organisms as well. The crystal structure of CksHs2 has been solved (Parge et al. 1993). Currently, there are no sequence motifs in the PROSITE database that match these three cell cycle proteins. The following motif, situated close to the C-terminus of the p13 family, is noteworthy: P-E-P-H-I-L-L-F. p21 and p16 are composed of ankyrin-like repeats – a review article about ankyrins was publishey by P. Bork (Bork 1993).

4 An Example for a MacPattern/PROSITE Session

Suppose you have found the following amino acid sequence and want to know if it contains a functional or structural motif found in other known proteins. Your sequence is:

SMRKIVATWMLEVCEEQKCEEEVFPLAMNYLDRFLSLEPVKKSRLQ
LLGATCMFVASK

You need: an Apple Macintosh computer, a connection to INTERNET, and a program for downloading files through INTERNET (Gopher, Fetch, or any program with FTP).

Search

1. Connect to "ftp.embl-heidelberg.de" (Address: 192.54.41.33) and login as anonymous. You will not need a password. Go into the "/pub/software/mac" directory and download "macpattern.hqx". (A list of programs that make use of PROSITE is in the "prosite.prg" file.)
2. Next, go into the "/pub/databases/prosite" directory and download prosite.dat, prosite.doc, prosite.lis, prosite.txt, experts.txt., and prosite.prg. This will take a little while.
3. On your Macintosh you will now find a file named "MacPattern.sea". The extension stands for self extracting archive. Clicking on the MacPattern.sea icon will activate the program and help you to create a folder for the MacPattern program.
4. Open the MacPattern program. This program provides a comprehensive "Help" menu, which is under the desk accessory menu. It is a good idea to take some time and browse through it.
5. Under the "File" menu choose "New Index" and then PROSITE. You will be asked to choose the ".dat file" – then click on "prosite.dat". You will see a message "Parsing PROSITE data file ..." and sometimes error messages. These messages are usually explained in the release information of the PROSITE database. Find out what they mean and act accordingly. Finally you will be asked if you want to sort the database alphabetically, which is a good idea, because it facilitates searching for motifs. Now you have an indexed PROSITE database on your computer called "prosite.inx".

6. Choose "Open Database File..." under the File menu and click on prosite.inx. A window with the database entries will appear on the screen.

7. Next, load your sequence with "Load Sequence File..." in the File menu. MacPattern can read a whole variety of different formats – consult the Help menu. A window will be displayed with the name of your sequence file and information on the format of your sequence file.

8. Click on the PROSITE database window and choose the motifs you want to search. Usually you will choose all of them in an initial run. Chosen motifs should be highlighted. Later on you might want to select subsets to speed up searches or to eliminate hits you are not interested in.

9. Under the PROSITE menu choose "Find". After the release of the mouse button the database will be screened for matches with your amino acid sequence. This is indicated by a growing bar in a communication window.

10. As soon as the search is complete, a results window will appear with the following results:

```
Sequence file: Your sequence
Pattern set:  Main Set

--------------------------------------

Sequence 1 (58 residues):

Matching pattern PS00292 CYCLINS:
   202: RKIVATWMLEVCEEQKCEEEVFPLAMNYLDRF
Total matches: 1

Total no of hits in this sequence: 1

========================================

1011 pattern(s) searched in 1 sequence(s), 58 residues.
Total no of hits in all sequences: 1.
Search time: 0:02 min.
```

Your sequence contains the cyclin motif. If you want to learn more about this motif, double click on it in the PROSITE database window. A new window will open with information about the motif. A list is given with all the sequences that were used to extract this motif followed by helpful remarks and literature references. The following paragraph shows an example:

ID CYCLINS; PATTERN.
AC PS00292;
DT APR-1990 (CREATED); JUN-1992 (DATA UPDATE); JUN-1994 (INFO UPDATE).
DE Cyclins signature.
PA R-x(2)-[LIVM]-x(2)-[FYW]-[LIVM]-x(8)-[LIVMC]-x(4)-[LIVMFY]-x(2)-[STAGC]-
PA [LIVMFYQ]-x-[LIVMFY](2)-D-[RK]-[LIVMFYW].
NR /RELEASE=26,38303;
NR /TOTAL=46(46); /POSITIVE=49(49); /UNKNOWN=0(0); /FALSE_POS=0(0);
NR /FALSE_NEG=10(10);
CC /TAXO-RANGE=??E?V; /MAX-REPEAT=1;
DR P30274, CG2A_BOVIN, T; P14785, CG2A_DROME, T; P20248, CG2A_HUMAN, T;
DR P24861, CG2A_PATVU, T; P04962, CG2A_SPISO, T; P18606, CG2A_XENLA, T;
DR P07818, CG2B_ARBPU, T; P18063, CG2B_ASTPE, T; P29332, CGB2_CHICK, T;
DR P20439, CG2B_DROME, T; P14635, CGB1_HUMAN, T; P15206, CG2B_MARGL, T;
DR P24860, CGB1_MOUSE, T; P30276, CGB2_MOUSE, T; P30277, CGB1_RAT , T;
DR P24862, CG2B_PATVU, T; P13952, CG2B_SPISO, T; P13350, CGB1_XENLA, T;
DR P13351, CGB2_XENLA, T; P25010, CG2A_DAUCA, T; P334800, CG21_ANTMA, T;
DR P34801, CG22_ANTMA, T; CG2B_ARATH,T;P30278, CG2B_MEDSA,T
DR P25011, CG21_SOYBN, T; P25012, CG22_SOYBN, T; P24865, CG21_SCHPO,T
DR P36630, CG22_SCHPO, T; P10815, CG23_SCHPO, T; P24868, CG21_YEAST, T;
DR P24869, CG22_YEAST, T; P24870, CG23_YEAST, T; P24871, CG24_YEAST, T;
DR P30283, CGS5_YEAST, T; P32943, CGS6_YEAST, T; P30284, CG21_EMENI, T;
DR P20437, CG11_YEAST, T; P20438, CG12_YEAST, T; P13365, CG13_YEAST, T;
DR P24866, CG11_CANAL, T; P24385, CGD1_HUMAN, T; P25322, CGD1_MOUSE, T;
DR P30279, CGD2_HUMAN, T; P30280, CGD2_MOUSE, T; Q04827, CGD2_RAT, T;
DR P30281, CGD3_HUMAN, T; P30282, CGD3_MOUSE, T; P24864, CG1E_HUMAN, T;
DR Q01043, CGH2_HSVSA, T;
DR P34638, CG2A_CAEEL, N; P25009, CG1P_SCHPO, N; P25008, CG1C_DROME, N;

DR P24863, CG1C_HUMAN, N; P24867, CG16_YEAST, N; P25693, CG17_YEAST, N;
DR P35190, CLG1_YEAST, N; P36613, CGM2_SCHPO, N;
DR P30286, CG1_MEDSA, P;
DO PDOC00264;
//
{PDOC00264}
{PS00292; CYCLINS}
{BEGIN}
****
* Cyclins signature *
****

Cyclins [1,2,3] are eukaryotic proteins which play an active role in controlling nuclear cell division cycles. Cyclins, together with the p34 (cdc2) or cdk2 kinases, form the Maturation Promoting Factor (MPF). There are two main groups of cyclins:

- G2/M cyclins, essential for the control of the cell cycle at the G2/M (mitosis) transition. G2/M cyclins accumulate steadily during G2 and are abruptly destroyed as cells exit from mitosis (at the end of the M-phase).
- G1/S cyclins, essential for the control of the cell cycle at the G1/S (start) transition.

In most species, there are multiple forms of G1 and G2 cyclins. For example, in vertebrates, there are two G2 cyclins, A and B, and at least three G1 cyclins, C, D, and E.

A cyclin homolog has also been found in herpesvirus saimiri [4].

The best conserved region is in the central part of the cyclins' sequences, known as the "cyclin-box", from which we have derived a 32 residue pattern.

- Consensus pattern: R-x(2)-[LIVM]-x(2)-[FYW]-[LIVM]-x(8)-[LIVMC]-x(4)-[LIVMFY]-x(2)-[STAGC]-[LIVMFYQ]-x-[LIVMFY](2)-D-[RK]-[LIVMFYW]
- Sequences known to belong to this class detected by the pattern: ALL, except for G1/S cyclins C from Human and Drosophila, puc1 from fission east and HCS26 from yeast.
- Other sequence(s) detected in SWISS-PROT: NONE.
- Last update: June 1994 / Text revised.

[1] Nurse P.
Nature 344:503–508(1990).
[2] Norbury C., Nurse P.
Curr. Biol. 1:23–24(1991).
[3] Lew D.J., Reed S.I.
Trends Cell Biol. 2:77–81(1992).
[4] Nicholas J., Cameron K.R., Honess R.W.
Nature 355:362–365(1992).

{END}

The nomenclature and symbols used are described in the "prosite.txt" file, and it is advisable to read through this information.

After downloading and indexing your PROSITE database, motif searches like the example above take only a few minutes. The information that comes with each entry is extremely helpful and precise. The references and the name of an "expert" ("experts.txt" file) can provide all the information you need to learn more about the function of your new protein.

The PROSITE database uses a format that can be used to build up a customized satellite database. A PROSITE-like database for cell cycle proteins is currently being created by the author. Readers interested in this database are welcome to send a note via INTERNET to "eckstein@mitotix.com". Please do not edit or change the original PROSITE database if you are planning to create your own database, as requested in the release information!

References

Attwood TK, Beck ME (1994) PRINTS – a protein motif fingerprint database. Protein Engineering 7(7):841–848

Bairoch A (1991) PROSITE: a dictionary of sites and patterns in proteins. Nucleic Acids Res 19:2241–2245

Barford D, Flint AJ, Tonks NK (1994) Crystal structure of human protein tyrosine phosphatase 1B. Science 263:1397–1404

Bork P (1993) Hundreds of ankryin-like repeats in functionally diverse proteins: Mobile modules that cross phyla horizontally. Proteins: Structure, Function and Genetics 17: 363–374

Brizuela L, Draetta G, Beach D (1987) p13suc1 acts in the fission yeast cell division cycle as a component of p34cdc2 protein kinase. EMBO Journal, 6:3507–3514

DeBondt HL, Rosenblatt J, Jancarik J, Jones HD, Morgan DO, Kim S-H (1993) Crystal structure of cyclin-dependent kinase 2. Nature 363:595–602

Draetta G, Brizuela L, Potashkin J, Beach D (1987) Identification of p34 and p13, human homologs of the cell cycle regulators of fission yeast encoded by cdc+ and suc1+. Cell 50:319–325

Ducommun B, Brambilla P, Draetta G (1991a) Mutations at sites involved in Suc1 binding inactivate Cdc2. Mol Cell Biol 11(12):6177–6184

Ducommun B, Brambilla P, Felix M-A, Franza BR, Karsenti E, Draetta G (1991b) Cdc2 phosphorylation is required for its interaction with cyclin. EMBO J 10(11):3311–3319

Dunphy WG, Kumagai A (1991) The cdc25 protein contains an intrinsic phosphatase activity. Cell 67(1):189–96

Fesquet D, Labbe JC, Derancourt J, Capony J-P, Galas S, Girard F, Lorca T, Shuttleworth J, Doree M, Cavadore J-C (1993) The MO15 gene encodes the catalytic subunit of a protein kinase that activates cdc2 and other cyclin-dependent kinases (Cdks) through phosphorylation of Thr161 and its homologues. EMBO J 12(8):3111–3121

Fuchs R (1991) MacPattern: protein pattern searching on the Apple Macintosh. Comput Applic Biosci, 7:105–106

Galaktinov K, Beach D (1991) Specific activation of cdc25 tyrosine phosphatases by B-type cyclins: evidence for multiple roles of mitotic cyclins. Cell 67(6):1181–1194

Gautier J, Solomon MJ, Booher RN, Bazan JF, Kirschner MW (1991) cdc25 is a specific tyrosine phosphatase that directly activates p34cdc2. Cell 67(1):197–211

Glotzer M, Murray AW, Kirschner MW (1991) Cyclin is degraded by the ubiquitin pathway. Nature 349:132–138

Guan KL, Dixon JE (1991) Evidence for protein-tyrosine-phosphatase catalysis proceeding via a cysteine-phosphate intermediate. J Biol Chem 266(26):17026–17030

Gyuris J, Golemis E, Chertkov H, Brent R (1993) Cdi1, a human G1 and S phase protein phosphatase that associates with Cdk2. Cell 75:791–803.

Harper JW, Adami GR, Wei N, Keyomarsi K, Elledge SJ (1993) The p21 Cdk-interacting protein Cip1 is a potent inhibitor of G1 cyclin-dependent kinases. Cell 75:805–816

Henikoff S (1993) Sequence analysis by electronic mail servers. Trends Biochem Sci 18: 267–268

Hoffmann I, Clarke PR, Marcote MJ, Karsenti E, Draetta G (1993) Phosphorylation and activation of human cdc25-C by cdc2-cyclin B and its involvement in the self-amplification of MPF at mitosis. EMBO J 12(1):53–63

Igarashi M, Nagata A, Jinno S, Suto K, Okayama H (1991) Wee1+-like gene in human cells. Science 353:80–83

Kirschner M (1992) The cell cycle then and now. TIBS 17(8):281–285

Knighton DR, Zheng J, Ten Eyck LF, Ashford VA, Xuoang N, Taylor SS, Sowadski JM (1991) Crystal structure of the catalytic subunit of cAMP-dependent protein kinase. Science 253:407–414

Kumagai A, Dunphy WG (1991) The cdc25 protein controls tyrosine dephosphorylation of the cdc2 protein in a cell-free system. Cell 64(5):903–914.

Lees EM, Harlow E (1993). Sequences within the conserved cyclin box of human cyclin A are sufficient for binding to and activation of Cdc2 kinase. Mol Cell Biol 13(2): 1194–1201

Marcote MJ, Knighton DR, Basi G, Sowadski JM, Brambilla P, Draetta G, Taylor SS (1993) A three-dimensional model of the Cdc2 protein kinase: localization of cyclin- and Suc1 binding regions and phosphorylation sites. Mol Cell Biol 13:5122–5131

McGowan CH, Russell P (1993) Human Wee1 kinase inhibits cell division by phosphorylating p34cdc2 exclusively on Tyr15. EMBO J 12(1):75–85

Millar JB, McGowan CH, Lenaers G, Jones R, Russell P (1991) p80cdc25 mitotic inducer is the tyrosine phosphatase that activates p34cdc2 kinase in fission yeast. EMBO J 10(13):4301–4309

Parge HE, Arvai AS, Murtari DJ, Reed SI, Tainer JA (1993) Human CksHs2 atomic structure: a role for its hexameric assembly in cell cycle control. Science 262:387–395

Russell P, Nurse P (1986) cdc25+ functions as an inducer in the mitotic control of fission yeast. Cell 45(1):145–153

Serrano M, Hannon GJ, Beach D (1993) A new regulatory motif in cell-cycle control causing specific inhibition of Cyclin D/CDK4. Nature 366:704–707

Solomon MJ, Harper JW, Shuttleworth J (1993) CAK, the p34cdc2 activating kinase, contains a protein identical or closely related to p40 MO15. EMBO J 12(8):3133–3142

Xiong Y, Hannon GJ, Zhang H, Casso D, Kobayashi R, Beach D (1993). p21 is a universal inhibitor of cyclin kinases. Nature 366:701–704

18 Purification of Cell Cycle Kinases

DIDIER FESQUET, JEAN-CLAUDE LABBÉ AND JEAN-CLAUDE CAVADORE

1 Background

The progression of eukaryotic cells through the cell cycle is under the control of sequential activations and subsequent inactivations of various Cyclin-dependent kinases (Cdks) (Sherr 1993). Among these complexes, two have been extensively described. The first one was the Cdc2-Cyclin B complex, also known as MPF (M phase promoting factor) which controls entry to and exit from M phase (Dorée 1990; Nurse 1990). The second one is the Cdk2-Cyclin A complex necessary for DNA replication which acts as SPF (S phase promoting factor) (Reed et al. 1991). The Cdc2 or Cdk2 kinase activity depends both on their Cyclin association and post-translational phosphorylation/dephosphorylation reactions (Pines 1993).

Two types of phosphorylation modulate Cdc2 or Cdk2 kinase activity. Phosphorylations on threonine 14 and tyrosine 15 are inhibitory, in contrast to phosphorylation of threonine 161, which appears to be absolutely required for Cdc2 kinase activity (Clarke and Karsenti 1991). Threonine 161 residue or its equivalent (Thr 167 in fission yeast) is conserved in all described Cdks (Meyerson et al. 1992). This threonine is phosphorylated by a Cdc2-kinase homologue, the Cdk-activating kinase (Cak) (Solomon et al. 1992). This kinase has been purified in Starfish (Fesquet et al. 1993), *Xenopus* (Poon et al. 1993; Solomon et al. 1993) and identified as the product of MO15 gene (Shuttleworth et al. 1990).

Both Cdc2-Cyclin B kinase and Cak can be easily prepared in a state near to homogeneity starting from Starfish or *Xenopus* oocyte homogenates. This chapter describes the preparation of these two kinases but it should serve as guide for the purification of other cell cycle kinases.

2 Methods

2.1 Preparation of the Starfish Oocyte Homogenates

Starfish *Marthasterias glacialis* females captured during the breeding season contain about 10^6 transparent oocytes arrested at the first meiotic prophase. These oocytes can be driven synchronously into M-phase by adding

1-methyladenine which is the hormone responsible for meiotic maturation in starfish.

Stock solutions

- Artificial-calcium free seawater

Washing solution I	mM	g/l
NaCl	500	29.2
$MgCl_2$, 6 H_2O	30	6.09
$MgSO_4$, 7 H_2O	17	4.19
MES	2	0.39

This solution is adjusted to pH 5.5 with HCl.

Washing solution II	mM	g/l
NaCl	500	29.2
$MgCl_2$, 6 H_2O	30	6.09
$MgSO_4$, 7 H_2O	17	4.19
Tris	2	0.24

This solution is adjusted to pH 8.2 with HCl.

- Homogenization buffer A

	mM	g/l
β glycerophosphate	144	44
EGTA	36	13.7
DTT	1.8	0.278
$MgCl_2$, 6 H_2O	27	5.49
Sucrose	200	68.46
KCl	250	18.63

Preparation

1. Cut the five arms of the starfish with large scissors and open it. Each arm contains two ovaries which are placed on a beaker filled with 50 ml of ice-cold buffer washing solution I. The absence of Ca^{2+} ions in this washing solution releases the oocytes from follicle cells and its acidic pH solubilizes the jelly coat surrounding the oocytes. Macerate the ovaries with forceps and filter through fine cheese cloth in order to remove fragments of ovaries.
2. Centrifuge 5 min at low speed (less than 100 *g* with a swinging bucket rotor) at 4 °C and discard the supernatant. Resuspend the oocyte pellet in 10 vol of ice-cold washing solution I and centrifuge at low speed (repeat twice). Then resuspended the pellet in 10 vol of washing solution II.
3. Add 1-methyladenine (Sigma; M 4876) to the final concentration of 1 μM and incubate the oocytes at 20 °C. Germinal vesicle breakdown (GVBD) generally occurs within 20–25 min after hormone addition. Since starfish

oocytes are transparent, GVBD is easily observed with a binocular microscope.

4. Forty minutes after hormone addition (the oocytes are at first meiotic metaphase), pellet the oocytes by low speed centrifugation (5 min at 4 °C), discard the solution II containing 1-methyladenine, add 2 vol of ice-cooled homogenization buffer A and crush the oocytes in a Potter homogenizer equipped with a Teflon pestle. After three strokes, centrifuge the homogenate at 12 000 *g* for 15 min at 4 °C (Sorvall SS-34 rotor).
5. The supernatant is collected and centrifuged at 100 000 *g* for 1 h in a swinging bucket rotor (TST 28.38 Kontron or equivalent). The supernatant is collected and frozen in liquid nitrogen. These extracts can be kept at –80 °C for several years without appreciable loss of Cdc2 kinase activity.

2.2 Purification of Cdc2-Cyclin B and Cak from Starfish

All fractions from each purification step are assayed for Cdc2 or Cak kinase activity as follows:

Cdc2 kinase assay

Histone H1 is a very good substrate for mitotic Cdc2 kinase, and Cdc2-Cyclin B kinase contributes to the major H1 kinase in eukaryotic cells at M-phase. Thus, the activity of Cdc2-Cyclin B is measured by H1 histone phosphorylation.

Samples of 2 μl from each chromatographic steps are diluted to 30 μl with 20 mM Hepes pH 7.4 and then mixed with 10 μl of a reaction mixture [4 mg/ml histone H_1, 80 mM Hepes pH 7.4, 40 mM $MgCl_2$, and 0.8 mM [$\gamma^{32}P$] ATP (100 cpm/picomol)]. After 10 min reaction at 25 °C, the resulting mixture is spotted on P81 phosphocellulose square paper (Whatman), extensively washed in water, wash once in ethanol, dried, and counted in liquid scintillation solution (see also chapters by P.R. Clarke, Part Two, and by Connell-Crowley and Harper, Part Four).

Cdk activating kinase assay

This assay is based on the ability of Cak to activate in vitro a Cdc2-Cyclin A or B complex. This two step reaction needs Cdc2 and Cyclin in soluble forms. For technical reasons, we use *Xenopus* Cdc2 and human Cyclin A, since available in our laboratory, but any soluble Cdc2 and mitotic Cyclins can be used (baculovirus or *E. coli* expressed protein) (Desai et al. 1992; Poon et al. 1993).

In the first step, in vitro translated *Xenopus* Cdc2 mixed with bacterially produced human Cyclin A is added to fractions to be assayed for Cak activity in the presence of $MgCl_2$ and ATP. When Cak activity is present in the tested fractions, Cdc2-Cyclin A complex becomes active due to threonine

161 phosphorylation. In the second step, the resulting activation is measured by histone H_1 phosphorylation.

Samples from each chromatographic steps are diluted with 20 mM Hepes pH 7.4 (from 1/40 up to 1/200). Dilution avoids interference with endogenous Cdc2 kinase activity and makes it possible to determine the peak of Cak activity in the different steps of purification. Diluted samples (2 µl) are added to 1 µl of reticulocyte lysate containing in vitro translated *Xenopus* Cdc2 (1 ng/µl), 1 µl of a mixture 1 mM ATP, 50 mM $MgCl_2$ and 1 µl (0.1 µg) of bacterially produced human Cyclin A. After 15 min reaction at 25 °C, the samples are diluted to 30 µl with 20 mM Hepes pH 7.4 and assayed for H_1 histone kinase activity as reported above.

Purification chart

1. Centrifugation of the homogenate
2. Dialysis of the supernatant
3. Ion exchange chromatography (SP-Sepharose)

⇙ ⇘

Positive fractions for Cak activity (0.1 to 0.25 M NaCl)	Positive fractions for Cdc2 activity (0.3 to 0.5 M NaCl)
⇩	⇩
Purification of Cak	Purification of Cdc2-Cyclin B
1. Heparin-Sepharose chromatography	1. $p13^{suc1}$ affinity chromatography
2. Gel filtration chromatography	2. Ion exchange chromatography (Mono S)
3. Ion exchange chromatography (Mono Q)	
4. Ion exchange chromatography (Mono S)	

Purification protocol

Note. All the following steps are performed at 4 °C.

1. The starfish homogenate (300 ml) supplemented with protease inhibitors (leupeptin, chymostatin, aprotinin, pepstatin, 1 mg/l each, and 1 mM PMSF) is dialyzed against 6 l of buffer A (50 mM Tris/HCl pH 7.3, 1 mM EGTA, 1 mM DTT, 1 mM NaF, 1 mM orthovanadate).

2. Dialyzed extract is centrifuged at 100 000 *g* for 45 min (TFT 50-38 rotor Kontron or equivalent), the supernatant filtered through 0.45-µm filter (Millipore) and loaded onto a 100 ml SP-Sepharose Fast flow column (Pharmacia) equilibrated with buffer A.

3. Bound proteins are eluted at a flow rate of 100 ml/h by a 0–1 M linear gradient (total volume 500 ml) of NaCl in buffer A Fractions of 10 ml are collected and assayed for both Cak and Cdc2-Cyclin B kinase activity (Fig. 1). Positive fractions for Cak and Cdc2 activity are pooled separately and further purified.

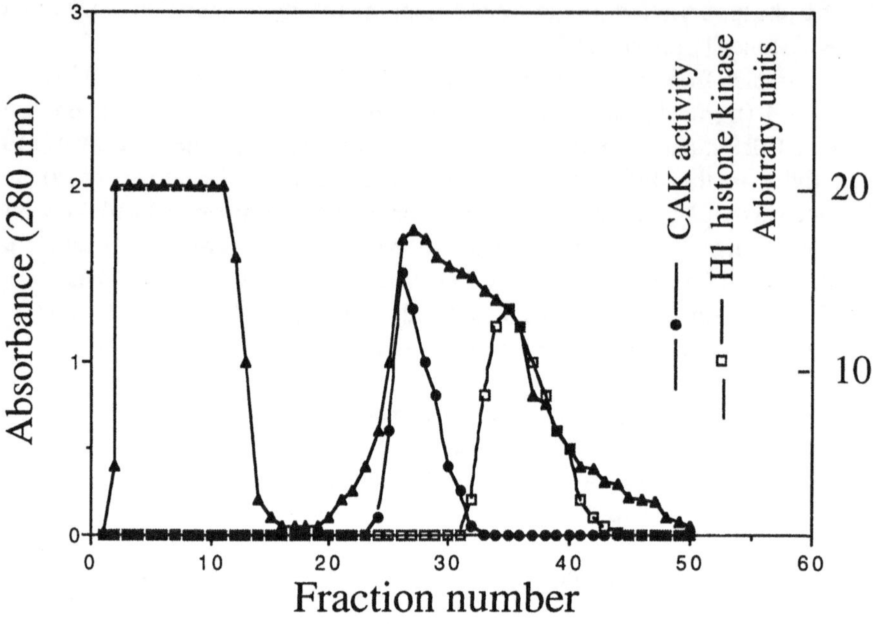

Fig 1. Representative chromatography of starfish homogenate on SP-Sepharose

Cak Purification

1. Positive fractions for Cak activity are pooled, diluted to a conductivity of 5 millimho with buffer A and loaded onto 10 ml heparin-Sepharose Fast flow (Pharmacia) column equilibrated in buffer A. After washing with 50 ml of buffer A, the retained proteins are eluted with a 0–1 M NaCl gradient in buffer A at a flow rate of 180 ml/h. Fractions of 3 ml are collected and tested for Cak activity after 1/40 dilution (Fig. 2A).

2. Positive fractions are pooled and proteins precipitated by adding 0.264 g/ml ammonium sulfate (45% saturation). The precipitate is collected by centrifugation, solubilized in 3 ml of buffer A, clarified by spinning in a TL 100 Beckman centrifuge (150 000 *g* for 15 min).

3. The supernatant is loaded onto a Superdex 200 HR 1.6/60 gel filtration column (Pharmacia) equilibrated in buffer A + 150 mM NaCl and eluted at a flow rate of 1 ml/min. Fractions of 2 ml are collected and tested for Cak activity after 1/80 dilution (Fig. 2B).

4. Fractions with Cak activity are pooled and diluted four times with buffer A and loaded onto a HR 5/5 Mono Q column (Pharmacia) equilibrated in buffer A. The bound proteins are eluted by a 0–0.4 M linear gradient of NaCl in buffer A at a flow rate of 1 ml/min (Fig. 2C). Fractions of 0.4 ml are collected and tested for Cak activity after 1/300 dilution.

5. Positive fractions are desalted by gel filtration on PD10 column (Pharmacia) equilibrated with buffer B (50 mM MES pH 6.5, 1 mM EGTA, 1 mM DTT) and applied onto HR 5/5 Mono S (Pharmacia) column equilibrated in buffer B. Bound proteins are eluted by a 0–0.4 M linear gradient of NaCl in buffer B and assayed for Cak activity after 1/200 dilution (Fig. 2D).

Analysis by SDS-PAGE (Fig. 3) of the most purified Cak shows the presence of three major copurifying proteins. Two of them peaking with the Cak activity correspond to 40 and 37 kDa polypeptides. The 37 kDa protein indicated by an arrow in Fig. 3 is the starfish homologue of the *Xenopus* MO15 gene product as determined by partial protein sequencing (Fesquet et al. 1993).

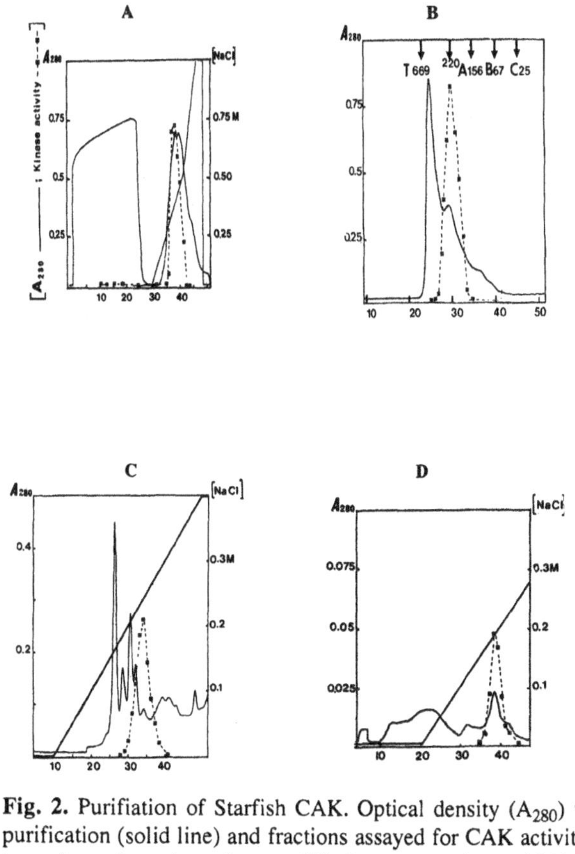

Fig. 2. Purifiation of Starfish CAK. Optical density (A_{280}) was recorded at each step of purification (solid line) and fractions assayed for CAK activity (dashed line). (**A**) Chromatography on heparin-sepharose; (**B**) chromatography on S200 Superdex: elution volumes of thryglobulin (T, 699 kDa), adolase (A, 156 kDa), bovine serum albumin (B, 67 kDa) and chymotrypsinogen (C, 25 kDa) are shown by arrows, as well as the elution volume of CAK activity (220 kDa); (**C**) chromatography an Mono.Q; (**D**) chromatography on Mono-S. See text for a complete description of the fractonation procedure

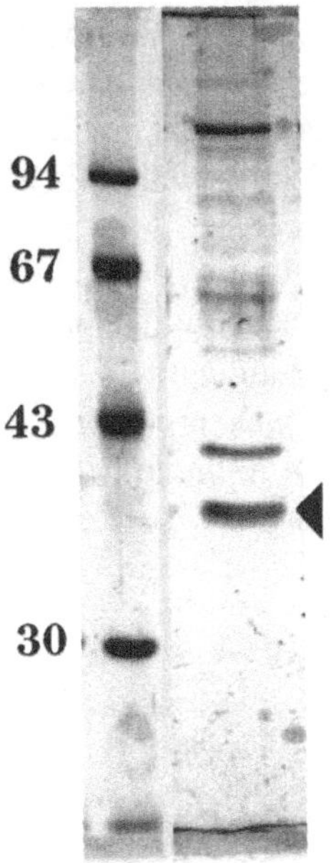

Fig. 3. Analysis by SDS-polyacrylamide gel electrophoresis and Coomassie blue staining of the pattern of proteins after the last step purification of Starfish CAK. Molecular weight markers are on the left side of the gel

Cdc2-Cyclin B purification

The most efficient step of this purification procedure is an affinity chromatography on immobilized yeast $p13^{suc1}$ protein which binds Cdc2 protein kinase. This affinity column can be prepared by immobilization of recombinant suc1 protein prepared as described in (Brizuela et al. 1987) on CNBr-activated Sepharose according to the manufacturer (Pharmacia) or purchased from Oncogene Science, Inc. (Cambridge MA USA).

1. Fractions with histone H_1 kinase activity resulting of the SP-Sepharose chromatography are loaded onto a 5-ml column of $p13^{suc1}$-Sepharose equilibrated with buffer A containing 150 mM NaCl.
2. After washing with 5 vol of equilibration buffer, wash with 5 vol of buffer C (50 mM Tris/HCl pH 6.8, 1 mM EGTA, 1 mM DTT).
3. Elute Cdc2-Cyclin B with 6 ml of buffer C containing 30 mg/ml of bacterially expressed p13suc1 protein, followed by 6 ml of buffer C.
4. Pool the 2 × 6 ml of buffer C containing the Cdc2 kinase and free $p13^{suc1}$, load onto a HR 5/5 Mono S column (Pharmacia) and elute the bound proteins by a 0–0.6 M NaCl gradient in buffer C at a flow rate of 0.8 ml/min. The Cdc2 kinase is eluted as a sharp peak around 0.3 M NaCl.

The $p13^{suc1}$ protein does not bind to the ion exchange chromatography column. Free $p13^{suc1}$ recovered in the flow through of this column can be re-used for other purifications. H_1 histone kinase activity corresponds to fractions containing only two polypeptides of 34 and 47 kD which have been identified by partial protein sequencing to Cdc2 and Cyclin B, respectively (Labbé et al. 1989).

2.3 Preparation of the *Xenopus* Oocyte Homogenate

Cak kinase can be prepared from unfertilized Xenopus eggs (see also chapter by P.R. Clarke, Part Two) arrested at metaphase II stage with modifications of the protocol used for starfish oocytes.

Stock solutions

- Washing solution I: 100 mM NaCl (5.84 g/l).
- Washing solution II: cysteine hydrochloride 2 % (*w*/*v*) in water. This solution is adjusted to pH 7.8 with NaOH.
- Homogenization buffer A.

	mM	g/l
β glycerophosphate, 2 H_2O·	80	18.72
EGTA	15	5.7
DTT	1	0.154
$MgCl_2$, $6H_2O$	10	2.033

Preparation

Day 1
Each female *Xenopus laevis* is injected with 125 units of pregnant mare serum gonadotropin (PMSG, Intervet, Boxmeer, Holland).

Day 3
Each female is injected with 1000 units of human chorionic gonadotropin (HCG, Sigma St Louis MO) and kept in solution I.

Day 4
Eggs laid during the night are collected in 50 ml round bottom tube with cap and washed three to four times with washing solution I. The jelly coat surrounding the eggs is removed by three to four washes with solution II. This treatment decrease the volume of the pelleted eggs by a factor of four. The eggs are again washed three times with solution I and once with ice-cold homogenization buffer.

The eggs (1 vol) are homogenized with 1 vol of homogenization buffer using a Potter equipped with a Teflon pestle. The homogenate is first centrifuged at 12 000 *g* for 15 min at 4 °C (SS-34 Sorvall 10 000 rpm) and the supernatant centrifuged again at 100 000 *g* for 45 min in a swinging bucket rotor. The top yellow layer is discarded and the soluble fraction is filtered under vacuum on 5.0 μm (Millipore SM, 5.0 μm), frozen in liquid nitrogen, and kept at –70 °C until use. From these extracts, it is possible to prepare both Cdc2-Cyclin B kinase and Cak. Only the purification of Cak is described here.

2.4 Purification of Cak from *Xenopus*

1. The homogenate (100 ml) diluted twofold with buffer A is loaded onto a 100 ml of Q Fast flow-Sepharose column (Pharmacia) equilibrated with buffer A, pH 8.0. Bound proteins are eluted at a flow rate of 100 ml/h by a 0-1 M linear gradient (total volume 500 ml) of NaCl in buffer A, pH 7.8. Fractions of 10 ml are collected and assayed for Cak activity. Fraction positive for Cak are pooled.
2. Positive fractions are loaded onto 10 ml heparin-Sepharose Fast flow column equilibrated in buffer A. After washing with 50 ml of buffer A, the retained proteins are eluted with a 0–1 M NaCl gradient in buffer A at a flow rate of 180 ml/h. Fractions of 3 ml are collected and tested for Cak activity. Positive fractions are pooled and precipitated by adding 0.264 g/ml ammonium sulfate. The resulting precipitate is collected by centrifugation, solubilized in 2.5 ml of buffer B (50 mM Tris/HCl pH 7.5, 300 mM NaCl) and loaded onto PD10 column (Pharmacia) equilibrated in buffer B. The first 2.5 ml are discarded and the eluate 3.5 ml are saved.

Affinity chromatography

3. An antibody affinity chromatography is performed on the PD 10 eluate. This affinity column is prepared by coupling an antibody directed against the C-terminus of *Xenopus* MO15 (EGMDQKDIAKKLSF) to protein A-Sepharose using dimethyl pimelimidate as the coupling agent (Pierce) (Schneider et al. 1982) (see also chapter by M. Pagano, Part Four).
   - The protein A-Sepharose (3 ml) is washed with 20 ml of 100 mM borate buffer (12.36 g/l) adjusted to pH 8.2.
   - The protein A-Sepharose and 4 ml of MO15 C-terminus antiserum are incubated for 3 h at 4 °C (rotation end over end),washed with 20 ml 100 mM borate buffer and then by 10 ml of 200 mM triethanolamine (26.6 ml/l).
   - Bound antibodies to protein A-Sepharose are crosslinked by incubation (45 min, room temperature) in the presence of 10 ml of 200 mM triethanolamine containing dimethyl pimelimidate (38.2 mg).
   - After this incubation period, the resin is collected by centrifugation and treated 10 min with 10 ml of ethanolamine (45 µl/50 ml of H_2O).
   - The resin is sequentially washed with 200 ml of 100 mM borate buffer, 100 ml of TBS (50 mM Tris/HCl pH 7.5, 150 mM NaCl). A blank cycle is performed in order to remove unbound antibodies by washing the

resin with 200 mM HCl/glycine pH 2.4 followed by extensive washing with TBS (50 ml).

- The saved fraction after PD10 chromatography is incubated 4 h at 4 °C with protein A-Sepharose antibody resin (rotation end over end) and washed with 20 ml of TBS supplemented with 0.05 % Tween 20.

Bound proteins to this affinity resin have Cak activity similar to that observed after direct immunoprecipitation with the anti C-terminus of Xenopus MO15 (Fesquet et al. 1993). The resin packed on a small Econo-column (Biorad) can be eluted alternatively with 6 M $MgCl_2$ or with 200 mM HCl/glycine pH 2.4 if SDS-PAGE analysis of bound proteins is required. When HCl/glycine elution is performed fractions of 150 µl are collected and neutralized by addition of 50 µl of 1 M Tris. Fraction aliquots (10 µl) are analyzed by SDS-PAGE (12.5 % polyacrylamide). Figure 4 shows the final preparation after affinity chromatography. Besides MO15 (M_r 40 000 indicated by an arrow) it contains only one additional polypeptide with an apparent M_r of 36 000.

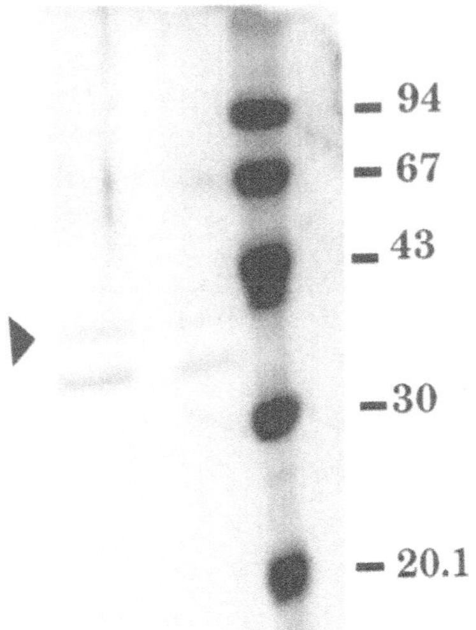

Fig. 4. Analysis by SDS-polyacrylamide gel electrophoresis and Coomassie blue staining of Xenopus CAK preparation. Molecular weight markers are on the right side of the gel

References

Brizuela L, Draetta G, Beach D (1987) p13^{suc1} acts in the fission yeast cell division cycle as a component of the p34^{cdc2} protein kinase. EMBO J 6:3507–3514

Clarke PR Karsenti E (1991) Regulation of p34^{Cdc2} protein kinase: new insights into protein phosphorylation and the cell cycle. J Cell Sci 100:409–414

Desai D, Gu Y, Morgan DO (1992) Activation of Human Cyclin-Dependent Kinases In vitro. Mol Biol Cell 3:571–582

Doree M (1990) Control of M-phase by maturation-promoting-factor. Curr Opin Cell Biol 2:269–273

Fesquet D, Labbé JC, Derancourt J, Capony JP, Galas S, Girard F, Lorca T, Shuttleworth J, Dorée M, Cavadore JC (1993) The MO15 gene encodes the catalytic subunit of a protein kinase that activates Cdc2 and other Cyclin-Dependent kinases (Cdks) through phosphorylation of Thr161 and its homologues. EMBO J 12:3111–3121

Labbe JC, Capony JP, Caput D, Cavadore JC, Derancourt J, Kaghad M, Lelias JM, Picard A, Doree M (1989) MPF from starfish oocytes at first meiotic metaphase is a heterodimer containing one molecule of Cdc2 and one molecule of Cyclin B. EMBO J 8: 3053–3058

Meyerson M, Enders GH, Wu CL, Su LK, Gorka C, Nelson C, Harlow E, Tsai LH (1992) A family of human Cdc2-related protein kinases. EMBO J 11:2909–2917

Nurse P (1990) Universal control mechanism regulating onset of M-phase. Nature. 344: 503–508

Pines J (1993) Cyclins and Cyclins dependent kinases: takes your partners. Trends Biochem. Sci 18:195–197

Poon R, Yamashita K, Adamczewski JP, Hunt T, Shuttleworth J (1993) The Cdc2-related protein p40^{MO15} is the catalytic subunit of a protein kinase that can activate p33^{Cdk2} and p34^{Cdc2}. EMBO J 12:3123–3132

Reed SI (1991) G1 specific Cyclins: in search of an S phase promoting factor. Trends Genet 7:95–99

Schneider C, Newman RA, Sutherland DR, Asser U, Greaves ML (1982) A one-step purification of membrane proteins using a high efficiency immunomatrix. J Biol Chem 257: 10766–10769

Sherr CJ (1993) Mammalian G1 Cyclins. Cell 73:1059–1065

Shuttleworth J, Godfrey R, Colman A (1990) p40^{MO15}, a Cdc2-related protein kinase involved in negative regulation of meiotic maturation of Xenopus oocytes. EMBO J 9: 3233–3240

Solomon MJ, Lee T, Kirschner MW (1992) Role of phosphorylation in p34Cdc2 activation: identification of an activating kinase. Mol Biol Cell 3:13–27

Solomon MJ, Harper JW, Shuttleworth J (1993) CAK, the p34^{Cdc2} activating kinase, contains a protein identical or closely related to p40^{MO15}. EMBO J 12:3133–3142

19 Interaction-Based Cloning of Novel Proteins

FRANCESCA FIORE, LOU LAMPHERE, AND JENO GYURIS

1 Background

It has become evident that interactions between proteins are the underlying mechanism of basic biological processes such as signal transduction, transcription, cell cycle progression, etc. For example, during the cell cycle, members of the Cdk family differentially interact with a number of cyclin regulatory subunits resulting in functionally distinct kinase complexes that regulate different stages of cell cycle progression (Draetta 1990; Sherr 1993). The cyclin/Cdk complexes contain other proteins that regulate their kinase activity (Xiong et al. 1993). Since these proteins form stable complexes, they are extremely amenable to interaction-based cloning using the two-hybrid system. The power of this approach has been well demonstrated recently by the interaction cloning of a battery of new cell cycle regulators such as $p16^{Ink4}$ (Serrano et al. 1993), $p21^{Cip1}$ (Harper et al. 1993), p130 (Hannon et al. 1993), and Cdi1/Kap (Gyuris et al. 1993; Hannon et al. 1994). In most cases, their presence in cyclin/cdk complexes had been detected previously by immunoprecipitation (Xiong et al. 1993).

The interaction-based cloning technique uses transcription as a phenotype to detect protein-protein interactions in yeast cells. It can be used to clone genes encoding proteins that bind to another known protein. It also offers an alternative approach to immunoprecipitation to study interactions between known protein partners. The two-hybrid system takes advantage of the modular structure of transcriptional activators that consist of at least two functionally distinct, separable domains: the DNA-binding domain and the transcriptional activator domain. Due to this modular structure, DNA-binding domains and transcriptional activator domains from different transcriptional factors can be combined to generate transcription factors with new properties. The development of the two-hybrid system derives from two key experimental observations.

- Ma and Ptashne showed that the transcriptional activator domain of Gal4 is able to activate transcription through Gal4 binding sites when it is brought into complex with the DNA-binding domain by noncovalent interaction (Ma and Ptashne 1988).
- Fields and Song demonstrated that Gal4 activates transcription through Gal4 binding sites when the Gal4 DNA-binding and activator domains are

separately fused to the yeast proteins Snf1 and Snf4, and brought together by the interaction of Snf1 and Snf4 proteins (Fields and Song 1989). They suggested and later demonstrated the feasibility of cloning proteins from a yeast genomic DNA library which expressed the genomic fragment encoded proteins in fusion with the Gal4 activator domain that interacted with the Gal4 DNA-binding domain fused to Sir4. They called this cloning strategy the two-hybrid system (Chien et al. 1991).

2 Principle and Applications

Several groups have developed interaction cloning systems utilizing the same principle (Chien et al. 1991; Dalton and Treisman 1992; Gyuris et al. 1993; Harper et al. 1993; Vojtek et al. 1993). Each system has three basic components: plasmids that allow the expression of a known protein in fusion with a site-specific DNA-binding domain, reporter genes that are bound by the DNA-binding domain but have no basal transcription activity, and vectors that express the cDNA encoded open reading frames in fusion with an N-terminal transcriptional activator domain.

Two DNA-binding domains are used in all systems: the DNA-binding domain of the yeast transcriptional activator Gal4 (Chien et al. 1991; Harper et al. 1993) or the heterologous *E. coli* repressor protein LexA (Gyuris et al. 1993; Vojtek et al. 1993). The Gal4 DNA-binding domain effectively localizes to the nucleus and binds to a 17 bp binding site that can be placed upstream of a reporter gene. Since *GAL4* is a yeast transcriptional activator, experiments utilizing the Gal4 DNA-binding domain must be performed in *gal4*$^-$ yeast to avoid interference with the endogenous GAL4. *gal4*$^-$ yeast cells are usually less healthy than *GAL4*$^+$ yeast and give rise to a higher number of false positives. The heterologous LexA protein does not contain a nuclear localization signal but enters into the nucleus and efficiently occupies its operator sites that are inserted in front of a reporter gene. As opposed to Gal4, LexA has no effect on the growth of yeast cells nor is there an endogenous yeast protein that binds to either the LexA operators or to the LexA protein. The LexA DNA-binding domain can be used in *GAL4*$^+$ cells allowing cDNA library expression under the inducible *GAL1* promoter. This feature allows the quick elimination of false positives (see later). Since the DNA-binding domains may slightly alter the properties of the fused proteins, it is conceivable that certain proteins will work with one but not with the other DNA-binding domain.

The existing interaction cloning systems also utilize different reporters. In most systems the reporter gene is *LacZ,* whose expression is controlled by a basic yeast promoter derived either from the *CYC1* or *GAL1* gene, in which the upstream activating sequences have been replaced by binding sites for the

DNA-binding domain being used. These reporter constructs are either integrated into the yeast chromosome or, alternatively, replicated on yeast plasmids. The activity of the reporters can be assayed on X-gal (5-bromo-4-chloro-3-indolyl-b-D-galactoside) containing indicator medium. Cells that contain active reporter genes produce β-galactosidase and can be identified by visual screening as blue colonies. To eliminate false positives, some systems utilize a second reporter gene. The expression of these selectable reporter genes allows the yeast cells to grow on a particular minimal medium. The common characteristic of these reporters is that they encode enzymes required for the synthesis of an amino acid. So far, the *LEU2* and *HIS3* genes have been developed into reporters by replacing their upstream regulatory elements for either Gal4 or LexA binding sites. The modified genes are integrated back into the chromosome of appropriate yeast strains so that the expression of the reporter gene is essential to grow on media that lacks leucine or histidine, respectively (Gyuris et al. 1993; Harper et al. 1993; Vojtek et al. 1993). In general, the sensitivity of reporters is proportional with the number of inserted binding sites: reporters with more binding sites will pick up weaker or shorter-lived interactions than reporters with fewer ones.

To express activation domain tagged proteins, the existing interaction cloning systems employ different activation domains. These include the strong activation domains of Gal4 (Chien et al. 1991; Harper et al. 1993) or the *Herpes simplex* virus transcriptional activator VP16 (Triezenberg et al. 1988; Dalton and Treisman 1992; Vojtek et al. 1993), or the artificial, medium strong activator B42 that derives from *E. coli* (Ma and Ptashne 1987; Gyuris et al. 1993). In each case, the activation moiety is fused N-terminally to the cDNA encoded proteins. When VP16 or B42 is used in combination with the LexA DNA-binding domain, the cDNA library can be expressed under the control of the inducible GAL1 promoter, making it possible to quickly eliminate false positives whose expression is independent from the activity of the promoter (Gyuris et al. 1993).

A number of interaction cloning systems have been developed thus far and all provide a reasonable chance to recover proteins that specifically interact with the protein of interest. Because they are all based on the same principle, some of the components are interchangeable, although they utilize the yeast selectable markers differently. Recently, CLONTECH introduced a two-hybrid system cloning kit (MATCHMAKER) with a large number of compatible cDNA expression libraries, based on the system developed in S. Fields laboratory at Stony Brook (Chien et al. 1991). In this chapter we describe the Interaction trap that was developed in the laboratory of R. Brent at the Massachusetts General Hospital (Gyuris et al. 1993). The methods outlined here can be used with slight modifications with other systems as well.

In the Interaction trap system the protein of interest is fused to the full-length LexA protein (bait). The LexA fusion protein binds to the LexA

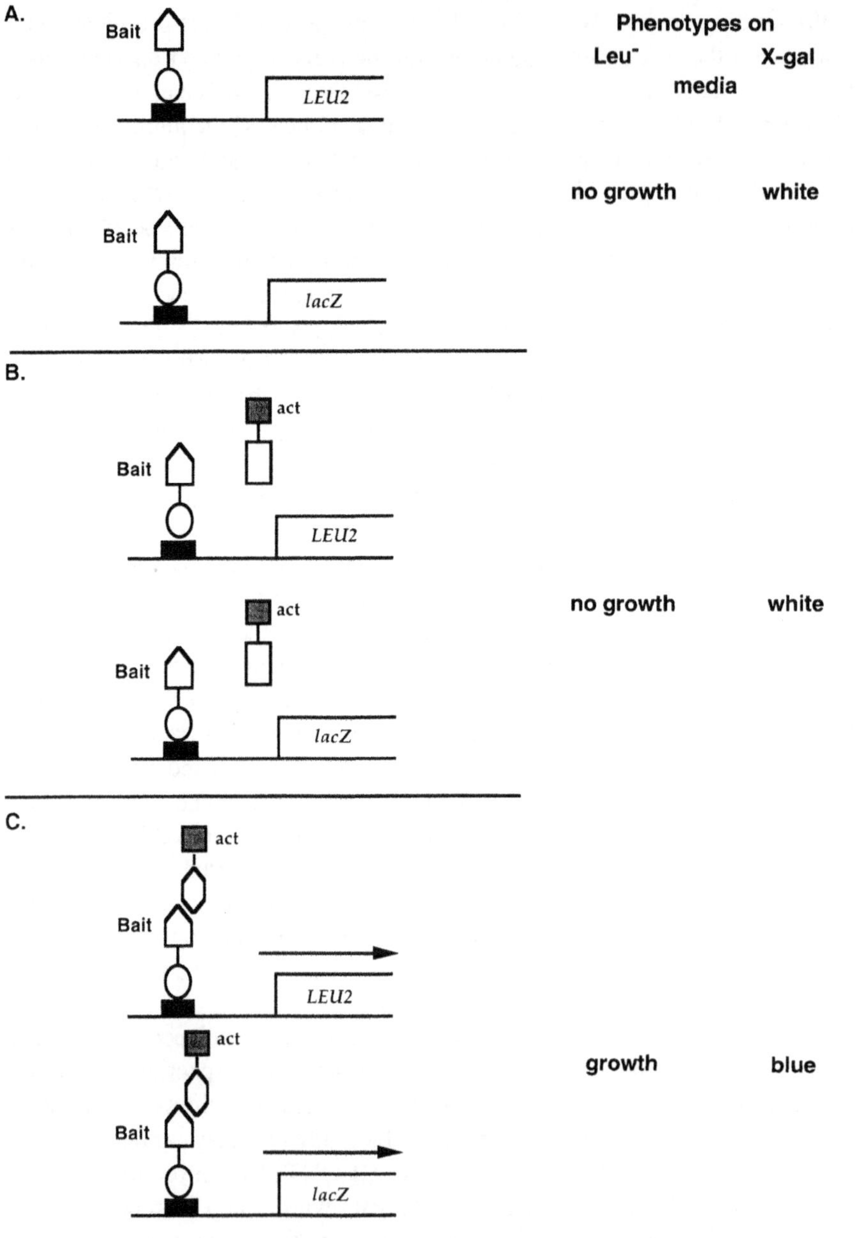

Fig. 1. The scheme of the Interaction trap. (**A**) The transcriptionally inert bait protein binds to the LexA operator sites in front of the LEU2 and LacZ reporter genes. The cells do not form colonies on Leu– media and do not turn blue on X-gal indicator plates because their transcription is not activated. (**B**) The expression of the library encoded proteins is induced in the presence of galactose. The reporter genes remain silent so that the cells can not form colonies on Leu– media and white on X-gal indicator plates, if the synthetised library protein is unable to interact with the bait. (**C**) In those cells in which the library encoded protein interacts with the bait the expression of the LexAop-LEU2 and LexAop-LacZ reporters is activated. The cells will form colonies on Leu– media and turn blue on X-gal indicator plates

binding sites (operators) inserted in front of two reporter genes. One reporter derives from the yeast *LEU2* gene in which the upstream activator sequences are replaced by LexA operator sites (*LexAop-LEU2*). The chromosomal, wild-type *LEU2* gene is substituted by the *LexAop-LEU2* construct. The *LexAop-LEU2* gene has no basal transcriptional activity, so the strain is unable to grow in the absence of leucine. The transcriptional activity of the *LexAop-LEU2* reporter can be detected by the ability of the cells to grow on medium lacking leucine, that requires the product of *LEU2* gene. The *LexAop-LacZ* gene serves as a second reporter and provides a secondary assay to validate and quantitate the activation that is presumably due to interaction between the bait and a member of the activation domain tagged cDNA library. The library vector, pJG4-5, is designed to express proteins as fusions with three protein moieties at the N-terminus: the SV40 nuclear localization signal (PPKKKRKVA) (Kalderon et al. 1984), the B42 transcriptional activation domain (Ma and Ptashne 1987); and the 12CA5 epitope tag (YPYDVPDYA) (Green et al. 1982) from the influenza virus hemagglutinin protein. The conditional expression of the library encoded proteins is directed by the *GAL1* promoter (Gyuris et al. 1993).

The cloning strategy using the Interaction trap is illustrated in Fig. 1. The constitutively expressed, transcriptionally inert bait protein binds to the LexA operator sites upstream of the *LEU2* and *Lac2* reporter genes but does not turn on their transcription (A). Synthesis of the library-encoded proteins is induced by growing the cells in the presence of galactose. If the library-encoded protein does not interact with the bait, the reporter genes are not active, the cells can not grow on media lacking leucine, and are white on X-gal containing indicator medium (B). Conversely, when the library-encoded protein interacts with the bait, then both reporter genes are active, the cells form colonies in the absence of leucine, and turn blue on X-gal indicator media (C).

3 Methods

3.1 Construction and Characterization of the Bait Protein

In order to start an interactor hunt, the protein of interest has to be fused to LexA and characterized in a series of control experiments. These controls are needed to establish that the LexA fusion protein is made in sufficient amount, localizes to the nucleus where it binds to the LexA operators, but does not activate the transcription of the reporter genes, so that the strain does not grow in the absence of leucine, and is white on X-gal indicator plates. The latter is the most important constraint on using interaction cloning systems.

Plasmids and Yeast Strains

LexA Fusion Plasmids. To construct the bait, the protein of interest has to be fused to the full-length LexA protein. The most commonly used LexA fusion plasmid, pLexA202+pl (Ruden et al. 1991) contains the entire LexA coding sequence (202 amino acids) including the DNA-binding and dimerization domains. The cloning sites downstream of the LexA are EcoRI, Bamhl, SalI, and the reading frame starts with the first G in the polylinker sequence: GAA TTC CCG GGG ATC CGT CGA CCT. The plasmid also contains the high-copy number 2 μ yeast replication origin and the *HIS3* selectable marker for propagation in yeast. The expression of the LexA fusion proteins is directed by the constitutive *ADH1* promoter/terminator to a level that generally sufficient for good operator occupancy upstream of the reporters. pEG202 is the same as pLexA202+pl except the unique cloning sites are EcoRI, BamHI, SalI, NcoI, NotI, XhoI (Ausubel et al. 1987–1994). The reading frame starts with the first G in the polylinker: GAA TTC CCG GGG ATC CGT CGA CCA TGG CGG CCG CTC GAG TCG ACC TGC AGC. Derivatives of the above plasmids with the SV40 nuclear localization signal exist for those rare proteins (for example membrane proteins) that are excluded from the nucleus.

Activator Domain Fusion Plasmids. pJG4-5 was developed to facilitate the expression of activator domain tagged cDNA encoded proteins. The selectable marker on this plasmid is *TRP1* an it also contains the 2 μ yeast replication origin. In pJG4-5 the fusion moiety contains, N to C-terminal, an ATG as a translational start, the SV40 nuclear localization signal to promote the nuclear localization of proteins that normally localize elsewhere in the cell, the artificial, medium strong transcriptional activator domain B42, and an epitope tag from the influenza virus hemagglutinin protein to allow the basic biochemical characterization of the fusion protein. The unique cloning sites, EcoRI and XhoI, are used for the insertion of cDNAs and allow the construction of directional cDNA libraries. The open reading frame starts with the first G in the polylinker: GAA TTC GGC CGA CTC GAG. The fusion proteins are expressed under the control of the inducible *GAL1* promoter. A number of cDNA libraries from different organisms are available on this vector (Gyuris et al. 1993).

LexAop-LacZ Reporter Plasmids. In addition to the plasmids expressing the LexA and activator domain fusions, the yeast cell contains another plasmid that carries the *LexAop-LacZ* reporter gene. The selectable marker on this plasmid is the *URA3* gene and multicopy 2 μ replication origin is present to ensure the maintenance of the plasmid in yeast cells. The *LacZ* gene is fused to the basic *GAL1* promoter in which the upstream activating sequence (*UAS*) is replaced by various number of LexA operators. There is one operator site in pRB1840 (Brent and Ptashne 1985), two in pJK103 (Kamens et al. 1990) and 8 in pSH18-34 (S. Hanes, personal communication). Reporters with more operator sites are more sensitive yet none of these reporter makes detectable amounts of β-galactosidase in the absence of a LexA fused activator, and they are all white on X-gal containing indicator plates.

LexAop-LEU2 Reporter Strains. A crucial part of the Interaction trap is a yeast strain that harbors the plasmids that express the bait, the activator domain tagged protein, as well as the *LacZ* reporter. In addition to these constructs, the strain also carries another reporter, *LexAop-LEU2*, which allows selection for cDNAs that encode interactors. The prototype of these reporter strains is EGY48 in which the chromosomal *LEU2* gene is replaced by *LexAop-LEU2* construct (Gyuris et al. 1993; Ausubel et al. 1987–1994). The genotype of EGY48 is *MATa* $ura3^{-}trp1^{-}his3^{-}$ *LexAop-LEU2*. Since the strain does not have a wild-type *LEU2* gene, whose product is essential for the biosynthesis of leucine, the strain does not grow in the absence of leucine. The *LEU2* reporter in EGY48 does not have residual transcriptional activity and is extremely sensitive. This sensitivity is probably due to the three LexA operators upstream to the *LEU2* gene that, in EGY48, can bind 6 molecules of the bait. The *LexAop-LEU2* reporter is activated by activator domain tagged proteins that only weakly interact with the bait allowing the cell to grow on leucine-deficient medium. Other, less sensitive, derivatives of EGY48 also exist in which one or two operator sites are placed upstream of the *LEU2* gene (E. Golemis, pers. comm.).

Additional Control Plasmids. In addition to the plasmids described above, a few other construct might help in the characterization of the bait. In plasmid pLexA-Cdk4 the entire human Cdk4 is expressed in fusion with LexA and can be used as a negative control because this construct alone does not activate the transcription of the reporters (F. Fiore and J. Gyuris, unpubl.). The pLexA202+pl or pEG202 parental plasmids are not ideal controls in these experiments because the encoded LexA protein includes several amino acids from the polylinker, making it a weak activator. Second, pJG4-5-Cyclin D1 (F. Fiore and J. Gyuris, unpubl.) which contains the full-length human Cyclin D1 protein in fusion with the B42 fusion moiety can be used in conjunction with pLexA-Cdk4 as a positive control, since these two proteins interact and strongly activate transcription of the reporters (F. Fiore and J. Gyuris, unpubl.). Whenever possible, it is also advisable to construct an activator fusion in pJG4-5 with a protein that is known to bind to the bait (pJG4-5-known interactor). This activator fusion protein might serve as a positive control to demonstrate that the LexA fused protein retained its ability to interact with other proteins. Two other plasmid may be useful during this experiments. pJK101 (J. Kamens, pers. comm.) is used in the "repression assay" (see below) and it contains two LexA operator sites that are placed between the UAS_{GAL} and basal *GAL* promoter element that are fused to the *LacZ* gene. The plasmid also carries the *URA3* yeast marker gene and the 2 μ replication origin. In the "repression assay" another plasmid, pJG4-4 (J. Gyuris, unpubl.) or any other plasmid, with the TRP1 selectable marker and 2 μ replication origin is needed as negative control.

Testing the Bait

All these experiments utilize a few basic yeast experimental techniques such as yeast transformation and media whose detailed description can be found elsewhere (Ausubel et al. 1987–1994). From the number of yeast transformation protocols, we suggest using the one developed by Schiestl and Gietz as described in (Gietz et al. 1992).

Testing

1. To construct yeast strains that carry both the *LexAop-LacZ* and *LexAop-LEU2* reporters, the bait, and the control pLexA-Cdk4, grow EGY48 in complete YPD media and transform with the following plasmid pairs:
pSH18-34 + pLexA-bait
pSH18-34 + pLexA-Cdk4.
Select for transformants on CM-Ura-His/Glu media at 30 °C for 3 days.
It is possible to transform yeast cells with more than one plasmid at a time, but the efficiency of transformation will be lower than with one plasmid. Usually this lower efficiency is not a problem because only a few transformants are needed to perform most of the experiments.

2. When the individual transformants appear, take one colony that it is transformed with pSH18-34 and pLexA-Cdk4 as well as three colonies in which the *LacZ* reporter, pSH18-34, and the bait is introduced and streak them on CM-Ura-His/Glu plates to make a master plate. If the bait protein is suitable for the interactor hunt, then one of these colonies will be used to introduce the library and to perform the selection for the interactors.
In order to verify that the transformants synthesize the expected size of bait protein, perform Western blot analysis using antibodies raised either against the LexA or the protein of interest. For the Western blot, inoculate the cells in 2 ml of liquid CM-Ura-His/Glu media and grow them at 30 °C with intensive shaking. When the density of the cultures reaches OD600 = 0.8, collect 1 ml of cells by centrifugation and resuspend the cell pellet in 50 μl of water. Add an equal volume of 2× Laemmli loading buffer and boil the samples for 3 min. Shear the DNA by sonication and load approximately 30 μl onto a SDS polyacrylamide gel. The running of the gel and the Western blot analysis are done according to the standard protocols (Ausubel et al. 1987–1994).

3. Grow colonies from the master plate that represent the three original bait transformants and the one control (pLexA-Cdk4) in liquid CM-Ura-His/Glu media. Transform the three independent bait transformants with:

p-JG4-5
pJG4-5-known interactor.

4. Transform the strain carrying the pLexA-Cdk4 bait with:
pJG4-5
pJG-4-5-Cyclin D1.

5. Select transformants at 30 °C on CM-Ura-His-Trp/Glu media for 3 days.

Assay for the Activity of the Bait on the *LexAop-LacZ* Reporter

Protocol

1. When the transformants have grown into large colonies, pick five independent colonies from each transformation and streak them on indicator plates:
CM-Ura-His-Trp/X-gal/Glu
CM-Ura-His-Trp/X-gal/Gal, Raf.

2. Incubate the plates at 30 °C for 3 days. Monitor for the developing blue color.

Expected results

pLexA-cdk4 + pJG4-5. White on both the glucose- and the galactose/raffinose-containing indicator plates because the LexA-Cdk4 fusion protein is transcriptionally inert.

pLexA-Cdk4 + pJG4-5-cyclin D1. White on the glucose-containing indicator media but it turns blue in less than 1 day on the galactose/raffinose indicator plate because the Cyclin D1 binds strongly to Cdk4.

pLexA-bait + pJG4-5. The color remains white on both kinds of indicator media if the bait protein is transcriptionally inert. Blue color will develop on both glucose and galactose plates if the bait activates transcription itself. The intensity of the color depends on how strong the intrinsic transcriptional activator domain of the bait is, and it can range from extremely pale blue after 3 days of incubation to intensive blue that develops in few hours. Often the appearing color is lighter in the presence of galactose/raffinose than glucose because the activity of the *ADH1* promoter, which drives the expression of the LexA fusion protein, drops by approximately 50% when galactose/raffinose is the only available carbon source.

pLexA-bait + pJG4-5-Known Interactor. If the bait protein is transcriptionally inert, the colonies stay white on the glucose indicator media but become blue on the galactose/raffinose-containing indicator plate where the known partner of the bait is expressed. The intensity of color depends on the strength of interaction between the bait protein and its partner. When the bait itself activates the *LacZ* reporter, the darker blue color on the galactose/raffinose plate is indicative of the interaction between the bait and the partner.

Assay for the Activity of the Bait on the *LexAop-LEU2* Reporter

It is also imperative to test the bait on the activation of the *LexAop-LEU2* reporter, since the selection for interactors is based on the ability of the interacting protein pair, but not the bait alone, to activate the *LexAop-LEU2* reporter and, consequently, to cause the yeast cell to grow in the absence of leucine. Often baits that do not activate the *LacZ* reporter but will activate the *LexAop-LEU2* reporter and allow EGY48 to grow in the absence leucine because the *LexAop-LEU2* is more sensitive than any of the *LacZ* reporters.

Protocol

1. Take independent colonies from each transformation and suspend them into 1 ml of water up to the density of $OD_{600} = 0.5$ (approx. 10^7 cells/ml). Make 10^2, 10^3, 10^4-fold dilutions in water and plate 100 µl from each dilution on
   CM-Ura-His-Trp/Glu
   CM-Ura-His-Trp/Gal, Raf
   CM-Ura-His-Trp-Leu/Glu
   CM-Ura-His-Trp-Leu/Gal, Raf plates.
2. Incubate these plates at 30 °C for 3–5 days and monitor the appearance of the colonies. If the growth rate of a given bait is significantly diminished on the CM-Ura-His-Trp/Glu and Gal, Raf plates as compared to the controls, it may indicate that its expression is toxic to the yeast cells.

Expected results

pLexA-cdk4 + pJG4-5. No colonies appear either on the glucose and the galactose/raffinose plates because the LexA-cdk4 bait is transcriptionally inert.

pLexA-Cdk4 + pJG4-5-cyclin D1. No colonies on glucose but colonies appear by the second day on the galactose/raffinose containing plates due to the binding of the activator tagged Cyclin D1 to the LexA operator bound LexA-Cdk4 bait.

pLexA-Bait + pJG4-5. No colonies should appear on either carbon source containing media in the first 3 days of incubation if the bait is inactive or only weakly activates transcription of the *LEU2* reporter. If colonies begin to appear by the fourth day, it indicates that the bait has some transcriptional activity but it will not interfere with the interactor hunt. If the bait activates the *LEU2* reporter, colonies will appear on both glucose and galactose/raffinose.

pLexA-Bait + pJG4-5-Known Interactor. If the bait itself does not activate *LEU2*, then no colonies will appear on glucose. Depending on the strength of interaction, colonies will appear from the second day of incubation on in the presence of galactose. If the bait itself activates the *LexAop-LEU2* reporter, then colonies will appear on both glucose and galactose and it is difficult to distinguish between the bait and the interaction-induced growth on the different plates.

Assay for the Nuclear Localization and Operator Binding of the Bait

The best result of the above experiments is that the bait protein alone is transcriptionally inert and does not activate the *LexAop-LacZ* and *LexAop-LEU2* reporters (i.e., white on X-gal-containing indicator plates, and does not grow in the absence of leucine) but it interacts with its partner (i.e., blue on the X-gal-containing indicator plates in the presence of galactose/raffinose, and it grows in the absence of leucine when galactose/raffinose are present). These experiments clearly demonstrate that the protein is made and localized to the nucleus in sufficient amount to occupy the LexA operator sites efficiently. When there is no known partner of the bait, the situation is more complicated. If the bait activates the reporters to various extents, then it is a good indication that the bait functions properly. For those baits that do not activate transcription, it is important to show that the LexA fusion protein is localized into the nucleus and occupies the LexA operator sites efficiently. This can be done by performing the "repression assay".

Protocol

1. For the repression assay, transform the following plasmid combinations:
   pJK101 + pJG4-4 (negative control)
   pJK101 + pLexA-Cdk4 (positive control)
   pJK101 + pLexA-bait.
   Grow the transformants on CM-Ura-His/Glu at 30 °C until colonies are obtained.
2. Streak five independent colonies from each transformation on CM-Ura-His/X-gal/Gal, Raf plates and incubate them overnight at 30 °C.
3. Examine the color of the colonies on the indicator plate.

Expected results

pJK101 + pJG4-4. The color of these colonies will be strong blue due to the activation of the *GAL1* promoter through the UAS_{GAL}, by the endogenous Gal4 transcriptional activator.

pJK101 + pLexA-Cdk4. These colonies will be lighter blue because the LexA-cdk4 protein enters into the nucleus and binds to the LexA operator sequences, inserted between the UAS_{GAL} and the *GAL1* TATA element, and interferes with the Gal4-mediated transcriptional activation of the promoter.

pJK101 + pLexA-Bait. If the bait protein enters into the nucleus and binds to the LexA operator sites of pJK101, then a decrease in β-galactosidase activity is detected. If the bait fails to enter into the nucleus and/or to bind the LexA operator sites of pJK101, then no change in the intensity of the developing blue color is detected as compared to that of the negative control.

Troubleshooting

- The bait alone activates the reporter genes.
  In this case there are three potential approaches to take. First, the sensitivity of the system can be attenuated by using yeast strains in which fewer LexA operators are placed in front of the *LEU2* gene and *LacZ* reporter plasmids (E. Golemis, pers. comm.; F. Fiore and J. Gyuris, unpubl.). Second, if the bait still activates the reduced sensitivity reporters, the threshold of the reporter constructs can be further decreased by using a diploid strain in which the LexA fusions activate the reporters less (E. Golemis, pers. comm.; J. Gyuris, unpubl.). The reason for this phenomenon is unknown. Third, a more extreme solution is to remove the transcriptionally active portion of the bait. If the active region is unknown, start first with the deletion of acidic stretches that are often responsible for the transcriptional activation in yeast (Ma and Ptashne 1987). If no stretches of acidic amino acids are found, a systematic deletion analysis is suggested. The apparent disadvantage of this approach is that deletions may remove regions of the protein that might be involved in the interaction with other proteins. Furthermore, the truncated derivatives may have slightly altered conformations that might also affect the ability of the bait to interact with other proteins.
- Bait does not repress in the "repression assay".
  Baits that do not activate the reporters and repress in the repression assay are ideal for use in the interaction hunt. Baits that neither activate nor repress are more problematic, especially when high level of expression is detected by Western blot analysis. In this case the bait might not localize to the nucleus efficiently or conversely; it is not capable of occupying the LexA operators effectively. The problem may be solved by the construction of a new bait in a LexA fusion vector that contains a nuclear localization signal.
- No interaction is detected with the known control protein.
  One explanation of the problem might be the exclusion of the bait from the nucleus (see above). The interaction should be tested again with a nuclear localization signal containing a version of the bait. If the bait with the nuclear localization signal is still unable to interact but represses in the "repression assay", then this might indicate that the conformation of the fused protein is impaired by the LexA fusion moiety. If the bait does not repress, it might mean that the conformation of the LexA is effected and, therefore, cannot bind to its operator sites. The distortion between the fused proteins might be relaxed by separating the two proteins by a flexible, glycine-rich hinge region (Chien et al. 1994).

3.2 Selection of Interactors

Once the bait is characterized, the hunt for the interactors begins. In the first step, the cDNA expression library is introduced into the strain that contains the bait, and a large number of cDNA transformants are collected. Every transformed yeast cell will carry one individual library plasmid. In the second step, those cells will be selected that express proteins that interact with the bait. The phenotype of the positives will be verified on X-gal containing indicator plates and the library plasmid from the true positives will be rescued. To prove the specificity of the interactors, the rescued library plasmids will be reintroduced into the strain containing the bait against which they were selected as well as into other strains that express different baits. When the interaction is reproduced and it is specific for the original bait the hunt for interactors is finished.

Library Transformation

The goal of this transformation step is to collect large numbers of library transformants, each of which will contain one member of the cDNA library. The library is made on the pJG4-5 vector that expresses the cDNA-encoded proteins under the control of the *GAL1* promoter. The selectable marker on pJG4-5 is *TRP1*. For the high efficiency transformation of the library, we suggest to use the protocol of Gietz et al. (Gietz et al. 1992) as described in (Ausubel et al. 1987–1994). This protocol utilize lithium-acetate to make the cells competent and single stranded carrier DNA to increase the transformation efficiency which is generally ~10^5 transformants/μg library DNA. Since the library DNA is usually valuable, it is advisable to perform pilot experiments and to start with the large scale library transformation when good transformation efficiency has been achieved.

The library transformants are plated onto media that permit the selection of cells that contain a library plasmid. The direct plating on media that also allows the selection for the cDNAs that encode interactor proteins (i.e. media that also lacks leucine and contains galactose as a carbon source) is disadvantageous from several reasons. First, the expression of potentially interesting cDNAs is sometimes toxic for the growth of the cell that carry them, so these cells (and cDNAs) would be underrepresented in the original pool if they were plated directly. Second, after simultaneous transformation and galactose induction, cells containing weaker interactors may not be able to support growth on medium lacking leucine, and consequently, would be lost.

Transformation protocol

1. Grow EGY48/pSH18-34 that carries the bait in CM-Ura-His/Glu media and make the cells competent according to (Gietz et al. 1992; Ausubel et al. 1987–1994).

2. Aliquot 200 µl of cell suspension into each of 20 Eppendorf tubes. Add 5 µg library DNA and 200 µg single-stranded salmon sperm DNA (Sigma) to each tube and proceed with the transformation as described in (Ausubel et al. 1987–1994). The volume of the library and the carrier DNA should not exceed 20 µl. It is important to perform a negative control of transformation using 200 µg single-stranded salmon sperm DNA only.

3. Plate the cells directly from the PEG solution onto 20 24 × 24 cm large plates (Nunc) containing CM-Ura-His-Trp/Glu media and incubate the plates for 2–3 days at 30 °C. To estimate the number of transformants obtained, sample 10 µl of cell suspension from one Eppendorf tube and plate onto CM-Ura-His-Trp/Glu media.

Expected results This transformation protocol will give ~2.5–5 × 10^5 transformants per large 24 × 24 cm plate, resulting in a total of 5–10 × 10^6 transformants.

Collecting the Primary Library Transformant Cells

Collecting cells

1. Cool the 20 24 × 24 plates with the transformants in the cold room (4 °C).

2. When the plates are cold and the agar has hardened, carefully scrape the transformants off the plates using a sterile glass microscope slide. Combine the cells in a 500-ml sterile Sorvall centrifuge tube, suspend them into sterile water, and collect the cells by centrifugation at 2000 *g* for 5 min. Be extremely careful to prevent contamination because the contaminating cells will come up as Leu^+ colonies during the selection process.

3. Resuspend the cells thoroughly in one pellet volume of buffer containing:
   65 % glycerol
   25 mM Tris/Cl pH 7.4
   100 mM $MgSo_4$.

4. Freeze 1-ml aliquots (approximately 20) in Eppendorf tubes at –80 °C.

5. Before the selection for interactors begins, thaw an aliquot and inoculate 500 µl of cell suspension into 4.5 µl of YPGal. Shake the cells at 30 °C for 5 h and plate dilutions on the following media:
   CM-Ura-His-Trp/Glu, Raf
   CM-Ura-His-Trp-Leu/Gal, Raf.

6. Incubate the plates at 30 °C for several days and count the colonies.

The number of colony-forming units in one tube can be calculated from the number of colonies on the CM-Ura-His-Trp/Glu, Raf plates. The number of colonies on the CM-Ura-His-Trp-Leu/Gal, Raf media will give the frequency at which the Leu^+ colonies appear. This will help to estimate the number of positives at the end of the large selection and design the selection experiment accordingly (see below).

Selection for Interactor Proteins

Selection protocol

1. Thaw an aliquot of cell suspension and add 500 μl to 4.5 ml YPGal as before. Shake it for 5 h at 30 °C to induce the expression of the *GAL1* promoter.

2. Plate the volume of cell suspension that is equivalent to ~5×10^6 colony-forming units onto an 150 mm CM-Ura-His-Trp-Leu/Gal, Raf plate. The number of colony-forming units to be plated depends on the number of the original, primary library transformants as well as on the frequency of Leu positives. Because not all cells that contain an interactor will plate with 100% efficiency on the Leu-deficient medium, it is advisable to plate five to ten times more colony-forming units than the original number of transformants, so that every primary transformant will be represented on the selection plate by five to ten individual cells. Incubate the plates at 30 °C for 2–4 days.
If the frequency of the occurrence of Leu-positive cells is too high, and large numbers of positives are expected (so that it is impossible to pick them by hand), cover the plate with a nitrocellulose (or equivalent) filter. Be careful not to leave air bubbles between the surface of the media and the filter. Plate the cells onto the filter directly and incubate the plates at 30 °C for 2–4 days. When Leu^+ colonies come up, subject them to a filter β-galactosidase assay (Breeden and Nasmyth 1985) to identify the Leu^+ and $LacZ^+$ blue colonies in which both reporter genes are active. Briefly, lift the filters from the plates and place them onto a Whatman 3 MM paper to dry for 5 min. To permeabilize the colonies, incubate the filters at −80 °C for 10 min, colony side up. Place the filters, colonies up, onto a Whatman 3 MM paper that is soaked in Z buffer (Ausubel et al. 1987–1994) that contains 1 mg/ml X-gal (excess of buffer removed) and incubate them at room temperature. Watch the color development. Strong interactions produce color in approx. 30 min, while the weaker ones develop in few hours. Pick the blue colonies and streak them on CM-Ura-His-Trp-Leu/Gal, Raf plates. Proceed with the analysis of these colonies âs described for the colonies that are isolated from the plates directly.

3. A second round of selection is necessary to purify the truly leu^+ colonies from the surrounding leu^- cells. Pick (and mark) the colonies that appeared on the second, third, and fourth days of selection with sterile toothpicks and streak them again on CM-Ura-His-Trp-Leu/Gal, Raf media.

Test for Galactose Dependence of the Leu^+ and $LacZ^+$ Phenotypes

Galactose dependence

1. Streak the colonies from the CM-Ura-His-Trp-Leu/Gal, Raf master plate onto CM-Ura-His-Trp/Glu media to shut off the expression of the activator-fused cDNA encoded protein. It is essential to shut off the expression of the cDNA encoded protein because its accumulated message and protein as well as the accumulated *LEU2* and *LacZ* gene products allow the cells to grow several generations and become pale blue without de novo expression of the cDNA fusion and the reporters. This would disguise the galactose dependence of the phenotypes.
2. Streak the colonies from the CM-Ura-His-Trp/Glu master plates on the following plates:
   CM-Ura-His-Trp/X-gal/Glu
   CM-Ura-His-Trp/X-gal/Gal, Raf
   CM-Ura-His-Trp-Leu/Glu
   CM-Ura-His-Trp-Leu/Gal, Raf.
3. Incubate the plates at 30 °C for 3–4 days and monitor the development of the color on the X-gal containing media and the growth of the colonies on the leucine-deficient plates.

Expected results The positive colonies are blue on X-gal plates and grow on media lacking leucine only if the expression of the activator-tagged, cDNA encoded protein is induced by galactose. Since the sensitivity of the *LEU2* reporters exceeds that of the *LacZ*, sometimes there are weak interactors that can induce growth in the presence of galactose but do not produce sufficient amount of β-galactosidase to develop color. These colonies should be considered as positive.

Specificity Test

The above experiments verified that the Leu^+, $LacZ^+$ phenotypes are dependent on the interaction between the bait and the activator-fused, cDNA encoded protein; but these experiments do not address the question of whether this interaction is specific for the bait and cDNA encoded protein pair, or whether it is due to the "stickiness" of the cDNA encoded protein that promiscuously binds to other proteins. To test this possibility, the cDNA expression plasmids have to be recovered from the positive colonies and reintroduced into a number of strains that carry the original and other unrelated baits. The true positives will interact with the original bait used for their selection, but not with the unrelated baits.

Specificity test

1. Grow up the positive colonies for the plasmid rescue in 2 ml of liquid CM-Trp/Glu media and isolate plasmid DNA as described in [26]. Transform an aliquot into a *E. coli* strain, for example KC8 (*pyrF::Tn5 hsdR*

leuB600 trpC9830 lacD74 strA galK hisB436), that contains a nonfunctional, mutant *trpC* gene. The lack of a functional *E. coli trpC* gene can be complemented by the introduction of the *S. cerevisiae TRP1* selectable marker gene that is carried by the library plasmid. Select the *E. coli* transformants on minimal media lacking tryptophane and containing ampicillin. This plasmid rescue protocol allows the selective rescue of the *TRP1*-containing library plasmid from a strain that contains two other plasmids that are also resistant for ampicillin but carry different yeast marker genes.

2. To reduce the work when large numbers of positives are obtained, it is important to sort the cDNAs into classes prior to the retransformation into yeast strains that carry the different baits. By the nature of the selection protocol, the true interactor cDNAs are usually selected in multiple isolates. By generating "fingerprints" of the cDNA inserts using EcoRI, XhoI, HaeIII triple digests, the identical cDNAs that produce identical restriction maps can be recognized.
   Cut approx. 1 μg of library plasmid miniprep DNA in 20 μl with EcoRI, XhoI, and HaeIII. Run the samples next to a similarly digested pJG4-5 vector control on 2.5% agarose 1/2 X TBE gel. Based on the restriction pattern, classify the cDNA clones.
3. Transform the strains that express the different baits and carry the pSH18-34 *LacZ* reporter with one representative plasmid of each class of cDNA isolates. Select for transformants on CM-Ura-His-Trp/Glu media.
4. Streak five independent transformants from each transformation on
   CM-Ura-His-Trp/X-gal/Glu
   CM-Ura-His-Trp/X-gal/Gal, Raf
   CM-Ura-His-Trp-Leu/Glu
   CM-Ura-His-Trp-Leu/Gal, Raf plates.
   Incubate the plates for up to 4 days at 30 °C and monitor the growth on the leucine-deficient media and the development of color on the X-gal indicator plates.

Expected results

The real interactors will reproduce the interaction with the bait that was used to select them, but will not interact with any of the unrelated, control baits.

3.3 Further Applications

Isolation of genes encoding proteins that touch another known protein is a relatively fast, straightforward process using interaction-based cloning. However, most of the time, the biological significance of the interaction is not revealed by these types of assays and has to be established using a variety of other approaches. Variations of the two-hybrid system may provide additional simple experimental tools to achieve this goal. Here, we overview a few modifications that significantly extend the power of this technology.

Since the interaction-based cloning technique identifies proteins through their ability to bind to other proteins, this system instantly provides information about the interaction domain or the position of the interaction domain within the cloned protein. One simple approach to find the interaction domain within a larger protein takes advantage of the random generation and expression of smaller pieces of the protein (Oliner et al. 1993). The fragments that potentially carry smaller domains can be generated by shearing the gene by sonication and cloning the sheared fragments into the activator fusion vector. This "domain library" is introduced into a yeast strain containing the bait. The positive colonies, in which the bait and the expressed protein domain interact, are selected using the protocol described above. After rescuing and analyzing the activator-fused domains, the smallest fragment will identify the minimum interaction domain that is necessary to bind to the bait (Oliner et al. 1993). The same approach can be taken to identify the complementary domain of the bait that interacts with the cDNA encoded protein.

Using PCR-generated mutagenesis, the amino acid residues of the new cDNA encoded protein that interact with the bait may be identified. In this approach, a library of mutant cDNA encoded proteins is generated by random misincorporation of nucleotides into the newly synthesized DNA strand by PCR (Cadwell and Joyce 1992). The mutagenized fragments are cloned into the activator fusion vector and introduced into a yeast strain that contains the bait. Colonies in which the interaction between the bait and the activator tagged protein is impaired by a mutation are identified as white or pale blue colonies on X-gal indicator plate by visual screening. The exact position of the mutation(s) is determined by sequencing the rescued plasmid(s). These mutations identify the amino acids that contact critical amino acids of the LexA-fused protein moiety. The contacted amino acids on the bait side can be determined, using the same technique. In this scenario, mutant versions of the LexA-fused protein are screened to identify those ones that restore the binding between the bait and a particular mutant version of the activator-tagged protein. This suppressor mutation identifies the amino acid residue in the bait that is contacted by the particular amino acid residue in the wild-type activator domain tagged protein. The phenotype of these mutants can be further studied in the authentic experimental system.

The interaction of the new cDNA encoded protein can be analyzed using a panel of mutant versions of the original bait with known phenotypes. If the cloned protein interacts with the wild type, but not with a mutant version of the bait, it suggests that the loss of interaction between the two proteins might cause the mutant phenotype. When naturally occurring mutants associated with known pathological syndromes and diseases are used, the loss of interaction suggests a mechanism by which the mutant allele might contribute to the development of the disease.

Certain important interactions, such as those between particular enzymes and substrates, may be weaker than the limit set by the reporter genes. These interactions may be studied using this system by incorporating a particular, catalytically inactive version of the enzyme that stabilizes the enzyme-substrate interaction.

The interaction of the newly isolated protein can also be tested with a panel of proteins whose function in the process is fairly well understood and which are known to be part of the same regulatory network as the original bait. The combination of the interaction data with the known role of the individual proteins in the particular biological process may suggest a testable hypothesis of the function of the new protein.

In certain interaction cloning experiments, it is assumed that the interaction between the protein of interest and its real partner depends on the presence of a small molecule, that could be a hormone or a drug (Chien et al. 1994; Lee et al. 1994, see also the chapter by V. Berlin and I. Chiu, Part Four). In these "ligand-dependent interaction" or "three-parts interaction" experiments, cDNA encoded proteins that interact in the presence or the absence of the particular small molecule can be isolated.

Furthermore, the interactor that is selected by its ability to bind to the bait protein can be used as a bait in another interaction cloning experiment allowing the identification of proteins that physically interact and may form regulatory pathways.

Finally, the most intriguing and potentially the most significant application of this technique is to isolate small peptides that bind to a particular domain of the protein of interest and thus prevent the complex formation between known protein partners. The use of peptide libraries in interaction-based cloning systems will create a battery of new experimental approaches that will help to refine structural and functional relationships within and between proteins.

References

Ausubel FM, Brent R, Kingston R, Moore D, Seidman JS, Struhl K (1987–1994) Current protocols in molecular biology. John Wiley & Sons New York

Breeden L, Nasmyth K (1985) Regulation of the yeast HO gene. Cold Spring Harbor Symp Quant Biol 50:643–650

Brent R, Ptashne M (1985) A eukaryotic transcriptional activator bearing the DNA specificity of a prokaryotic repressor. Cell 43:729–736

Cadwell C, Joyce G (1992) Randomization of genes by PCR mutagenesis. PCR Methods and Applications 2:38–33

Chien CT, Bartel PL, Sternglanz R, Fields S (1991) The two-hybrid system: a method to identify and clone genes for proteins that interact with a protein of interest. Proc Natl Acad Sci USA 88:9578–9582

Chiu I, Katz H, Berlin V (1994) RAPT1, a mammalian homolog of the yeast TOR, interacts directly with the FKB12/rapamycin complex. Proc Natl Acad Sci 91:12574–12578

Dalton S, Treisman R (1992) Characterization of SAP-1, a protein recruited by serum response factor to the c-fos serum response element. Cell 68:597–612

Draetta G (1990) Cell cycle control in eukaryotes: molecular mechanisms of cdc2 activation. Trends Biochem Sci 15:378–382

Fields S, Song O (1989) A novel genetic system to detect protein-protein interaction. Nature 340:245–246

Gietz D, St Jean A, Woods RA, Schiestl RH (1992) Improved method for high efficiency transformation of intact yeast cells. Nucl Acid Res 20:1425

Green N, Alexander H, Olson A, Alexander S, Shinnick TM, Sutcliffe JG, Lerner RA (1982) Immunogenic structure of the influenza virus hemagglutinin. Cell 28:477–487

Gyuris J. Golemis E, Chertkov H, Brent R (1993) Cdi1, a human G1 and S phase protein phosphatase that associates with Cdk2. Cell 75:791–803

Hannon G, Demetrick D, Beach D (1993) Isolation of the Rb-related p130 through its interaction with Cdk2 and cyclins. Genes & Development 7:2378–2391

Hannon G, Casso D, Beach D (1994) KAP: a dual specificity phosphatase that interacts with cyclin-dependent kinases. Proc Natl Acad Sci 91:1731–1735

Harper JW, Adami GR, Wei N, Keyomarsi K, Elledge SJ (1993) The p21 Cdk-interacting protein Cip1 is a potent inhibitor of G1 Cyclin-dependent kinases. Cell 75:805–816

Hoffman CS, Winston F (1987) A ten minute DNA preparation from yeast efficiently releases autonomous plasmids for transformation of E. coli. Gene 57:267–272

Kalderon D, Roberts BL, Richardson WD, Smith AE (1984) A short amino acid sequence able to specify nuclear localization. Cell 39:499–509

Kamens J, Richardson P, Mosialos G, Brent R, Gilmore T (1990) Oncogenic transformation by vRel requires an amino-terminal activation domain. Mol Cell Biol 10:2840–2847

Lee JW, Ryan F, Swaffield JC, Johston SA, Moore DD (1994) Ligand-dependent interaction of a thyroid and retinoic X receptors with a new member of a conserved family of ATPases. Nature 374:91–94

Ma J, Ptashne M (1987) A new class of yeast transcriptional activators. Cell 51:113–119

Ma J, Ptashne M (1988) Converting a eukaryotic transcriptional inhibitor into an activator. Cell 55:443–446

Oliner, JD, Pietenpol JA, Thiagalingam S, Gyuris J, Kinzler KW, Vogelstein B (1993) Oncoprotein MDM2 conceals the activation domain of tumor suppressor of p53. Nature 362:857–860

Ruden DM, Ma J, Li Y, Wood K, Ptashne M (1991) Generating yeast transcriptional activators containing no yeast protein sequences. Nature 350:426–430

Schiestl RH, Gietz RD (1989) High efficiency transformation of intact yeast cells using single-stranded nucleic acids as a carrier. Curr Genet 16:339–346

Serrano M, Hannon G, Beach D (1993) A new regulatory motif in cell cycle control causing specific inhibition of cyclin D/CDK4. Nature 366:704–707

Sherr CJ (1993) Mammalian G1 cyclins. Cell 73:1059–1065

Triezenberg SJ, LaMarco KL, McKnight SL (1988) Evidence of DNA:protein interactions that mediate HSV1-immediate early gene acitivation by VP16. Genes Dev 2:730–742

Vojtek AB, Hollenberg SM, Cooper JA (1993) Mammalian Ras interacts directly with the serine/threonine kinase Raf. Cell 74:205–214

Xiong Y, Zhang H, Beach D (1993) Subunit rearrangement of the cyclin-dependent kinasesis is associated with cellular transformation. Genes Dev 7:1572–1583

20 Expression of Cell Cycle Proteins Using Recombinant Baculoviruses

GREGORY J. HANNON

1 Background, Principle, and Applications

The ability to express proteins in heterologous systems has been crucial to the biochemical analysis of many cell cycle regulators. Until recently, bacterial expression systems have reigned supreme due to their ease of use and their ability to produce vast quantities of a desired protein. However, many proteins are recovered from bacteria in an inactive form due either to improper folding or to lack of critical covalent modifications (e.g., phosphorylation or glycosylation).

This chapter discusses an alternative expression system in which recombinant baculoviruses are used to direct protein expression in cultured insect cells. This system shares with bacteria the ability to express foreign proteins to high levels with final yields ranging from 0.1 to 50% of total cellular protein. However, the baculovirus expression systems far exceed bacterial systems in the proportion of proteins which are expressed in a native conformation. Many post-translational modifications including phosphorylation, N- and O-linked glycosylation and specific proteolytic cleavages are properly executed in infected insect cells. Furthermore, undesired proteolytic cleavages that are so often a problem in bacterial expression are minimized. In bacterial expression systems, proteins of more than 100 kDa are rarely produced intact. However, baculovirus systems are not subject to similar size limitations. Finally, multiple genes can be expressed simultaneously to form multiprotein complexes within infected insect cells. This can be accomplished either through coinfection with several different recombinant viruses (allowing easy alteration of individual components) or through the use of vectors which direct the expression of multiple proteins from a single recombinant virus (giving maximal yield of a particular complex).

Baculovirus expression systems also have several drawbacks in comparison with bacterial systems. The first is cost. Culture of insect cells is equivalent in cost to the maintenance of mammalian cells which, while not prohibitive, is certainly greater than the cost of growing bacterial cultures. The second disadvantage of the baculoviral systems is speed. After construction of the baculovirus transfer vector, it requires roughly 2 weeks to generate a sufficient stock of high titer baculovirus for protein production, whereas 1 day is sufficient to generate the equivalent bacterial expression strain. For these reasons, baculovirus systems will probably never supplant bacterial expression systems but will instead serve as the source for many proteins or protein complexes for which bacterial expression is unsuitable.

Although baculoviruses form a large family of double-stranded DNA viruses, the term baculovirus, for the purposes of this chapter, refers to the best studied strain, *Autographa californica* nuclear polyhedrosis virus (AcNPV). Infectious AcNPV particles enter susceptible host cells, and viral DNA is un-coated in the nucleus where all viral transcription and replication occur. At roughly 6 h postinfection, viral DNA synthesis begins. Initially (approx. 10 h postinfection), viral particles are released into the culture media by budding from the host cell membrane. Later in the course of infection (after about 26 hours), viral particles accumulate in the nucleus, embedded within a matrix consisting predominantly of a single protein, the polyhedrin protein. This aggregate of polyhedrin protein and viral particles is visible in the light microscope and is known as an occlusion body. Disruption of the polyhedrin gene which is not essential for virus propagation in tissue culture originally served as a marker for distinguishing "occlusion-negative" recombinant viruses from "occlusion-positive" nonrecombinants. Fortunately, this somewhat cumbersome method has been greatly simplified by recent advances in vector design (see below).

Recombinant baculoviruses have proven essential for the biochemical and physical analysis of many cell cycle regulators (see for example Desai et al. 1992; DeBondt et al. 1993; Parker et al. 1993; Rosenblatt et al. 1993). For example, expression of Cyclin D/Cdk4 complexes in insect cells resulted in the first convincing demonstration of Cdk4 kinase activity and provided the first evidence that the retinoblastoma gene product (Rb) might be an in vivo substrate of these kinases (Matsushime et al. 1992). In addition, experiments in insect cells provided a guide to determining conditions for assay of Cdk4 kinase activity in mammalian cells (Matsushime et al. 1994). Recombinant baculoviruses have also been useful for analyzing the properties of Cdk activating phosphatases and inhibitory kinases (see for example Desai et al. 1992; Parker et al. 1993). Recently, insect cell systems have proven critical to the analysis of a two distinct families of cyclin kinase inhibitors (Xiong et al. 1993; Serrano et al. 1993; Zhang et al. 1994). Finally, interactions between the Rb family of pocket proteins and various cyclins and cdk kinases have been analyzed using baculoviral systems (Hannon et al. 1993; Dowdy et al. 1993; Ewen et al. 1993; Kato et al. 1993). In this chapter, I will provide an overview of the methods used in the construction and maintenance of recombinant baculoviruses. More detailed information can be obtained in a recent manual devoted solely to baculoviral expression systems (O'Reilly et al. 1992). I will also provided sample protein expression protocols that will hopefully serve as a useful guide in designing protocols specific to individual needs.

2 Methods

Materials

The facilities needed for the production and propagation of recombinant baculoviruses are essentially similar to those needed for the propagation of mammalian cells (e.g., laminar flow hood, standard tissue culture plastic ware, etc.). The following additional materials are necessary:

- Refrigerated low temperature incubator. Insect cells are maintained at 27 °C. Although this is higher than room temperature, standard incubators cannot be reliably maintained below 28.5–29 °C. An electrical outlet inside the incubator is also useful (for suspension culture), but CO_2 is unnecessary.
- Baculovirus expression kit. Although it is possible to create recombinant viruses from publicly available materials, I strongly recommend the purchase of a kit which supplies the expression vectors and, most importantly, the linear baculovirus DNA for use in transfections (see below). We routinely use the kit from Pharmingen (San Diego, CA). As of the writing of this chapter, the kit costs approximately $ 300 and contains reagents sufficient for the production of about 50 different recombinant viruses. Similar kits are also available from a number of other manufacturers.

2.1 Growth and Maintenance of Insect Cells

A number of insect cell lines are available for use with recombinant baculoviruses. The most common, derived from *Spodoptera frugiperda*, are Sf9 and Sf21. Other proprietary cell lines are available (e.g., Hy5 and tNiSel). These claim large increases in protein production, and in some cases these claims have been verified. However, I recommend that virus production and initial experiments be done in the more common cell lines and that proprietary lines be tested only if increased yields of protein are essential. Sf9 and Sf21 must be maintained in a medium designed for insect cell culture. We routinely use Graces' insect media from Gibco BRL that can be purchased as a liquid or in dry form.

If media is reconstituted from powder, particular attention should be paid to its pH, since media components precipitate and cells fail to grow if the media deviates from pH 6.2. Graces' media must be supplemented prior to use with 10% heat inactivated fetal bovine serum, lactalbumin hydrolysate, yeastolate ultrafiltrate, and 50 µg/ml gentamicin sulfate. All of these components are also available from Gibco. Equivalent insect cell media preparations are available from a number of sources. It should be noted that defined, protein free culture media have recently been developed. These are particularly useful for the production and purification of secreted proteins. For all operations, insect cell media should be equilibrated to room temperature prior to use.

Monolayer culture After first reviving insect cells from long term storage, they should be maintained in monolayer culture until sufficient cells are available for transition to suspension culture. Insect cell maintenance is similar to mammalian cell culture with a few important exceptions. Insect cells adhere loosely to plastic dishes, and care must be taken to avoid dislodging cells during routine operations. Liquid should be added to a monolayer culture by tipping the dish and pipetting onto the lower edge of the plate. Cells should be subcultured at a 1 : 3 to 1 : 5 split ratio several times per week (the doubling time of Sf9 cells is approximately 18–24 h). Cells can be removed from the plate by simply pipetting a direct stream of media onto the cell monolayer or by scraping cells from the plate with a disposable cell scraper. Trypsin is unnecessary and should be avoided. One should bear in mind that insect cells are fragile and that the healthiest cell population will result from a minimum of handling. Healthy cells have a uniform, rounded appearance. Presence of some cells in the culture media is normal, but these cells should not exceed 2–5% of the total cell population. Unhealthy cultures are marked by reduction in growth rate and by cells which flatten and attach firmly to the plate.

Suspension culture Once suitable numbers are achieved, cells should be transferred to suspension culture for maintenance. Suspension cultures are not only easier to maintain than monolayer cultures but also contain generally healthier cells. Suspension cultures should be maintained at 27 °C in spinner flasks with the stirrer rotating at approx. 50 rpm. Initial seeding density should be 1–5×10^5 cells/ml, and mature cultures should be maintained at densities between 0.5 and 2×10^6 cells/ml. Cultures should be transferred to new flasks approximately once per month.

Long-Term storage Insect cells can be stored frozen in liquid nitrogen exactly as mammalian cells. A mixture of 90% heat inactivated fetal calf serum and 10% DMSO is a suitable freezing medium.

2.2 Production of a Recombinant Baculovirus

Transfer vectors Generation of a recombinant baculovirus involves a recombination event between a linear baculoviral genome and a baculoviral transfer vector containing the target gene. Available baculoviral transfer vectors are nearly as varied as bacterial expression plasmids. In general, two major choices must be made. The first is whether the protein should be produced in its native form or as a fusion protein. Second, one must choose the promoter for protein production.

Using commercially available vectors, protein fusions with glutathione-S-transferase or a poly-histidine tag can be constructed. These allow essentially single-step purification on glutathione Sepharose or Ni+-NTA-agarose, respectively. In most cases, an engineered protease cleavage site allows removal of the tag from which the native protein can then be purified. Vectors which direct secretion of recombinant proteins are also available. This can provide a particularly effective purification scheme when coupled with the protein free insect cell media which have recently become available.

By far the most common promoter used in baculoviral expression systems is the polyhedrin promoter. This promoter is activated very late in the infection cycle and tends to produce the highest yield of recombinant protein. While suitable for most purposes, polyhedrin based vectors sometimes fail to produce active protein if extensive posttranslational modification is required. In these cases, a promoter which is active earlier in the infection cycle (e.g., the 39 K promoter) may be a better choice. Several multiple promoter vectors are also available for use in the production of multicomponent complexes. For example, up to three subunits of a complex may be expressed from a single virus, eliminating the need for careful titration of three independent viruses for multiple infection.

We have mainly utilized baculovirus transfer vectors from Pharmingen (San Diego, CA), but a variety of vectors are also available from other sources (e.g., Clontech, La Jolla, CA; Kodak, Rochester, NY). Specific information concerning transfer vectors should be obtained from their manufacturers.

Transfection of insect cells

Creation of a recombinant baculovirus requires two components. A baculoviral transfer vector containing the gene of interest and a linear baculoviral genome. These are cotransfected into insect cells, and viruses which direct expression of the gene of interest are created through a recombination between the two DNAs. The most convenient systems (e.g., those from Pharmingen and Clontech) utilize linear baculoviral DNA which contains a lethal deletion. Therefore viral particles can only be produced following successful recombination. These systems claim (and our experience supports their claims) that >99.9% of viral particles produced in the transfection direct the expression of the desired protein. This innovation greatly reduces the labor needed to produce a recombinant virus and has essentially eliminated the need for plaque purification of viruses in many cases. Other systems make use of a beta-galactosidase marker for identifying recombinant viruses which then require plaque purification.

Transfection of insect cells is done in 12-well tissue culture dishes by a calcium phosphate procedure. Lipofection procedures should also be suitable for virus production. The protocol listed below allows the production of approx. 50 viruses from a single baculovirus transfection kit, rather than the 5 viruses which can be produced using the manufacturers' instruction set.

Creation of a recombinant virus

1. Seed 1 × 10^6 Sf9 cells from a suspension culture into one well of a 12-well tissue culture dish.

2. Allow the cells to attach for ca. 15 min to 1 h at room temperature (if cells do not attach in this period of time, they are probably unhealthy and should not be used.

3. During this period, prepare the DNA for transfection.
   - Dilute 10 µg of transfer vector into sterile water to a final volume of 50 µl (the transfer vector should be equivalent to CsCl-gradient purity).
   - Add 50 µl of chloroform and vortex.
   - Spin for ca. 1 min at top speed in a microcentrifuge. *Note.* Do not open the tube outside the tissue culture hood.

   These steps eliminate many contamination problems by sterilizing the transfer vector preparation. Other sterilization methods (e.g., ethanol precipitation and resuspension in the hood in sterile water) are equally effective.

4. Mix 1 µl (200 ng) of the transfer vector with 50 ng of linear baculoviral DNA in 100 µl of transfection buffer B [25 mM Hepes, pH 7.1 (+/ 0.05), 125 mM $CaCl_2$, 140 mM NaCl].

5. Remove media from the attached cells and replace with 100 µl of Graces' insect media containing 10 % heat inactivated FBS (transfection buffer A).

6. Wait 5 min, then add the DNA solution dropwise to the cells while rocking the plate. The fine calcium phosphate precipitate which forms at this step may make the solution appear cloudy.

7. Place the 12-well plate inside a zip-lock (or equivalent bag) with several moist paper towels. This creates a humidified atmosphere and prevents the transfection from drying.

8. Incubate the bag at 27 °C for 4 h.

9. Gently remove the transfection solution by aspiration and replace it with 1 ml of complete Graces' media.

10. Replace the plate in the humidified bag and incubate for 4–5 days at 27 °C.

11. Collect the culture media and clarify by centrifugation at 1000 *g* for 5 min.

Amplification of viral stocks

Amplification of virus occurs in two stages. The first amplification is performed in a 60-mm tissue culture dish and yields approximately 4 ml of ~2–5 × 10^6 plaque forming unit (pfu)/ml virus. The second amplification takes place in a 10 cm dish and generally results in the production of 10 ml of 1–2 × 10^8 pfu/ml virus. Subsequent amplifications for propagation of high titer viral stocks can be performed essentially as described below, scaling cell numbers and media volumes to the surface area of the amplification plate. It is essential to begin viral amplifications at a multiplicity of infection (m.o.i.) of less than 1. High m.o.i. amplifications tend to select for defective deletion viruses which fail to produce recombinant protein and thus reduce overall yield. If this problem is suspected, then re-amplification at low m.o.i. should eliminate the problem.

Primary amplification

1. Seed 5 × 10^6 insect cells into a 60 mm tissue culture plate.
2. After cells have attached (~15 min) remove the media and add ~4–5 ml of fresh media.
3. Add half of the supernatant removed from the original transfection.
4. Incubate for 4 days at 27 °C.
5. Collect the culture media and clarify by centrifugation. Viral titer should now be ~2–5 × 10^6 pfu/ml.

Secondary amplification

1. Seed 8 × 10^6 insect cells into a 10-cm tissue culture dish.
2. After cells have attached, add half of the primary viral amplification.
3. Incubate at 27 °C for 4 days.
4. Clarify the final viral stock. The titer should be ~1 x 10^8 pfu/ml.

Determination of viral titers and plaque purification

Since the baculoviral transfection systems are so efficient and so reproducible, titering and plaque purification can generally be skipped until viruses are tested using an assumed titer of 1 × 10^8 pfu/ml. However, maximal reproducibility and optimal protein production often require titering and plaque purification.

1. Seed ca. 7 × 10^6 cells onto a 10-cm plate (ca. 5 plates per virus to be titered).
2. Change to fresh media and incubate for 2 h at 27 °C.
3. Prepare dilutions of the virus in Graces' medium. About 200 plaques per plate is a convenient number for counting, so dilutions ranging from 10^{-3} to 10^{-6} are common.
4. Add an appropriate amount of virus to each plate and incubate for a further 1 h while rocking at room temperature.

5. Prepare an agarose overlay which prevents viral spread and allows visualization of plaques: dissolve (by microwave or boiling) low melting point agarose in protein-free medium to a final concentration of 1.6–2%. Cool to 45 °C in a water bath. Add an equal volume of prewarmed complete medium to the agar solution. The final concentration of low melting point agarose in the overlay should be between 0.8 and 1%.
6. Remove the media from each titer plate and gently overlay 10 ml of the agarose solution by pipetting into the bottom of a tipped dish.
7. Incubate plates undisturbed on a level surface for ~20–30 min while the agar hardens.
8. Incubate at 27 °C in a humidified environment. Visible plaques become apparent in ca. 6 days. Plaques are sometimes more visible if plates are held against a dark background.
9. If plaque purification is desired, individual plaques can be isolated using a sterile pipette. Virus can be eluted from the agar by incubating the plaque in 500 µl of medium for several hours at room temperature. Amplification of the purified plaque to create a high titer virus stock can be accomplished as described above.

Storage of virus Viral supernatants can be stored for about 6 months at 4 °C if protected from light. For longer term storage, supernatants can be frozen directly at 80 °C.

2.3 Production of Recombinant Protein

Production of recombinant protein is accomplished by infection of an insect cell host with the recombinant virus. Optimal multiplicity of infection may differ for individual viruses, but a good starting point is an m.o.i. of ca. 5–10. Time of incubation following infection can also dramatically affect protein yield, and several different time points should be examined. Generally, maximal production is reached at 48–72 h postinfection. Protein production can be measured in any of a number of ways depending upon the availability of reagents. Proteins that are produced at high levels can generally be visualized in whole cell lysates prepared derived from ^{35}S-labeled cells. In many cases, the protein band can even be visualized by Coomassie staining.

One great advantage of baculoviral systems is the ability to express multiple proteins in the same cell by coinfection with several viruses. Simultaneous infection with multiple viruses is usually sufficient for producing moderate amounts of multicomponent complexes. However, multiple infections generally reduce overall protein yield, and the optimal m.o.i. ratios may need to be determined if large quantities of specific complexes are required. In these cases the use of multi promoter vectors should also be considered.

2.4 Sample Protocols

Small-Scale Production of Labeled Proteins in Insect Cells (e.g., for Immunoprecipitation)

Protocol A

1. Seed 1 × 10^6 Sf9 cells in one well of a 12-well tissue culture plate.
2. After the cells have attached, replace the media with 1 ml of fresh Graces' media.
3. Add viral stock(s) to the media to achieve a m.o.i. of ca. 5–10 per virus, or the optimal m.o.i. for each specific virus if one has been determined.
4. Incubate at 27 °C for 48–72 h.
5. Aspirate media and GENTLY rinse once with 1 ml of labeling media Labeling media: Graces' insect cells media minus methionine (Gibco BRL), dialyzed fetal bovine serum (Gibco BRL). Do not add lactalbumin hydrolysate or yeastolate ultrafiltrate. Antibiotics are optional but recommended.
6. GENTLY overlay cells with ca. 1 ml labeling media.
7. Incubate at 27 °C for 30 min.
8. Add ^{35}S-methionine to ~0.5 mCi/ml and incubate at 27 °C for 2–6 h.
9. Aspirate labeling media and GENTLY rinse once with PBS.
10. Lyse cells as required for the next step of the procedure (see chapter by Jenkins and Xiong, Part Four).

Large-Scale Production of a Poly-Histidine Tagged Protein in Suspension Culture

Large-scale production of recombinant proteins can be done either in monolayer or in suspension culture, and each method has its proponents. The best approach is to determine which works best for a specific protein. Monolayer cultures, while more work, are more forgiving. Suspension cultures have the advantages of lower cost and ease of handling. However, for protein production, cells in suspension culture must be in top condition. Lysis and purification protocols are specific to the desired product and purification method. The example protocol given below was suitable for the expression and purification of a poly-histidine tagged fusion version of p16^{Ink4} protein for biochemical analysis (Serrano et al. 1993).

Protocol B

1. Collect insect cells from 1 l of suspension culture (2.5 × 10^6 cells/ml) by centrifugation at 800 *g* (insect cells are sensitive to centrifugation and should be handled carefully).

2. Resuspend gently in 250 ml of fresh media. Use protein-free media if a secreted protein is being purified.

3. Add virus to the optimal m.o.i. Determination of optimal multiplicity is crucial for obtaining the highest levels of expression.

4. Incubate at 27 °C, stirring at 50–60 rpm, for 48–72 h. Again, determination of optimal expression time is worthwhile.

5. Harvest the cell pellet by centrifugation at 800 *g*.

6. Gently resuspend the cell pellet in PBS and again collect by centrifugation.

7. Resuspend the cell pellet in hypotonic lysis buffer (ca. four times the volume of the pellet).

Hypotonic lysis buffer
20 mM Tris, pH 8.0
1 mM EDTA
5 mM ß-mercaptoethanol
Protease inhibitors

8. Incubate on ice for 10 min.

9. Lyse the cells in a dounce homogenizer using the tight pestle. Approximately 5–15 strokes should be sufficient. Cell lysis should be >90% and can be monitored microscopically.

10. Add NaCl to 300 mM and remove debris by centrifugation at 20 000 *g* for 10 min.

11. Remove the clear supernatant and pass over a 1–2-ml column of Ni+-NTA agarose (available from Qiagen) which has been prewashed in loading buffer.

12. Wash with 10 column volumes of loading buffer.

13. Wash with 10 column volumes of loading buffer containing 25 mM imidazole.

14. Elute with ten 1-ml steps of loading buffer containing 100 mM imidazole. The peak of protein is generally in fractions 2–5.

Substantial purification should be obtained by this procedure, depending on the expression level of the fusion protein. Purity of ~50% is common. The yield from this purification will vary from several hundred micrograms to several milligrams.

Active Cyclin D1/Cdk4 Kinase from Insect Cells

Insect cells have been particularly useful for the biochemical analysis of Cyclin dependent kinases and their regulators (Desai et al. 1992; Matsushime et al. 1992; Serrano et al. 1993; Xiong et al. 1993, Zhang et al. 1994). Particularly in the case of Cyclin D/Cdk4, complexes baculoviral systems were crucial for identifying potential kinase substrates and for determining conditions for assaying complexes from mammalian cells (Matsushime et al. 1992, 1994). The protocol given has been useful for all Cyclin/Cdk pairs tested to date.

Protocol C

1. Seed 3 x 10^7 insect cells in a 15-cm tissue culture plate.
2. After cells have attached, replace the media with 25 ml of fresh media.
3. Add Cyclin D1 virus at an m.o.i. of ca. 5 and Cdk4 virus at an m.o.i. of ca. 10. Other Cyclin/Cdk pairs will show different optimal ratios of virus.
4. Incubate at 27 °C for 60 h.
5. Cells can be labeled with ^{35}S-methionine at this point or can remain unlabeled (see above for the labeling protocol).
6. GENTLY rinse the monolayer twice with kinase buffer.

Kinase buffer
20 mM Tris pH 8.0
10 mM $MgCl^2$
1 mM EGTA

7. Tilt the plate and drain for ca. 1 min. Remove all residual buffer by aspirating.
8. Add 1 ml of kinase buffer + protease inhibitors to the plate.
9. Suspend the cells in the 1 ml of buffer using a disposable cell scraper.
10. Collect the suspension into a 3-ml disposable plastic tube.
11. Lyse the cells by repeated passage through a 26-g needle (about 15–20 strokes is sufficient; lysis can be monitored microscopically).
12. Split the lysate into two microcentrifuge tubes and clear the supernatant by two sequential 10 min centrifugation at top speed.
13. Add glycerol to ~10% and store at 80°C.
14. Assay for Cyclin kinase activity can be done in whole lysate or in immunoprecipitates according to protocols described in the chapters by Connell-Crowley and Harper, Part Four, P. R. Clarke, Part Two, and Fesquet et al., Part Four (see also: Matsushime et al. 1992; Xiong et al. 1993, Serrano et al. 1993).

References

DeBondt HL, Rosenblatt J, Jancarik J, Jones HD, Morgan DO, Kim SH (1993) Crystal structure of Cyclin dependent kinase 2. Nature 363:595–602

Desai D, Gu Y, Morgan DO (1992) Activation of human Cyclin dependent kinases in vitro. Mol Biol Cell 3:571–582

Dowdy SF, Hinds PW, Louie K, Reed SI, Arnold A, Weinberg RA (1993) Physical interaction of the retinoblastoma protein with human D-type Cyclins. Cell 73:499–511

Ewen ME, Sluss HK, Sherr CJ, Matsushime H, Kato J, Livingston DM (1993) Functional interactions of the retinoblastoma protein with mammalian D-type Cyclins. Cell 73: 487–497

Hannon GJ, Demetrick D, Beach D (1993) Isolation of the Rb-related p130 through its interaction with Cdk2 and Cyclins. Genes Dev 7:2378–2391

Kato J, Matsushime H, Hiebert SW, Ewen ME, Sherr CJ (1993) Direct binding of Cyclin D to the retinoblastoma gene product (pRb) and pRb phosphorylation by the Cyclin D dependent kinase Cdk4. Genes Dev 7:331–341

Matsushime J, Ewen ME, Stom DK, Kato JY, Hanks SK, Roussel MF, Sherr CJ (1992) Identification and properties of an atypical catalytic subunit ($p34^{J3/Cdk4}$) for mammalian D type G1 Cyclins. Cell 71:323–334

Matsushime H, Quelle DE, Shurtleff SA, Shibuya M, Sherr CJ, Kato JY (1994) D type Cyclin dependent kinase activity in mammalian cells. Mol Cell Biol 14:2066–2076

O'Reilly D, Miller L K, Luckow VA (1992) Baculovirus expression vectors, a laboratory manual. W.H. Freeman, New York, NY

Parker LL, Walter SA, Young PG, Piwnica-Worms H (1993) Phosphorylation and inactivation of the mitotic inhibitor Wee1 by the Nim1/Cdr1 kinase. Nature 363:736–738

Rosenblatt J, DeBondt H, Jancarik J, Morgan DO, Kim SH (1993) Purification and crystallization of human Cyclin dependent kinase 2. J Mol Biol 230:1317–1319

Serrano M, Hannon GJ, Beach D (1993) A new regulatory motif in cell cycle control causing specific inhibition of Cyclin D/Cdk4. Nature 366:704–707

Xiong Y, Hannon GJ, Zhang H, Casso D, Kobyashi R, Beach D (1993) p21 is a universal inhibitor of Cyclin kinases. Nature 366:701–704

Zhang H, Hannon GJ, Beach D (1994) p21-containing Cyclin kinases exist in active and inactive conformations. Genes Dev 8:1750–1758

21 Cell Cycle-Regulated Transcription Factors: Bandshift Assays

PIDDER JANSEN-DÜRR

1 Background

Transcriptional control plays an important role in the regulation of cell cycle progression in eukaryotes. Studies in yeast have shown that a family of transcription factors which controls the expression of several cell cycle regulated genes, in particular genes involved in DNA synthesis, tightly regulates the commitment of a cell to enter into S phase (reviewed by Andrews and Mason 1993). Although in higher eukaryotes, similar control mechanisms are just beginning to be discovered and understood, several cell cycle-regulated transcription factors have been already described. This chapter presents as an example the bandshift analysis of the transcription factor E2F, which appears to be centrally involved in control of the G1/S transition in mammalian cells (Nevins 1992) and may serve as a paradigm for cell cycle regulated transcription factors. Indeed it has been shown that:

- Expression of E2F in quiescent cells induces S phase entry.
- Different forms of E2F are present at various stages of the cell cycle (Mudryj et al. 1991; Pagano et al. 1992a).
- E2F associates physically to several proteins of proven capacity as regulators of the cell cycle, like Cyclins, Cdks, the product of the retinoblastoma gene, p107, p130, etc. (Lees et al. 1992; Cobrinik et al. 1993; Schwarz et al. 1993).
- E2F is a target for several viral oncogenes that are known to override growth control mechanisms acting at the G1/S boundary (Phelps et al. 1991).

Bandshift assays have played an important role in most of the above listed discoveries. Furthermore, due to the relatively easy experimentation required, this technique has proved useful to carry out large series of experiments addressing other specific questions like posttranslational modifications of DNA binding proteins (Bagchi et al. 1989); footprint, and methylation interference experiments can be carried out to delineate the details of a given DNA-protein interaction (Jansen-Dürr et al. 1989). In combination with protocols for the preparation of protein extracts from very low quantities of cells (or tissue samples), this technique allows a molecular analysis of many physiological processes that may be very difficult to study by other biochemical methods. In particular, the common need to perform long time course experiments, familiar to most, if not all, cell cycle researchers, makes this technique an attractive alternative to other techniques demanding higher amounts of starting material.

2 Principle and Applications

The basic principle of the method exploits changes of the electrophoretic mobility of a DNA fragment in a nondenaturing gel, that are observed when the fragment is bound by a protein(s) with sufficient affinity to stick to the DNA fragment during electrophoresis. Since only the DNA fragment is radioactively labeled in such an experiment, the presence of a binding protein is deduced from a change of the mobility of the labeled fragment in the gel. DNA-protein complexes display lower mobility than the unbound ("free") probe; thus, the DNA band appears shifted up in the gel as a result of protein binding ("bandshift").

In general, proteins can bind to DNA fragments in two different modes:

- *"Nonspecific" interactions*, which would occur with any DNA fragment irrespective of particular nucleotide sequences:
  Proteins with such properties, usually characterized by a high proportion of basic amino acids, may actually have very high affinity to DNA (e.g., histones) and give rise to very stable complexes in bandshift experiments. However, in most cases, such complexes can be eliminated by the addition of defined quantities of "random" DNA. This excess random DNA, which carries no radioactive label, titrates the nonspecific DNA binding proteins, thereby removing such proteins from the labeled probe. Several types of random DNA have been used; for most applications it may be sufficient to use double-stranded salmon sperm DNA, physically sheared to an average fragment length of a few kilobases. Alternatively, different poly-desoxyribonucleotides can be used, poly dI.dC and poly dA.dT being most frequently applied. The sequence composition of this "random" DNA may be important and its influence should be evaluated in every particular case, since experimental results may vary considerably with different preparations of "random" DNA. It has been shown, for example, that detection of E2F complexes is straightforward in the presence of poly dA.dT, while the use of poly dI.dC nearly precludes detection of this factor. Most likely, this is due to the fact that poly dI.dC shows sufficient homology to the CGCGCG motif inherent to the E2F binding site.

- *Sequence-specific binding of proteins to particular DNA motifs.*
  This is the property of cellular transcription factors which accounts for the high versatility of transcriptional control of gene expression in all living organisms. Proteins involved in transcription control, but also in DNA replication and other DNA-mediated processes, "recognize" a particular nucleotide sequence (e.g., TTTCGCGCG in the case of E2F) and bind to it with high affinity. In contrast, they have much reduced affinity to other sequences (including "random" DNA) or even to the point mutated motif (e.g., TTTATCGCG in the case of E2F). Since it has been often reported that binding sites for distinct factors are in close vicinity or even physically overlap, the use of defined point mutations is mandatory to identify the actual DNA binding protein being studied in a particular experiment.

The first step of a bandshift experiment is the identification of possible specific complexes among the huge number of bands seen under varying experimental conditions. To this end, increasing amounts of "random" competitor DNA should be added to a fixed amount of protein extract (usually we start with 5-10 μg of whole cell extract or 1–5 μg of nuclear extract) prior to gel electrophoresis. Several conditions, each displaying a distinct banding pattern, are then selected for the follow-up experiment.

In a second step, selectivity of binding to a particular DNA sequence motif is assayed by addition to the binding reaction of increasing amounts of the unlabeled DNA fragment used to generate the labeled probe (homologous competitor). In a parallel experiment, the same quantities of a similar DNA fragment of divergent sequence (heterologous competitor; the use of defined point mutations is advisable here, see above) are added. Bands which disappear at a given concentration of the homologous but not the heterologous competitor are due to specific binding to the motif under investigation. To account for possible variations in the concentration of DNA solutions and other experimental variations, I recommend considering only bands whose sensitivity to competition by the heterologous DNA fragment is at least ten times lower than sensitivity to competition by the homologous DNA fragment.

In a third step, the protein composition of specific bands, identified as described above, can be analyzed by the addition of specific antibodies to proteins of interest. In most cases, recognition by the antibody of a protein in a nucleoprotein complex will result in a further reduction of the electrophoretic mobility (supershift), since the antibody associates with the multiprotein complex and increases its molecular weight. In other cases, addition of an antibody leads to the disappearance of a given complex. It is assumed that the antibody interacts with an epitope of the antigen that is involved in complex formation, (thereby disrupting the interaction of the antigen with the other components of the complex) or it forms an immunocomplex that precipitates (cf. Pagano et al. 1992b)

3 Methods

3.1 Preparation of Labeled Probe

The best results for the analysis of single transcription factors are obtained with short synthetic oligonucleotides (around 25 bp) used as labeled probes. Such fragments are of excellent purity and usually can be labeled to very high specific activity. Labeling can be performed by either polynucleotide kinase and γ-[^{32}P]ATP (5' end labeling) or by Klenow DNA polymerase and any suitable α-[^{32}P]dNTP (3' end labeling). There are several advantages for labeling with Klenow polymerase, as compared to labeling with polynucleotide kinase:

- Oligonucleotides can be designed in a such way that several radioactive nucleotides can be incorporated into a single DNA molecule, increasing the specific activity.
- It is possible to design double-stranded oligonucleotides that will be labeled selectively at one strand; this option is important, if "footprinting" or methylation interference studies are planned (see below).
- We found that in some cases crude cellular extracts display considerable phosphatase activity, which may lead to a rapid loss of label from a 5' end-labeled probe.

When 5'-end labeling is used, it is also possible to selectively label one strand; in this case, just label the appropriate single strand of a double-stranded oligonucleotide, heat inactivate the polynucleotide kinase, and anneal with the unlabeled complementary strand.

Gel-purification of the double-stranded DNA fragment, subsequent to the labeling reaction, is recommended. In addition to providing a reliable measurement of labeling efficiency, this technique eliminates any contaminating radioactivity and undesired DNA that might interfere with the binding reaction, e.g., single-stranded oligonucleotides. After elution from the gel, the labeled probe can be stored at –20 °C in 50% glycerol, at a radioactive concentration of 5000–10 000 cpm/µl.

In some cases, especially when a larger promoter fragment is to be analyzed by bandshift techniques, it may be necessary to use restriction fragments as radioactive probes. Again, it is possible to obtain fragments labeled at either one or both single strands, depending on the application. A combination of restriction enzyme cleavage and Klenow labeling should be designed for every particular case. 5'end-labeling is another possibility; however, in this case the DNA ends have to be dephosphorylated. Usually, much lower specific activity was obtained with restriction fragments, as compared to synthetic oligonucleotides; thus, the use of such probes is not recommended.

3.2 Preparation of Extracts

High salt whole cell extracts

The following protocol works well for preparation of extract from cultured cells. A slight modification of the protocol, involving ultrasonic disruption of solid tissue samples as an additional step prior to the standard protocol, was successfully used to extract DNA binding proteins from tissue samples.

1. Wash cells in PBS on the dishes.
2. Scrape cells from the dishes with a rubber policeman and collect them in PBS; finally, collect cells by centrifugation for 1 min at 13 000 rpm in an Eppendorf centrifuge.
3. Remove the supernatant carefully; keeping the cell pellet on ice, add extraction buffer (1.5 times the volume of the pellet).
4. Suspend cells in the buffer by pipetting.
5. Keep 20 min on ice.
6. Freeze in –80 °C refrigerator for at least 20 min. Pellets can be kept at –80 °C for longer periods of time.
7. Let the cells thaw on ice for 20 min.
8. Rigorously vortex the suspension for 10 s; spin in Eppendorf centrifuge at 4 °C (10 min, 13 000 rpm), recover the supernatant (extract).
9. Measure protein concentration and freeze in liquid N_2 in small aliquots; 5–10 µg should be sufficient for one bandshift experiment.

Buffer

Extract Buffer
20 mM HEPES pH 7.9
25 % glycerol
0.4 M NaCl
1 mM EDTA
1 mM EGTA
1 mM DTT
1.2 mM PMSF

Phosphatase inhibitors where appropriate

Other extracts

In some cases, it may be preferable to work with nuclear rather than whole cell extracts. In principle, the methods described here are suitable for such extracts as well. However, we found that a distinct advantage of the extraction procedure described above may be the good conservation of protein-protein interactions. In our hands, such interactions are much less preserved in extracts prepared according to several protocols described by others, most likely because our protocol is designed to keep the extraction procedure as short as possible.

3.3 E2F Bandshift Experiment

Preparation of the gel

We use gel plates of the size 30 cm × 20 cm and 1.5-mm spacers, together with a 28-well comb (1-cm wells). Prepare about 80 ml of a solution containing 4.5% acrylamide/bis (29 : 1) and 0.3 × TBE. Add 500 μl of 10% ammonium persulfate and 50 μl of TEMED.

Pour the gel, insert the comb and keep the two glass plates tightly together (use clamps, alternatively use a full 2-l bottle to press glass plates together).

After 1 h at horizontal position, the gel (should be polymerized) is transferred to the cold room (+ 4 °C) and put in the gel apparatus, containing precooled 0.3 × TBE buffer (45 mM Tris-borate, 1 mM EDTA).

Wait for at least 1 more h to allow temperature adaptation. Remove the comb just prior to loading the samples, rinse the slots with 0.3 × TBE.

Binding reaction

Incubate cell extract (0.5–4 μl, to be titrated in each case) together with 2 mM $MgCl_2$ and sonicated salmon sperm DNA (100–2000 ng/reaction; to be titrated in each case) for 10 min at room temperature with 10 000 cpm of hot probe, contained in 2 μl of 50% glycerol (Ficoll is not recommended here, as it impairs the resolution of high MW complexes), in a total volume of 10 μl. Where appropriate, 0.5–2 μl of specific antibodies are added. In this case, continue the incubation 50 more min on ice.

Load the reaction to the **bottom** of the gel slot by using a flattened tip (e.g., from Stratagene). Avoid any perturbations of the loaded material at this stage; you should see the sample as a small layer, due to its high glycerol content (addition of dyes is not recommended, because they were found to perturb the banding pattern).

Electrophoresis

We found that the electrical parameters are absolutely critical, in order to obtain clear and reproducible banding patterns. Before loading the samples, check current. In our system, a voltage of 280 V should yield a current between 45 and 50 mA just prior to loading. (**Note** that the current will be higher if checked before the gel has cooled down!) We run the gels at 280 V for 30 min, followed by 320 V for 60 min. During electrophoresis, current decreases due to exhaustion of ions; at the end of the run it is usually between 20 and 30 mA. The gel should never warm up! (Check by touching with your hand.)

We follow electrophoresis by running bromophenol blue solution (0.5% in 50 % glycerol) in a free slot. At the end of electrophoresis, the dye (comigrating approximately with a labeled 25-bp fragment) is about 1 cm within the lower edge of the glass plate. The gel is fixed for 10 min in ethanol/acetic acid (10%/10%), transferred to Whatman DE 81 paper and dried on a gel dryer. The gel is then exposed for autoradiography.

References

Andrews BJ, Mason SW (1993) Gene expression and the cell cycle – a family affair. Science 261:1543–1544

Bagchi S, Raychaudhuri P, Nevins J (1989) Phosphorylation-dependent activation of the adenovirus-inducible E2F transcription factor in a cell-free system. Proc. Natl. Acad Sci USA 86:4352–4356.

Cobrinik D, Whyte P, Peeper DS, Jacks T, Weinberg RA (1993) Cell cycle-specific association of E2F with the p130 E1A-binding protein. Genes Dev 7:2392–2404

Jansen-Dürr P, Mondesert G, Kedinger C (1989) Replication-dependent activation of the adenovirus major late promoter is mediated by the increased binding of a transcription factor to sequences in the first intron. J Virol 63:5124–5132

Johnson D, Schwarz J, Cress D, Nevins J (1992) Expression of transcription factor E2F1 induces quiescent cells to enter S phase. Nature 365:349–352

Lees E, Faha B, Dulic V, Reed S, Harlow E (1992) Cyclin E/Cdk2 and Cyclin A/Cdk2 kinase associate with p107 and E2F in a temporally distinct manner. Genes Dev 6: 1874–1885

Mudryj M, Devoto SH, Hiebert SW, Hunter T, Pines J, Nevins JR (1991) Cell cycle regulation of the E2F transcription factor involves an interaction with Cyclin A. Cell 65: 1243–1253

Nevins JR (1992) E2F: A link between the Rb tumor suppressor protein and viral oncoproteins. Science 258:424–429

Pagano M, Dürst, M, Joswig S, Draetta G, Jansen-Dürr P (1992b). Binding of the human E2F transcription factor to the retinoblastoma protein but not to Cyclin A is abolished in HPV-16-immortalized cells. Oncogene 7:1681–1686

Pagano M, Draetta G, Jansen-Dürr P (1992a). Association of Cdk2 kinase with the transcription factor E2F during S phase. Science 255:1144–1147

Phelps WC, Bagchi S, Barnes JA, Raychaudhuri P, Kraus V, Münger K, Howley PM, Nevins JR.(1991) Analysis of trans activation by human papillomavirus type 16 E7 and adenovirus 12S E1A suggests a common mechanism. J Virol 65:6922–6930

Schwarz JK, Devoto SH, Smith EJ, Chellapan SP, Jakoi L, Nevins JR (1993) Interactions of the p107 and Rb proteins with E2F during the cell proliferation response. EMBO J 12:1013–1020

22 Immunoprecipitation and Immunoblotting in Cell Cycle Studies

CHRISTOPHER W. JENKINS AND YUE XIONG

1 Background, Principle, and Applications

Immunoprecipitation (IP) and immunoblotting (Western blotting) are potential tools for addressing questions about the expression of a protein and its interactions with other proteins. Immunoprecipitation and immunoblotting are, as their names imply, techniques dependent on immunological agents – namely antibodies. These techniques take advantage of the high specificity and tight binding that antibodies provide to separate the desired proteins out from the background of all the other cellular proteins present. Immunoprecipitation has long been used as a tool for assessing a protein's presence, while immunoblotting provides a more accurate means of quantitating the amount of the protein that is present. Immunoblotting also serves to establish or verify the identity of proteins isolated in immunoprecipitation. A vast number of other questions can also be addressed by these two related techniques.

- What is the abundance and the rate of protein synthesis?
- What is the relative molecular weight of the protein?
- Does the level of the protein change during the cell cycle?
- What other protein(s) interact with the antigen protein?
- Is the protein posttranslationally modified?
- If so, is the modification cell cycle dependent?

When combined with the use of different types of cells or tissues, with different conditions for culturing cells, or with the use of normal and transformed cells, the above questions can be extended to study the interactions between pathways controlling the cell cycle and pathways regulating cell differentiation, cell growth, and cell transformation.

When coupled with metabolic labeling and antigen competition, the immunoprecipitation technique is a potent tool for examining protein-protein interactions, provided an antibody that can co-immunoprecipitate proteins complexed with the antigen is available. The development of protein microsequencing techniques, requiring as little as 10 picomoles of protein to obtain sufficient peptide sequence information, has recently made it possible to isolate genes encoding proteins identified through co-immunoprecipitation by large-scale immunopurification (see for examples, Ewen et al. 1991; Settleman et al. 1992; Xiong et al. 1993b). Together with the yeast two-hybrid system (see the chapter by F. Fiore et al. Part Four), the immunoprecipitation-microsequencing method has become one of two powerful alternatives to classic yeast genetics for identifying genes involved in cell cycle control as well as other cellular pathways. This is especially the case in studies of mammalian cells, where conventional genetic isolation methods do not apply.

Using immunological methods, over a half dozen cell cycle regulatory proteins have been identified in mammalian cells in the last 2 years by virtue of their physical interaction with the primary cell cycle regulators: Cyclins and/or Cyclin-dependent kinases (Cdks). These include proliferating cell nuclear antigen (PCNA, Xiong et al. 1992, 1993a), the p53-regulated Cdk inhibitor p21 (Xiong et al. 1992, 1993a, 1993b; Zhang et al. 1993), the Cdk4 specific inhibitor $p16^{Ink4}$ (Serrano et al. 1993; Xiong et al. 1993a), the Cyclin A-associated p19 (Xiong et al. 1993a), the Cyclin E-Cdk2 inhibitor $p27^{Kip1}$ that mediates cellular response to treatment with the growth inhibitor TGF-β and to contact inhibition (Polyak et al. 1994), and a 28-kDa (p28) protein specifically associate with Cyclin D-Cdk6 complexes in human T lymphocytes following mitogenic activation by phytohemagglutinin (Meyerson and Harlow 1994). As more novel Cyclins and Cdks are discovered, more mitogenic pathways are characterized, and more cell lines and cell types are analyzed, the number of novel cell cycle regulatory proteins identified by virtue of their ability to physically interact with a Cyclin, a Cdk, or a Cyclin-Cdk complex will undoubtedly continue to rise.

A prerequisite for applying these techniques is the availability of antibodies suitable for immunoprecipitation and immunoblotting. Generation of antibodies against the C-terminal residues of proteins is described in the chapter by M. Pagano (Part Four). For the preparation of other kinds of antibodies, see Harlow and Lane (1988) or Asai (1993). The choice of antibodies for immunoprecipitation and immunoblotting can often be critical to obtaining good results, and the two procedures in fact are quite different in their requirements for an optimally useful antibody. Immunoblotting requires one good antibody that will recognize the denatured form of a protein, whereas immunoprecipitation usually requires an antibody which recognizes the native form of its antigen. An antibody that can form immunocomplexes with its native antigen without dissociating other protein(s) associated with the antigen would be the best choice for the studies of protein-protein interaction. Unfortunately, there is not a clear procedure to follow in generating such an ideal antibody and it is still largely an empirical practice. One tip from studies over the last several years for generating antibodies capable of reacting with the denatured proteins in immunoblotting and of immunoprecipitating native Cyclin-Cdk complexes is that antibodies against C-terminal residues of the various individual Cdk proteins and antibodies against whole proteins of the various Cyclins expressed and purified from bacteria generally work very well (Draetta and Beach 1988; Pines and Hunter 1989, 1990;

Koff et al. 1991; Matsushime et al. 1991, 1994; Elledge et al. 1992; Dulic et al. 1992; Pagano et al. 1992; Rosenblatt et al. 1992; Xiong et al. 1992, 1993a; Bates et al. 1994; Meyerson and Harlow 1994).

This chapter describes a detailed protocol for coupled [^{35}S] methionine metabolic labeling and immunoprecipitation (^{35}S-IP), and coupled immunoprecipitation and immunoblotting (IP-Western). This procedure was modified from previous methods based on our experiences in last few years of studying mammalian Cyclins and Cdks in normal human fibroblasts and should need only slight modifications when applied to other cells (e.g., epithelial cells) or when using different radioactive precursors (e.g., ^{32}P-orthophosphate metabolic labeling for examining protein phosphorylation). Other related techniques involving the use of antibodies such as immunofluorescence and immunohistochemistry are described separately in this book.

2 Experimental Protocols

2.1 [^{35}S]-Metabolic Labeling Immunoprecipitation ([^{35}S]-IP)

! *Note:* Wear gloves for all the following [^{35}S]-IP procedures, beginning with step 4.

^{35}S-Methionine labeling of mammalian cells

A Monolayer Cells

1a. Chosen cell lines are grown to 60–90% confluence in an appropriate culture medium.

2a. Aspirate off the culture media and wash the cells once gently with 5 ml of 37 °C PBS.

3a. Add 1 ml/60 mm dish, 2.5 ml/100 mm dish, or 5 ml/150 mm dish of prewarmed labeling media [DMEM without methionine, glutamine, and cysteine (ICN or Gibco-BRL), supplemented with 10% dialyzed fetal bovine serum and antibiotics (Gibco-BRL)]. Incubate the plates at 37 °C for 30–40 min to allow depletion of the intracellular pools of sulfur-containing amino acids.

! **4a.** In a fume hood (^{35}S evaporates!), add [^{35}S] to the culture. The amount of ^{35}S-methionine used is a limiting factor and is largely determined by the number of immunoprecipitations to be carried out on a cell lysate. Use of 50 to 100 μCi ^{35}S-methionine for each immunoprecipitation is suitable for most proteins, and the upper limit can be as high as 2 mCi per 100 mm dish. Incubate the plates for 30 min to 6 h (usually 4 h) at 37 °C, taking care to place the plates in a shielded box in an incubator labeled for radioactive use. A piece of charcoal paper placed in the shield box can help absorb a large amount of the evaporating ^{35}S. About once an hour during labeling gently tilt the plate back and forth and left to right several times to redistribute the media evenly.

B Suspension Cells

1b. Cell cultures are grown to an appropriate cell density. After counting the cells with a hemocytometer and determining the number of cells per milliliter, calculate the volume containing 1×10^7 cells and place in a 15 ml Falcon tube. Spin the cells down for 2 min at 3000 rpm.

2b. Aspirate off the supernatant. Resuspend the pellet in 10 ml prewarmed labeling media [DMEM without methionine, glutamine, and cysteine (ICN or Gibco-BRL), supplemented with 10% dialyzed fetal bovine serum and antibiotics (Gibco-BRL)].

3b. Spin down for 2 min at 3000 rpm. Aspirate the supernatant, and resuspend the pellet in 2.5 ml labeling media. Place the sample in a 60 mm tissue culture dish and incubate at 37 °C for 30 min to allow depletion of the intracellular pools of sulfur-containing amino acids.

4b. In a fume hood add the appropriate amount of [^{35}S] methionine (see step 4a for Monolayer Cells above). 6 h is a reasonable incubation for labeling suspension cells.

Cell lysis

A Monolayer Cells

5a. When the labeling is complete, take the plates to a fume hood and aspirate off the ^{35}S-containing media. Rinse each of the plates once with 5 ml (10 ml for a 150 mm plate, 3 ml for a 60-mm plate) of cold PBS, and aspirate it off.

6a. Place the plates on ice and add 1 ml of lysis buffer per 60- or 100-mm plate or 2 ml for a 150-mm plate. Some cell lines will require 20 to 30 min on ice with occasional shaking for lysis; however, most should be fairly rapidly lysed. When lysis becomes apparent (check under a microscope if it is unclear), keeping the lid on, hold the plate partially open with one hand while tilting the plate to one side. The buffer should be pooled on one side of the plate. Using a P1000 adjustable pipette (set for 600 µl to help prevent contamination of the pipette) take up some of the buffer and re-release it at the top edge of the plate to allow the lysed cells and buffer to re-accumulate in the pool. Repeat this several times until confident that the (sometimes visible) particulate cellular matter has accumulated in the pool. Take care, in pipetting, not to blow bubbles with the buffer as it increases the risk of ^{35}S contamination. Carefully collect each of the lysates into 1.5 ml Eppendorf centrifuge tubes (*Note:* do not freeze the lysates at this stage if the intent is to see protein-protein complexes).

A number of different lysis buffers for cultured mammalian cells have been developed in the past and choice of a lysis buffer depends on the purpose of the experiments and type of cells used. Two common lysis buffers for cultured mammalian cells are NP-40 and RIPA buffer. The standard lysis buffer we use for immunoprecipitation aimed at identifying protein-protein interactions in human fibroblast cells is Nonidet P-40 lysis buffer (NP-40 LB). NP-40 LB appears to work well for many different types of cells including human keratinocytes and lymphocytes. The major advantage of NP-40 LB is that it appears to preserve most, if not all, of the Cyclin-Cdk complexes when an appropriate antibody is used. The only exception is the Cyclin D1-Cdk4 kinase, whose activity is only preserved when Tween-20 is used as detergent (Matsushime et al. 1994). RIPA buffer, on the other hand, is a good choice for initial characterization of a newly produced antibody. Quite often, although not always, RIPA buffer dissociates protein complexes. By providing more stringent conditions of immunoprecipitation, RIPA buffer removes many of the antigen-associated specific bands isolated with NP-40 LB unless they are simply cross-reactive.

NP-40 LB (Nonidet P-40 lysis buffer)

Final concentration	For 500 ml stock (keep at 4 °C)
	452.5 ml dH_2O
50 mM Tris-HCl (pH 7.5)	25 ml 1M Tris-HCl (pH 7.5)
150 mM NaCl	15 ml 5M NaCl
0.5% NP-40	2.5 ml NP-40
50 mM NaF	1.05g NaF

Immediately before use add the following 100× stocks to the lysis buffer:

<u>100× stocks</u>	
100 mM Na_3VO_4	(1.83 g sodium orthovanadate in 100 ml dd H_2O, kept at 4 °C; add 10 µl per ml of lysis buffer)
100 mM DTT	(154 mg in 10 ml dd H_2O, kept at –20 °C; add 10 µl per ml of lysis buffer)
100 mM PMSF	(174 mg in 10 ml 100% ethanol, kept at 4 °C; add 10 µl per ml of lysis buffer)
100× Protease inhibitor	(2.5 mg/ml leupeptin, 2.5 mg/ml aprotinin, 100 mM (= 15 mg/ml) benzamidine, and 1 mg/ml trypsin inhibitor in dd H_2O, kept at –20 °C); add 10 µl per ml of lysis buffer

Note. All the –20 °C stocks can be thawed and frozen several times with the exception of the Na_3VO_4. Also, PMSF, phenylmethylsulfonyl fluoride, has a half-life of approximately 30 min in aqueous solution. Add PMSF last, and keep it on ice until use.

RIPA buffer

Final concentration	For 100 ml stock
	80 ml dH_2O
50 mM Tris-HCl (pH 7.5)	5 ml 1 M Tris-HCl (pH 7.5)
150 mM NaCl	3 ml 5 M NaCl
1% NP-40	1 ml NP-40
0.1% SDS	1 ml 10% SDS
1% sodium deoxycholate	10 ml 10% sodium deoxycholate

The same additions as for NP-40 LB (10 µl per ml lysis buffer of Na_3VO_4, protease inhibitor, PMSF, and 100 mM DDT) should be made prior to use.

7a. Wrap the tube(s) containing the cell lysate in saran wrap and place on a rotator in the cold for 10 min to continue cell lysis.

B Suspension Cells

5b. When labeling is complete, take the plate to the fume hood and using a p1000 (per step 6a above) place the cells in a 15 ml Falcon tube. Rinse the plate once with 2 ml cold PBS and add this wash to the 15 ml tube containing the cells. Spin the cells down at 3000 rpm for 2 min in a centrifuge designated for radioactive use, and place on ice.

6b. Solutions for lysis are the same as for monolayer cells (see step 6a above). Aspirate off the supernatant and lyse the cell pellet with 1 ml of NP-40 LB, pipetting up and down several times. Collect the lysate into a 1.5 ml Eppendorf tube, and place on ice.

7b. Wrap the tube containing the lysate in Saran wrap and place on a rotator in the cold for 30 min to continue cell lysis. The remaining procedures for suspension cells are identical to those for monolayer cells beginning with preclearing – step 8.

8. Centrifuge (in a microcentrifuge labeled for radioactive use, preferably in a fume hood) at maximum speed for 5–10 min at 4 °C to pellet the cell nuclei. This step removes chromatin, which otherwise causes a high background.

9. Preclear to remove bead-specific precipitates. Add 50 to 100 µl of Protein A beads (in a 50% slurry, prewashed with PBS, see below) to new 1.5-ml Eppendorf centrifuge tubes. Transfer the cell lysate supernatants to the new tubes. Wrap the tubes in Saran wrap and place on a rotator at 4 °C for 20 to 30 min. This simple preclearing step is sufficient when a high titer or high affinity antibody is used (so that only small volumes of antibody and beads are needed), but it may be necessary to carry out additional preclearing steps when a high background is encountered. Preimmune sera or normal sera may be used to further preclear the cell lysate as necessary. Fixed *S. aureus* Cowan I (SAC) is also often used as substitute for Protein A beads in preclearing because of its availability in large quantities and tight binding to immunoglobulins. However, we have repeatedly found that the use of crude preparations of SAC in preclearing produces considerably higher background than Protein A beads.

Prewashing of Protein A beads: Aliquots of Protein A beads should be centrifuged at low speed (high speeds can damage the beads), and the supernatant aspirated off. The beads should then be washed three times with 0.1% NP-40 in PBS (making sure to adequately resuspend the beads each time), re-centrifuged at low speed, and the supernatants aspirated off. After the final wash, the beads should be resuspended in PBS in a 1 to 1, vol/vol ratio, and stored at 4 °C.

10. After the preclearing step, centrifuge the lysate in a microcentrifuge (designated for use with radioactive materials) for 5 min at low speed (4000–5000 rpm) to pellet the beads. The supernatant is now ready for immunoprecipitation.

Immunoprecipitation

11. Add the desired antibodies (with or without competing peptides) as appropriate to fresh 1.5-ml Eppendorf tubes. Add pre-cleared cell lysate supernatant (from step 10) to the appropriate tubes. Take extreme care not to carry over any of the beads from preclearing – they are an obvious source of high background. Wrap the tubes in Saran wrap and place them on a rotator in the cold for 2 h to overnight.

12. Briefly spin (5–10 s) the tubes in the radioactive-use microcentrifuge, to bring any radioactive material away from the tube lids before they are opened.

13. In a fume hood, open the tubes, and add prewashed Protein A beads to each of the tubes. Different brands of immobilized Protein A beads have different capacities for IgG (e.g., ca. 12–15 mg IgG per ml of gel for PIERCE Protein A agarose beads). Based on this approximation, we routinely use 5 to 10 μl of prewashed Pierce Protein A agarose beads (in a 50% slurry) per microliter of unpurified rabbit sera. The smaller the amount of beads used, the lower the background will be. Make sure the beads in the stock tube are adequately resuspended before each addition, and take care that the outside of the pipette tips used are free of additional beads when they are added to the tubes. Close the tubes, wrap them in Saran wrap, and place on a rotator in the cold for 1 h.

14. In a fume hood, initially spin the tubes for 2 min at 2000 rpm to pellet the beads. Then, aspirate off the supernatants. Be careful to use as fine a pipette tip as possible (a gel-loading tip is a good choice) to prevent accidental removal of the beads. Do not touch the tip to the beads, but aspirate the supernatant off as cleanly as possible (aspirating the bubbles first helps maintain a clear and clean interface). Also be aware that the beads will shift their position if the tube is tilted, so it is often best to aspirate from the top side of the tube to avoid their accidental aspiration. Changing the tip between each wash is not necessary.

15. Wash the pellets with a large volume (1 ml) of lysis buffer per tube. The wash buffer is identical to the lysis buffer used, except that protease inhibitors are not added. Close the tubes, gently invert several times, and tap the ends to disperse the pellets. Spin the tubes for 2 min at 2000 rpm, and then aspirate off the supernatant as before. Repeat for a total of three washes.
Note. If the background is found to be too high, transferring the last 1 ml of wash solution into a new microcentrifuge tube may help to reduce some of the background, which is presumably sticking to the walls of tube.

16. After the final aspiration, add 10 µl of 2× SDS sample buffer to the lid of each tube, close, and briefly spin the tubes to draw the sample buffer down to the pellet.

17. Open the tubes to prevent popping, and boil the samples on a 100 °C heater for 4 min. (Subsequently spin down for 30 s if necessitated by the contents being disturbed.)

18. Load *all* of the supernatant (beads can be loaded as well, but they tend to clog the pipet tips) onto a SDS gel.

19. Run the gel for an appropriate length of time. Mark the upper left side (as viewed when loading) of the plates. Take the gel off and, in the hood, open the plates. Cut off the stacking gel and the upper left corner and discard them in the radioactive waste. (Remember to carefully wash out all of the gel apparatus).

20. In the hood, place the remaining gel in a baking dish containing enough Enhance solution (NEN) for the gel to move about freely. Cover the dish with Saran wrap (the fumes are both noxious and hazardous), and place it on a shaker for 30 min at a setting that just ensures gentle motion of the gel about the dish.

21. At the end of the 30 min, carefully holding onto the gel, empty the Enhance into an appropriate waste container. Taking care to avoid the remaining fumes and also taking care not to pour water directly onto the gel, add enough distilled water to the dish to cover the gel. Cover the dish with Saran wrap, and return it to the shaker for 15 to 20 min (prolonged soaking in water will result in the gel becoming too fragile).

22. The gel should now be ready for drying. Rinse the gel several times in dH_2O to remove precipitates. Then, wet two pieces of Whatman paper slightly larger than the gel with dH_2O and place them on the gel dryer. Carefully placing three fingers under each side of the gel, thumbs in front, lift it up– taking care that the cut upper left corner remains on the upper left. then place the gel on the paper: holding the gel vertical over the paper with the bottom edge of the gel just touching the paper, so it forms as straight a line of contact as possible, hold the top edge closer to you than the top-must edge contacting the paper, and gently roll the gel forward onto the paper as you lower and move your hands forward. Not distorting the form of the gel is important, and it can be lifted back off the paper and repositioned several times of necessary, as long as damaging the gel is avoided. Wet glove fingertips with dH_2O before making any further adjustments, as dry gloves can easily tear the gel. Smooth out any air bubbles, overlay the gel with saran wrap, and further overlay it with a mylar sheet. Dry the gel on the gel dryer at 80 °C for 1 h.

23. Proceed to develop the dried gel through phospho-imaging (2 h to an overnight exposure) and/or autoradiography (2 to 5 day exposure while stored at –80 °C). The dried gel must be discarded in the radioactive waste when finished.

2.2 Immunoblotting (Western Blot)

All steps for Western blotting are included in the IP-Western protocol.

2.3 Coupled Immunoprecipitation and Western Blotting (IP-Western)

Cell lysis All steps for cell lysis are the same as for the ^{35}S-IP, except that no precautions against radioactive contamination are needed, since there is no intermediate labeling step, and cells are lysed directly after reaching the desired level of confluency.

Immunoprecipitation Before preceding to add antibody after cell lysis (preclearing is not required) it may be necessary to determine the protein concentrations of the cell lysates. Prepare samples in duplicate, adding 5 µl from each lysate (save some lysis buffer for use as a blank) to 1 ml 1x Bradford Assay buffer, and read the OD at 595 nm for the samples on a photospectrometer. Based on the results, the amount of lysate added to each tube containing antibody can be adjusted to an equivalent protein concentration.

All subsequent steps including the use of Protein A beads, the post-immunoprecipitation wash, and SDS-PAGE are the same as described for the [^{35}S-IP].

Western blotting

The following steps are designed for use of a BIO-RAD semidry trans-blot SD, if a different apparatus is used, follow the manufacturer's instructions for buffer conditions and voltage and current settings.

!

1. Wear gloves to work on the transfer from the SDS-PAGE gel. Cut six pieces of 3-mm Whatman paper slightly larger than the size of the gel, and one piece of nitrocellulose larger than the gel but smaller than the Whatman paper. Cut the upper left corner of the nitrocellulose.
2. Wet three pieces of the Whatman paper in Western transfer buffer, remove excess buffer and place them onto the platinum anode. Roll a pipet over the surface of the paper to exclude all air bubbles.

   Western transfer buffer

Final concentration	For 1 l stock
48 mM Tris base	5.82 g Tris base
39 mM glycine	2.93 g glycine
0.0375% SDS	3.75 ml 10% SDS
20% methanol	200 ml methanol

   Adjust to a final 1 l volume with dH_2O
3. Wet the nitrocellulose in the same buffer and place it onto the Whatman paper, again excluding all air bubbles.
4. Next, wet the gel in the transfer buffer for 5 to 10 s and place it onto the nitrocellulose.
5. Finally, wet the other three pieces of Whatman paper and, removing excess buffer, place them onto the gel, and exclude air bubbles. If air bubbles are not adequately removed, the transfer will be neither even, nor complete.
6. Place the trans-blot cathode onto the stack and then the safety cover.
7. Turn on the power supply. Transfer the gel at a constant voltage between 15 to 25 V for 30 min. Do not exceed 25 V with the BIORAD apparatus. With other equipment, the following formula may be of use: 2.5 mA per each cm^2 of gel, setting a voltage of 40 V, but see the manufacturer's instructions.
8. While the gel is being transfered, make up 100 ml of blocking solution. For alternative blocking and wash solutions, see Harlow and Lane (1988) or Assai (1993).

Blocking solution:	98 ml 1× PBS
	2 ml 10% Tween-20
	5 g nonfat dry milk

   Continue stirring until use.

9. Add 100 ml dH_2O and 1 ml Ponceau S stock to a glass dish. Open the trans-blot apparatus and remove the sandwich of papers. Remove the nitrocellulose from the sandwich with forceps (which should be used to handle the nitrocellulose by the edges for the remainder of the procedure), and place it in the Ponceau S solution to stain for 2 to 5 min.

Ponceau S stock solution:
2 g Ponceau S
30 ml trichloroacetic acid
100 ml dH_2O

Wrap the solution bottle with aluminum foil. The solution is stable at room temperature.
Note. 2% Ponceau S is used as 10× (Sambrook et al. 1989), but it can be used as 50 or even 100×.

10. Pour out the Ponceau S staining solution, and rinse twice with dH_2O. Use a pencil to mark the upper left corner of the transferred gel, each lane, and the molecular weight markers. Cut the filter to a more appropriate size if necessary. The molecular weight size markers and the 54-kDa heavy chains of the immunoglobulins should be easily visible.
Note. If the protein of interest has an apparent M_r of 54 000, avoid boiling the sample (step 17 of the immunoprecipitation procedure). The antigen will be eluted anyway because of the SDS present in the sample buffer. In these conditions, the two heavy chains of the immunoglobulins will not be dissociated – they will appear at M_r 96 000 and will not interfere with the visualization of the studied protein.

11. Pour out the H_2O and add some PBS to destain the nitrocellulose, with slow shaking until the protein bands disappear (about 5 to 10 min).

12. Pour off the PBS, and add the blocking solution to the nitrocellulose (30 to 50 ml per filter). Place the dish on a shaker (slow) for at least 1 h to overnight. Add 5 µl/ml 10% sodium azide to the blocking solution if it is going to be left overnight.

13. Pour out the blocking solution and rinse the blot once with PBS.

14. Pour off the PBS, and add enough blocking solution to cover the nitrocellulose (approximately 0.1 ml per 1 mm^2 of nitrocellulose). Also add the primary antibody. The dilution of antibody used is entirely antibody-dependent. The range of dilution runs from 1 : 5 (for some low titer monoclonal antibodies, it is possible to directly incubate the filter in the hybridoma supernatant without any dilution) to 1 : 5000. Place the dish on a shaker (slow) for at least 1 h to overnight at room temperature (adding sodium azide per step 12), or on a rocker if the volume of solution is very small.

15. Transfer the blot to a new dish, saving the blocking solution containing the antibody at 4 °C after adding sodium azide, if the antibody is precious. The blocking solution can be reused up to four times within 2 weeks.

16. Wash the blot five to six times, each time for 10 min with 0.2% Tween-20 in PBS (be sure to have a few liters on hand) and slow shaking.

17. Add some blocking solution (0.1 ml per 1 mm^2 of nitrocellulose) and secondary antibody (e.g., horseradish peroxidase-conjugated mouse anti-rabbit antibody in a 1 : 10 000 dilution if the primary antibody is of rabbit origin and the signal is to be detected by luminescence) to the nitrocellulose. Continue with slow shaking for 1 h at room temperature. Remember that horseradish peroxidase is inhibited by sodium azide!

18. Wash the blot five times for 10 min each with 0.2% Tween-20 in PBS and slow shaking as before.

19. Wash the blot once for 5 min with PBS and slow shaking.

20. The blot is now ready for developing with the detection reagents. The remaining steps must proceed rapidly, as the detectable signal will essentially be lost after 30 to 45 min. Make sure the darkroom/X-ray developer is available before proceeding, and have all reagents and equipment ready to go. Using ECL detection reagents (RPN 2106, Amersham), mix equal volumes of each of the reagents in a fresh dish. Place the blot in the mixture for exactly 1 min, with frequent agitation, making sure all blot surfaces receive sufficient contact with the reagents.

21. Pencil marks up, remove the filter from the ECL detection reagents, and place it on a piece of Whatman paper, to partially blot it dry. Then, quickly place the nitrocellulose filter on another piece of Whatman paper, orienting it with the pencil marks up and the proper corner on the upper left. Secure its position with a small piece of tape on the upper left corner. Wrap the blot/Whatman paper in Saran-wrap, and proceed to the dark room. Place the Saran wrapped blot in an X-ray cassette, and using a timer, expose the blot to X-ray film for between 30 s and 5 min, after first placing a slight bend in the left upper corner of the film for later ease of orientation. Develop the films, and if available time remains, adjust the exposure times as necessary. If the signal is very strong, taping the blot to the inside of the X-ray cassette lid should aid in avoiding overexposure and double-images.

References

Asai DJ (ed) (1993) Methods in cell biology volume 37: antibodies in cell biology. Academic Press, New York

Bates S, Bonetta L, MacAllan D, Parry D, Holder A, Dickson C, Peters G (1994) CDK6(PLSTIRE) and CDK4(PSK-J3) are distinct subset of the Cyclin-dependent kinases that associate with Cyclin D1. Oncogene 9:71–79

Draetta G, Beach D (1988) Activation of cdc2 protein kinase during mitosis in human cells: cell cycle-dependent phosphorylation and subunit rearrangement. Cell 54:17–26

Dulic V, Lees E, Reed SI (1992) Association of human Cyclin E with a periodic G1-S phase protein kinase. Science 257:1958–1961

Elledge SJ, Richman R, Hall FL, Williams RT, Lodgson N, Harper JW (1992) CDK2 encodes a 33-kDa Cyclin A-associated protein kinase and is expressed before CDC2 in the cell cycle. Proc. Natl Acad Sci USA 89:2907–2911

Ewen ME, Xing YG, Lawrence JB, Livingston DM (1991) Molecular cloning, chromosomal mapping, and expression of the cDNA for p107, a retinoblastoma gene product-related protein. Cell 66:1155–1164

Harlow E, Lane D (1988) Antibodies: a laboratory manual. Cold Spring Harbor Laboratory, Cold Spring Harbor

Koff A, Cross F, Fisher A, Schumacher J, Leguellec K, Philippe M, Roberts JM (1991) Human Cyclin E, a new Cyclin that interacts with two members of the CDC2 gene family. Cell 66:1217–1228

Matsushime H, Quelle DE, Shurtleff SA, Shibuya M, Sherr CJ, Kato JY (1994) D-type Cyclin-dependent kinase activity in mammalian cells. Mol Cell Biol 14:2066–2076

Matsushime H, Roussel MF, Ashmun RA, Sherr CJ (1991) Colony-stimulating factor 1 regulates novel Cyclins during the G1 phase of the cell cycle. Cell 65:701–713

Meyerson M, Harlow E (1994) Identification of G1 kinase activity for Cdk6, a novel Cyclin D partner. Mol Cell Biol 14:2077–2086

Pagano M, Pepperkok R, Verde F, Ansorge W, Draetta G. (1992) Cyclin A is required at two points in the human cell cycle. EMBO J 11:961–971

Pines J, Hunter T (1989) Isolation of a human Cyclin cDNA: evidence for Cyclin mRNA and protein regulation in the cell cycle and for interaction with $p34^{cdc2}$. Cell 58:833–846

Pines J, Hunter T (1990) Human Cyclin A is adenovirus E1A-associated protein p60, and behaves differently from Cyclin B. Nature 346:760–763

Polyak K, Kato J, Solomon MJ, Sherr CJ, Massague J, Roberts JM, Koff A (1994) $p27^{Kip1}$, a Cyclin-Cdk inhibitor, links transforming growth factor-β and contact inhibition to cell cycle arrest. Genes Dev 8:9–22

Rosenblatt J, Gu Y, Morgan D (1992) Human Cyclin-dependent kinase 2 is activated during the S and G2 phase of the cell cycle and associates with Cyclin A. Proc Natl Acad Sci USA 89:2824–2828

Sambrook J, Fritsch EF, Maniatis T (1989) Molecular cloning: a laboratory manual. Cold Spring Harbor Laboratory, Cold Spring Harbor

Serrano M, Hannon GJ, beach D (1993) A new regulatory motif in cell cycle control causing specific inhibition of cyclin D/CDK4. Nature 366:704–707

Settleman J, Narasimhan V, Foster LC, Weinberg RA (1992) Molecular cloning of cDNAs encoding the GAP-associated protein p190: implications for a signaling pathway from ras to the nucleus. Cell 69:539–549

Xiong Y, Zhang H, Beach D (1992) D type Cyclins associated with multiple protein kinases and the DNA replication and repair factor PCNA. Cell 71:505–514

Xiong Y, Zhang H, Beach D (1993a) Subunit rearrangement of Cyclin-dependent kinases is associated with cellular transformation. Genes Dev 7:1572–1583

Xiong Y, Hannon G, Zhang H, Casso D, Kobayashi R, Beach D (1993b) p21 is a universal inhibitor of the Cyclin kinases. Nature 366:701–704

Zhang H, Xiong Y, Beach D (1993) Proliferating cell nuclear antigen and p21 are components of multiple cell cycle kinase complexes. Mol Biol Cell 4:897–906

23 Protein Mapping in Cell Cycle Studies

EMMA LEES

1 Background

Immunoprecipitations are used frequently in cell cycle studies to study protein levels, kinase activities, and protein-protein interactions. The ability to coprecipitate components of the cell cycle machinery has greatly facilitated our understanding of the cell cycle by revealing who interacts with whom. It was early immunoprecipitations of Cyclin A that suggested that there may be other Cdc2-related kinases (Faha et al. 1992), these observations ultimately leading to the cloning of *cdk2*. Co-immunoprecipitations have allowed the identification of most Cyclin/Cdk interactions, and also the identification of other regulatory proteins that interact with these complexes. A critical element in these discoveries was the ability to identify coprecipitating proteins either using antibodies or by peptide mapping.

Immune complexes can be resolved by polyacrylamide gel electrophoresis in SDS-containing buffers (SDS-PAGE). However, since the separation is based on molecular size only, it is not possible to deduce that two polypeptides of identical mobility are related. In addition, two related proteins may migrate differently as a result of posttranslational modification or degradation. Many different techniques have been employed to establish the true relationship between proteins, such as immunological characterization and amino acid composition. Possibly one of the most stringent is that of peptide mapping and in this chapter I will describe in detail how to perform such analysis using partial proteolysis modified from the method of Cleveland (Cleveland et al. 1977).

2 Principle and Applications

In the Cleveland method, the polypeptides of interest are located after SDS-PAGE by a brief period of staining and destaining. In cell cycle studies it is rare to be blessed with that much protein so the protocol detailed here uses proteins metabolically labeled with ^{35}S methionine. ^{32}P-orthophosphate can also be used, but since the polypeptide is likely to have only a few phosphorylation sites, the peptide pattern generated is much less complex, therefore potentially less informative. Once visualized, the polypeptides are cut out and the gel slices placed in sample wells of a second polyacrylamide gel slab. These sample wells also receive known amounts of a specific protease which is active in the presence of SDS. During the second electrophoretic separation, the protease digests the polypeptide and the resulting peptides are separated according to their molecular weight. Several dilutions of enzyme are used to ensure partial proteolysis and multiple peptide species. The peptide banding pattern generated by such a procedure is characteristic of the protein substrate. Although this method is particularly useful for mapping of polypeptides fractionated first by SDS-PAGE, it is equally applicable to a pure protein from any source.

There are several different proteases that can be used in this method. The most commonly used, because of its unique specificity, is *Staphylococcus aureus* V8 protease. This V8 protease specifically cleaves peptide bonds on the carboxy-terminal side of either aspartic or glutamic acids. This rather unique specificity makes it a useful tool for protein chemistry and peptide mapping studies. It has a molecular weight of 27 000 Da and optimum pH of 4.0 and 7.8 with hemoglobin as the substrate. The enzyme is inhibited by di-isopropyl flurophosphate (DFP) and monovalent anions such as F-, Cl-, CH3COO-, and N03-. Other enzymes that can be used for peptide mapping include papain, subtilisin, chymotrypsin, elastase, and ficin.

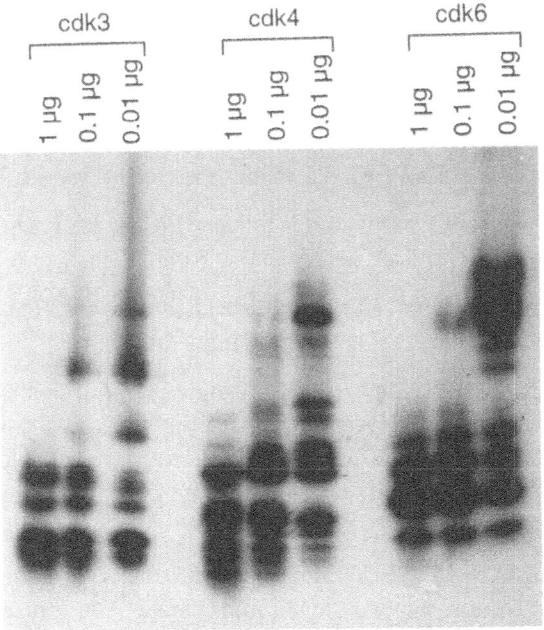

Fig. 1. Peptide maps of Cdk3, Cdk4 and Cdk6 using *Staphylococcus aureus* V8 protease. ^{35}S methionine labeled in vitro translation products of Cdk3, Cdk4, and Cdk6 were treated with dilutions of V8 protease as shown. The proteolytic products were resolved on a 15% polyacrylamide gel, and exposed to X-ray film after fluorography. (Courtery W. Seghezzi, DNAX Research Institute)

Peptide mapping by two-dimensional methods is still a more stringent test of identity, but the value of this one-dimensional method is its speed and simplicity in comparing multiple polypeptides. The specificity of the V8 protease allows discrimination of closely related polypeptides. Figure 1 shows partial proteolytic mapping of Cdk3, Cdk4, and Cdk6 using V8 protease. You will be able to see that the patterns are quite distinct even between quite related proteins (71% of identity between Cdk4 and Cdk6, 48% of identity between Cdk3 and Cdk4, and 48% of identity between Cdk3 and Cdk6). Partial proteolytic mapping has become a widely used technique in cell cycle studies, the results generally considered definitive. A few examples include the cloning of human *cdk2* (Tsai et al. 1991), the demonstration that Cyclin A and Cyclin E associate with p107 and p130 (Faha et al. 1993), and the characterization of kinases bound to Cyclin D1 (Xiong et al. 1992).

3 Methods

3.1 Preparation of Polypeptide Bands for Analysis

Metabolic labeling The amount of protein and radioactivity required for a good signal will depend on the abundance of the protein within the cell and its half-life, and will have to be deduced empirically for each application. However, as a guideline, in the analysis of Cdks that coprecipitate with Cyclin E we used the following procedure:

1. Label three 9-cm dishes of cells with 1 mCi ^{35}S methionine (each) for 4 h.
2. Prepare a detergent lysate and then immunoprecipitate with Cyclin E antisera using standard techniques (see Chapter by Jenkins and Xiong).
3. Collect immune complexes on Protein-A Sepharose beads, wash, then dissociate complexes by the addition of 1× Laemmli's sample buffer.
4. Load the sample into three lanes of a 10% polyacrylamide gel. In order to allow the gel slices to fit easily into the sample wells of the second gel (in which the proteolytic digestion will occur) it is advisable to use a smaller well dimension for the preparative gel. Use a 20-well comb for this first gel, then an 18-well comb for the second gel.
5. Run the gel as normal but DO NOT FIX OR PROCESS this gel or the V8 protease will not work. Simply remove the gel from the gel plate, rinse with water and dry down. Mark three corners of the gel with radioactive ink, to enable correct orientation of the autoradiograph, and expose to film overnight at –70 °C.

Since this gel is not fluorographed you need very hot samples to enable overnight visualization (12 h). If this preparative gel needs a longer exposure, then your final V8 gel will take a minimum of 1 week to develop, usually longer, and is not recommended. Your protein band should ideally be visible with an overnight exposure since this level of signal reproducibly gives the best results. A peptide map should then be obtainable with a 1–3-day exposure to film. If your protein is not labeled, you will need a minimum of 10 μg in the initial band to obtain peptides that are easily visible by Coomassie staining.

In vitro translation products

In vitro translated proteins are frequently used as a source of protein for comparison with the in vivo metabolically labeled polypeptide whose identity is sought. The same polypeptide labeled with ^{35}S-methionine either in vitro or in vivo tends to give very similar if not identical peptide maps so comparisons can safely be made.

For the best results, it is important to try and have the two polypeptides for comparison labeled at similar intensities. Generally this requires less than 1 μl of protein translated in vitro using a standard protocol. It is useful to run a preparative gel with several dilutions of in vitro translated material so that protein bands of equal intensity to the in vivo labeled polypeptide can be used in the analysis.

On the rare occasion that a gene translates very poorly, it is advisable to first immunoprecipitate the in vitro translated protein, to get rid of all the other proteins in the reticulocyte lysate, and then load that immunoprecipitated material on a gel. Do not attempt to load more than 5 μl of translation product directly!

3.2 Partial Proteolytic Mapping

Protocol

1. Outline the polypeptides of interest on the reverse side of the X-ray film using a china pencil. Turn the film over and line up the film with the gel using the radioactive marks. Rub the film to transfer the china pencil markings to the gel itself. In this way you should be able to transfer the outline of the bands to the gel so that you can accurately remove gel slices containing the polypeptides of interest. Cut out the bands carefully and store in a microcentrifuge tube until use.
2. Prepare a 15% polyacrylamide gel according to the SDS-discontinuous system, but with a stacking gel about twice the normal depth (i.e., about 5 cm is suitable).
3. Place the gel in a gel rig and fill the wells with 1× Cleveland's soak buffer.

Cleveland's soak (5 ×)
0.5% SDS
5 mM EDTA
0.625 M Tris pH 6.8.

Dilute to 1 × and add DTT to final concentration 2.5 mM before use.

4. Push the dried gel slices (paper included) into three adjacent wells and align along the bottom of the well. Millipore forceps and a fine spatula are invaluable tools for this task. Try to avoid air bubbles between the gel slice and the base of the well. It is preferable to leave a gap between each set of three digests. Allow gel slices to swell for about 10 min in the Cleveland's soak buffer.

5. Overlay the slice with 20 μl of 20% glycerol in Cleveland's soak.

6. Overlay this with 10 μl of 10% glycerol/ Cleveland's soak containing a dilution of V8 protease. (Endoproteinase Glu-C" Boehringer Mannheim, Cat No. # 791 156 or Protease, *S. aureus* V8 Worthington Enzymes Cat No. # LS03605; Stock 10 mg/ml in water and aliquot. Do not refreeze once thawed!)
   To ensure that partial proteolysis is achieved and that an informative peptide map is obtained it is advisable to use more than one dilution of protease. It is standard to use three. The dilutions I use are from a 10 mg/ml stock: 1/10, 1/100, 1/1000. It is best to add the most dilute enzyme first to minimize digestion due to spillover into adjacent wells. Load a lane of markers for reference. Also load 1 μg BSA in sample buffer to all spacer lanes to keep lanes straight.

7. Carefully fill the lower and upper reservoir of the gel rig with standard SDS-PAGE running buffer. Electrophoresis is begun at 20 mA until the dye front reaches the separation gel, then it can be continued at 50 mA until the dye front reaches the bottom of the gel. During the initial electrophoresis the protease and polypeptide substrate are costacked and digestion takes place. For larger polypeptides, this initial step can be adjusted to allow the protease more time for digestion. For example, for the protein product of the retinoblastoma gene product, which is 105 kDa in size, the initial electrophoresis is performed at 15–20 mA. Once the protein has migrated about half way through the stacking gel, the power is shut off for 30 min. Electrophoresis is then continued at 20 mA until the dye reaches the resolving gel and then the power is turned up to 50 mA.

8. Process gel as normal, taking care when fluorographing and drying because of the high polyacrylamide percentage.

4 Special Hints for Application and Troubleshooting

The most important determinants for a successful V8 peptide map are the amount of protein or level of radioactive incorporation into the protein, and homogeneity of the polypeptide in the gel slice. Other problems will be minor and are easily solved.

- The pattern of peptide bands looks smeared.
  This is usually caused by the presence of more than one polypeptide species in the original gel slice. Repeat preparative gel and possibly try to change gel acrylamide percentages to get better resolution in the region of interest. Be careful when cutting gel slices: cut them thin, risking losing some, rather than cutting fat bands that are more likely to be contaminated.
- On exposure of final autoradiogram, black gel slices are visible in the wells and the peptides generated are weak.
  This usually happens with larger proteins when you have incomplete digestion. Electrophoresis more slowly and allow digestion to proceed for longer by shutting of power for a period of time. Also check enzyme stocks.
- On exposure of the final autoradiogram, all three lanes look similar or all peptide species very small.
  With smaller proteins electrophoresis conditions or dilutions of protease may need to be altered to give you only partial proteolysis.

5 Interpretation of Results

If two proteins have completely dissimilar peptide maps by V8 partial proteolysis, it is very unlikely that they are related. Conversely, if the peptide profile is very similar or identical then they can be considered as the same protein with reasonable certainty. It should be added that although V8 peptide maps can be reproducible, it is NOT recommended that you attempt to compare maps prepared from different experiments. Day to day variations in the procedure and the amount of protein obtained from the preparative gel can alter the digestion profile enough to mislead the interpretation. You can only obtain reliable comparisons from the same experiment.

6 References

Cleveland DW, Fischer SG, Kirschner MW, Laemmli UK (1977) Peptide mapping by limited proteolysis in sodium dodecyl sulfate and analysis by gel electrophoresis. J Biol Chem 252:1102–1106

Faha B, Meyerson M, Tsai L-H (1992) Adenovirus E1A targets one of two Cyclin A-kinase complexes. Pezcoller Foundation Symposium, pp 107–120

Faha BF, Harlow E, Lees EM (1993) The adenovirus E1A-associated kinase consists of Cyclin E/p33Cdk2 and Cyclin A/p33Cdk2. J Virol 67:2456–2465

Tsai L-H, Harlow E, Meyerson M (1991) Isolation of the human Cdk2 gene that encodes the Cyclin A- and adenovirus E1A-associated p33 kinase. Nature 353:174–177

Xiong Y, Zhang H, Beach D (1992) D type Cyclins associate with multiple protein kinases and the DNA replication and repair factor PCNA. Cell 71:505–514

24 From Peptide to Purified-Antibody

MICHELE PAGANO

1 Background

Antibodies have played an important role in the study of the cell cycle. In particular, they enabled investigators to monitor fundamental aspects of the cell cycle such as:

- Protein expression.
- Protein-protein interactions (e. g., through coimmunoprecipitation, super-band shift assays, etc.).
- Enzymatic activity such as protein kinase or phosphatase activity.
- Protein subcellular localization.
- In vivo protein functions (e. g., though microinjection or electroporation).

In this chapter, I will describe a simple protocol for the preparation and purification of polyclonal antibodies using synthetic peptides. Although some of the steps described could be performed with commercially available kits, the latter generally do not give more reproducible results or save time.

Anti-peptide antibodies work excellently in Western blot, immunoprecipitation, indirect immunofluorescence, and other techniques. Importantly, by appropriately designing the peptide, one can obtain an anti-peptide antibody that either does or does not interfere with the folding and activity of the antigen (see Sect. 3, Methods).

Compared to monoclonals, polyclonal antibodies are easier and often faster to prepare. Probably the only disadvantage is that each injected rabbit provides a large but limited amount of polyclonal antibody, whereas a monoclonal hybridoma provides an unlimited source.

Compared to polyclonal antibodies raised against the full length protein, anti-peptide antibodies can be more specific, as there is less chance they will cross-react with other members of the same protein family, and they can be prepared even when the full-length protein is not immediately available. However, in order to prepare a peptide, one needs to know at least a partial primary sequence of the protein, whereas upon purification of the full length protein, antibodies to the native protein can be prepared in the absence of sequence information. Even so, today it is easy enough to obtain microsequence from a few micrograms of purified protein and use such information to prepare an anti-peptide antibody that in turn can be used to clone the relevant gene.

2 Principle and Applications

Step one is the choice of a peptide (6 to 12 amino acids) from the protein of interest. Ideally, the peptide should not be masked in the folded protein, since, in such an instance, the antibody would work only with the denatured protein. C-terminal and N-terminal peptides are often exposed in the native protein and are therefore often the obvious choice. When using internal peptides, it is better to choose one containing hydrophilic amino acids, which has more probability of being exposed (see: Kyte and Doolittle 1982, for the hydropathic index, and Hopp and Woods 1981, for the hydrophobic index of amino acids). Internal peptides can be synthesized with their N-terminus acetylated and with their C-terminus amidated to mimic their respective status in the native protein.

The choice of the peptide will depend upon several other considerations:

- Peptides that are conserved in a whole family of proteins should be used when the aim is to produce an antibody able to recognize all the members of the family. Unique peptides should be used when the aim is an absolute specificity. For example, an antibody to the PSTAIRE peptide recognizes many Cdk protein kinases, since the PSTAIRE motif or a related one is present in several Cdk kinases (Meyerson et al. 1992). By contrast, the C-terminal peptide of the human Cdk2 kinase is unique and the antibodies prepared against it do not cross react with other Cdk kinases (Elledge et al. 1992; Pagano et al. 1992; Rosenblatt et al. 1992; Tsai et al. 1993).
- If a protein binding site for a second protein or for DNA is known, the information can be used to prepare antibodies that inhibit such binding or, alternatively, that do not interfere with this interaction. The PSTAIRE motif in Cdc2 is involved in the interaction with Cyclin A and Cyclin B (De Bondt et al. 1993; Marcote et al. 1993). Antibodies against this region inhibit the Cdc2-Cyclin complex formation (J. Lukas, pers. comm.).
- The choice of a terminal peptide can be ideal when a particular activity must be preserved. For example, to measure the activity of Cdc2 kinase in a small amount of protein extract (10–100 µg) for the first time, it was necessary to prepare an antibody to its C-terminal residue (Draetta and Beach 1988), since all the antibodies previously produced against the full-length protein immunoprecipitated the protein but not its activity. Currently the use of anti-C-terminal peptide antibodies is generally considered the simplest way to assay Cdk kinase activities in small samples originat-

ing from cells in different phases of the cell cycle. Anti-terminal peptide antibodies have also been used to assay Cdc25C phosphatase activity in similar samples (Hoffmann et al. 1993). Finally, although the anti C-terminal peptide antibodies to the human Cdc2 and Cdk2 kinases do not inhibit their activity in vitro, they do inhibit their functions in vivo (Pagano et al. 1993; Tsai et al. 1993). The reasons for the latter effect remain speculative.

- When it is known that a protein undergoes posttranslational modification in vivo, it is useful to prepare antibodies against unmodified and/or modified peptides (e.g., the N-terminus can be synthesized acetylated or it is possible to synthesize phosphorylated peptides). With luck, one can obtain a polyclonal antibody able to recognize only a specific form. For example, an antibody prepared against the unphosphorylated Cdc2 VEKI-GEGTYGVVYKARHKLS peptide only recognizes the dephosphorylated form of Cdc2 but is unable to react with the phosphorylated form of tyrosine 15 (Baratte et al. 1992).

In order to obtain a good antibody, the purity of the peptide should be at least 80%.

3 Methods

3.1 Preparation of Antigen

MBS-Coupling procedure

This method is the most commonly used method for coupling a peptide to a carrier. It is suitable for peptides having a cysteine at their N- or C-terminus. If more than one cysteine is present, avoid this method because all the cysteine residues will be coupled. If no cysteine residue is present in the chosen peptide sequence, you should add one at the amino-terminus for C-terminal peptide or at the carboxy-terminus for N-terminal peptide. In this way, after coupling, the peptide will be oriented as in the native protein. For internal peptides the position of the added cysteine is unimportant.

1. Prepare a 10 mM KPO_4 (pH 7.2) solution adding 73 µl of 1 M K_2HPO_4 and 27 µl of 1 M KH_2PO_4 to 10 ml distilled water.
2. Prepare a 50 mM KPO_4 (pH 6.0) solution adding 515 µl of 1 M K_2HPO_4 and 4485 µl of 1 M KH_2PO_4 to 100 ml distilled water.
3. Mix 0.25 ml of a 20 mg/ml solution of keyhole limpet hemocyanin (KLH, Calbiochem) (dissolved in 10 mM KPO4 pH 7.2) with 35 µl of a 20 mg/ml solution of m-Maleimidobenzoyl-N-hydroxysuccinimide ester (MBS, Pierce) (dissolved in dimethyl formamide). Add MBS very slowly (5 µl at

a time) to ensure that the local concentration of formamide is not too high, since KLH is insoluble in >30% formamide. Stir or incubate on a rotator 30 min at room temperature.

4. Filter on a PD10 Sephadex G-25 column (Pharmacia) pre-equilibrated with 50 ml of 50 mM KPO_4 pH 6.0 to remove free MBS. Collect ten fractions of 0.5 ml. Assay protein concentration. Pool fractions containing proteins.

5. Dissolve 5 mg peptide in 1 ml PBS. If the peptide is insoluble add a few microliters of diluted acetic acid (if the pH is above 7) or few microliters of diluted sodium hydroxide (if the pH is below 7). If necessary, sonicate mildly.

6. Mix the dissolved peptide with the KLH solution. Check to ensure the pH is approximately 7–7.5. Neutralize if necessary. Incubate on a rotator at room temperature for 3 h to overnight.

7. Divide into 500-µl aliquots (ca. 0.8 mg/ml, adjusting volume with PBS), freeze, keep at –70 °C.

Glutaraldehyde-Coupling procedure

This method cross-links amino groups, and is therefore only suitable for peptides with few or no lysines or arginines.

1. Weigh out 5 mg of peptide and an equal weight of carrier (keyhole limpet hemocyanin, KLH from Calbiochem).

2. Dissolve the peptide and carrier protein in 0.1 M $NaHCO_3$ at a concentration of 2 mg/ml. (If the peptide is difficult to dissolve, try making a more concentrated solution in 50% DMSO, and then gradually diluting into coupling buffer KLH. A maximum of 10% DMSO final concentration can be tolerated).

3. Add glutaraldehyde (Sigma ultra pure grade 1; 25% stock stored at – 20 °C) to the peptide/carrier solution to a final concentration of 0.05% (v/v). Check pH and if necessary raise to 8.4 by adding NaOH. Mix in a glass tube at room temperature 2 h to overnight. The solution will usually go a pale yellow color.

4. Add 1 M glycine ethyl ester (adjust pH to 8.0 with NaOH) to a final concentration of 0.1 M to block unreacted glutaraldehyde. Leave for 30 min at room temperature.

5. Precipitate the coupled carrier with 4–5 volumes of ice cold acetone at – 70 °C for 30 min. Pellet protein at 10 000 *g* for 10 min, pour off acetone and air-dry pellet. Redisperse the pellet in saline (0.9% NaCl) at 0.8 mg

carrier/ml, using a Dounce homogenizer. Alternatively, dialyze overnight at 4 °C in 2 l (with two to three changes) of PBS.

6. Divide into 500 µl aliquots, freeze, keep at –70 °C.

3.2 Rabbit Immunization Schedule

Immunization

Day 0

1. Collect about 5 ml of serum for a preimmune bleeding.
   The dose suggested is 0.4 mg in 0.5 ml per injection. However, lower doses can be used in case of scarcity of antigen (e.g., 0.1 mg in 0.5 ml for the first injection and 0.05 mg for the follow ones). The minimum dose is 0.01 mg in 0.5 ml but the efficiency of immunization will be lower.
   If necessary, dilute antigen with PBS. Mix with the same volume of complete Freund's adjuvant and sonicate for a few seconds to homogeneity (do overheat during sonication). Inject intra-lymphnode, subcutaneously, or by both routes.

Day 14

2. Prepare the sample as above but use incomplete Freund's adjuvant (IFA). Inject intradermally (better) or subcutaneously.

Day 21

3. Same as day 14 (boost).

Day 28

4. First bleed (20 ml blood).

Day 49

5. Boost.

Day 63

6. Second bleed (30–50 ml of blood).

Then boost every month and bleed 14 days later (approximately 50 ml of blood) until about eight bleeds have been collected. Alternatively, boost every month and bleed weekly (approximately 20 ml of blood) for 4–6 months.

3.3 Isolation of Serum from Rabbit Blood

Isolation of Serum

1. Bleed the rabbits from the ear vein. Let the blood sit undisturbed in a sterile plastic disposable 50-ml capped test tube for at least 1 h at 37 °C. This allows clotting to occur.
2. Separate blood from the side of the test tube by rimming the test tube with a wooden stick.
3. Let the blood sit for several hours to overnight at 4 °C.
4. Remove as much serum as possible into a clean tube. Now centrifuge the clot for 10 min at 2500 *g* (preferably in the cold). Combine the serum which has been centrifuged out of the clot with the rest of the serum, and centrifuge for 15 min at 2500 *g*. Carefully remove the serum without disturbing the pellet.
5. Add NaN_3 as preservative to a final concentration of 0.02%
6. Aliquot into sterile Eppendorf test tubes and freeze at –20 °C. Small aliquots for testing can be stored at 4 °C.

Alternative method

1. Collect rabbit blood in vacutainer tubes (Becton/Dickinson, brand 55T) for serum separation.
2. Let blood stand undisturbed at 37 °C for at least 1 h. Continue with steps 4 and 5 above.

3.4 Peptide Coupling to Sepharose for Antibody Purification

Coupling protocol

The gel used is cyanogen bromide (CNBr) activated Sepharose 4B (Pharmacia). One g Sepharose powder yields approximately about 3.5 ml swollen gel. Do not centrifuge >2000 rpm, which can damage the structure of the gel. Activated Sepharose is coupled to the peptide through the peptide's free amino group. For this reason, do not use Tris- or NH_4-containing buffers which can block the protein binding sites of the gel.

1. To prepare one column, place 2 g (dry weight) of CNBr-activated Sepharose 4B in a test tube with 50 ml of 1 mM HCl. Leave to mix on a rotator wheel for 15 min at room temperature.
2. Wash gel four to five times with 50 ml of 1 mM HCl and two to three times with coupling buffer (0.1 M $NaHCO_3$, 0.5 M NaCl pH 8.3).
3. To prepare the peptide, dissolve 10 mg of peptide in 10 ml of coupling buffer.

4. To the swollen gel add 10 ml coupling buffer containing peptide in a test tube and leave on a rotating wheel overnight at 4 °C (do not use magnetic stirring). Keep all buffers cold after this stage.

5. Centrifuge at 2000 rpm for 5 min. Check the yield of coupling, assaying the protein concentration in the supernatant (usually more than 90% should be coupled). Resuspend pellet in 50 ml coupling buffer.

6. Centrifuge at 2000 rpm for 5 min and resuspend pellet in 50 ml 1 M ethanolamine pH 8.4–9.0 (use HCl to adjust pH) to remove the coupling buffer. Handle the ethanolamine in a fume hood, since it is harmful when inhaled.

7. Centrifuge at 2000 rpm for 5 min and resuspend pellet in 50 ml 1 M ethanolamine. Place on the mixer wheel for 2 h at room temperature or overnight in the cold room to block any remaining active protein binding sites on the gel.

8. To remove any loosely bound peptide from the gel, wash gel alternately three times with:

   - 0.1 M NaAc, 0.5 M NaCl pH 4.0 (use acetic acid to adjust pH to 4).
   - 0.1 M Tris, 0.5 M NaCl pH 8.0.

9. Store the gel in PBS/azide at 4 °C.

3.5 Affinity-Purification of Antibodies

Purification of antibodies

1. Equilibrate the column matrix (to which the antigen had been bound) with 10 bed-volumes of 50 mM Tris-HCl, pH 7.4, containing 0.4 M NaCl, 0.1% NP40.

2. Add the serum sample diluted 1 : 1 with equilibration buffer. The amount will depend on the antibody titer. Generally, 10 ml of serum is appropriate for 2 ml of resin and should yield between 0.5 and 1.5 mg of affinity-purified antibody contained in the most enriched two to three fractions. Incubate overnight at 4 °C on a rotator.

3. Pour on a Bio-Rad Econo column. Collect flowthrough and save (check that you have actually depleted it of specific antibodies).

4. Wash with 10–15 column volumes of equilibration buffer.

5. Elute with 2–2.5 column volumes in 0.43 ml aliquots of 100 mM glycine, pH 2.5, in tubes containing 70 µl 1 M Tris, pH 7.0.

6. Measure the protein concentration. Pool accordingly. It is advisable to test fractions with the highest protein concentration separately, since fractions can vary in specificity and level of background.

7. To the affinity purified antibody add 0.02% azide. Add BSA to 0.5 mg/ml if the concentration is very low (≤0.2 mg/ml). Alternatively, concentrate and desalt the preparation either by centrifugation using Centricon 10 tubes (Amicon), or by dialysis and subsequent lyophilization. Storage at high protein concentration increases stability. Aliquot and store one vial at 4 °C for further testing and store remaining vials at –70 °C.
8. Wash the column extensively with equilibration buffer and then store it in PBS in presence of 0.02% azide.

Alternative method

1. Tilt a Bio-Rad Econo column on a support at 45° to the bench, pour gel slurry into the column and allow to settle.
2. In a cold cabinet equilibrate the column with 0.1% NP40 in PBS at a flow rate 12 ml/h for at least 2 h until the baseline settles.
3. Load the column with crude serum a rate ≤5 ml/h or recycle the serum overnight.
4. Wash the column with PBS containing 0.4 M NaCl and 0.1% NP40 at a flow rate of 10 ml/h to remove all proteins which have not specifically bound to the column.
5. Elute the antibody specifically bound to the column using 2–2.5 column volumes with 0.5 ml fractions of 1 M propionic acid. Be careful at this stage to elute into tubes containing 0.5 ml of 0.5 M Na_2HPO_4 and to mix immediately to neutralize the solution. Minimizing contact with the propionic acid is critical to avoid inactivation of antibody.
6. Using the chart recorder or the Bio-Rad dye reagent, select the fractions containing the highest protein concentration and pool. Alternatively test these fractions separately (see preceding section no. 6).
7. Depending on the purity of the preparation, an additional step of ammonium sulfate precipitation may be required. If so, add dropwise an equal quantity of 80% ammonium sulfate solution to the antibody and stir in the cold room for 2 h.
8. Centrifuge 8000 rpm for 20 min and resuspend pellet in 5 ml PBS.
9. Dialyze overnight at 4 °C in 2 l of PBS (with two to three changes).
10. Following dialysis, add 0.02% azide. If the protein concentration is ≤0.2 mg/ml, add BSA to 0.5 mg/ml. Alternatively, concentrate and desalt it either by centrifugation using Centricon 10 tubes (Amicon), or by dialysis and subsequent lyophilization. Aliquot and store as described in the preceding section no. 7.

3.6 Cross-Linking Antibodies to Protein A-Sepharose

Antibody coupling

This is a simple method of purifying a cellular antigen free of immunoglobulins.

1. Wash 1 ml of swollen protein A-Sepharose twice with 5 ml of 0.1 M sodium borate pH 8.2.
2. Add 250 µl of crude anti-serum in 5 ml of the same buffer.
3. Incubate for 2 h or overnight on a rotator.
4. Wash five times by centrifugation in borate pH 8.2. Resuspend in 10 ml of borate pH 8.2. Take 50 µl for SDS-PAGE to check for IgG depletion.
5. Resuspend the resin pellet in 30 ml of 0.2 M sodium borate pH 9.0, 40 mM dimethyl pimelidate (addition of dimethyl pimelidate decreases the pH to about 7.8).
6. Incubate 1 h at room temperature. Take 50 µl for SDS-PAGE to check the yield of the coupling.
7. Wash in 0.1 M Tris pH 8.0 for 1 h hat room temperature on a rotator.
8. Keep gel in Tris buffer/azide at 4 °C.

Acknowledgments. My research is supported in part by the Human Science Frontier Program grant RG-496/93.

References

Baratte B, Meiyer L, Galaktinov K, Beach D (1992) Screening for antimitotic compounds using the Cdc25 tyrosine phosphatase, an activator of the mitosis-inducing p34Cdc2/Cyclin B protein kinase. Anticancer Res 12:873–880

DeBondt HL, Rosenblatt J, Jancarik J, Jones HD, Morgan DO, Kim SH (1993) Crystal structure of human Cdk2: implications for the regulation of Cyclin-dependent kinases by phosphorylation and Cyclin binding. Nature 363:595–602

Draetta G, Beach D (1988) Activation of Cdc2 protein kinase during mitosis in human cells: cell cycle-dependent phosphorylation and subunit rearrangement. Cell 54:17–26

Elledge SJ, Richman R, Hall FL, Williams RT, Lodgson N, Wade Harper J (1992) cdk2 encodes a 33-kDa Cyclin A-associated protein kinase and is expressed before Cdc2 in the cell cycle. Proc. Natl. Acad. Sci. USA 89:2907–2911

Hoffmann I, Clarke PR, Marcote MJ, Karsenti E, Draetta G (1993) Phosphorylation and activation of human Cdc25-C by Cdc2-Cyclin B and its involvement in the self-amplification of MPF at mitosis. EMBO J 12:53–63

Hopp TP, Woods KR (1981) Prediction of protein antigenic determinants from amino acid sequences. Proc Natl Acad Sci USA 78:3824-3828

Kyte J, Doolittle RF (1982) A simple method for displaying the hydropathic character of a protein. J Mol Biol 157:105–132

Marcote MJ, Knighton DR, Basi G, Sowadski JM, Brambilla P, Draetta G, Taylor SS (1993. A three-dimensional model of the Cdc2 protein kinase: localization of Cyclin- and suc1-binding regions and phosphorylation sites. Mol Cell Biol 13:5122–5131

Meyerson M, Enders GH, Wu CL, Su LK, Gorka C, Nelson C, Harlow E, Tsai LH (1992) A family of human Cdc2-related protein kinases. EMBO J 11:2909–2917

Pagano M, Draetta G, Jansen-Durr P (1992) Association of Cdk2 kinase with the transcription factor E2F during S phase. Science 255:1144–1147

Pagano M, Pepperkok R, Lukas J, Baldin V, Ansorge W, Bartek J, Draetta G (1993) Regulation of the human cell cycle by the Cdk2 protein kinase. J Cell Biol 121:101–111

Rosenblatt J, Gu Y, Morgan DO (1992) Human Cyclin dependent kinase 2 is activated during the S and G2 phases of the cell cycle and associates with Cyclin A. Proc Natl Acad Sci USA 89:2824–2828

Tsai L, Lees E, Faha B, Harlow E, Riabowol K (1993) The Cdk2 kinase is required for the G1 to S transition in mammalian cells. Oncogene 8:1593–1602

Index

Actinomycin D 43
Activation domain
- B42 149
- GAL4 146, 147, 148, 149, 211, 212
- VP16 147, 149
Additional DNAse I 38
Amino acid pattern 188
Amino acid starvation 66–67
2-Aminopurine 94
3-Aminotriazole 152, 154
Ankyrin 194
Antibody
- antigen competition 250
- metabolic labeling 250
Antigen unmasking 7
Anti-fading agent 22
- DABCO 22
- n-propyl gallate 22
- p-phenylene-diamine 22
Anti-peptide antibody 271–276
- choice of the peptide 272
- modified peptides 273
- purification 276–279
- terminal peptides 272
AOX genes 170, 172
Aphidicolin 68
Apoptosis 94–95
ATP 190–193
- binding motif 193
- dependent kinases 190
Automatic injection system (AIS) 81

Baculovirus 234–238
- amplification of viral stock 236
- determination of viral titers 237
- plaque purification 237
- transfection of insect cells 235
Bandshift assay 243
Beta-galactosidase activity (β-galactosidase activity) 148
Blocking agents 20
Bromodeoxyuridine (BrdU) incorporation 24, 49
- acid denaturation method 25, 50
- enzyme digestion-based method 25
- immunodetection 24, 51
- thermal denaturation method 51

Caffeine 93, 96
Calcineurin 145, 147, 148, 153
$CaPO_4$ transfection 89, 91
Carnoy's fixative 96
Cdc2-Cyclin B 200
- activation by cyclins 110
- activity 202
- fluorescence in situ hybridization 30–32
- kinase assay 111, 202
- phosphorylation by Wee1 167
- purification 203–206
Cdc25
- active site motiv 192
- anti-peptide antibodies 273
- catalytic nucleophile 192
- chromosomal anomalies 31
- fluorescence in situ hybridization 30–32
Cdk activating kinase (Cak) 200, 207
- activity 204, 205, 207–209
- expression in E. coli 163
- kinase assay 202
- purification from extract 203–206, 208
- purification of recombinant Cak 163, 164
Cdk inhibitors (Cdkis)
- Cdki in DNA replication assay 178
- chromosomal anomalies 31
- expression in insect cells 239–240
- fluorescence in situ hybridization 30–32
Cell division cycle (Cdc) mutants 119, 125
Cell-surface antigens 18
Centrifugal elutriation of
- mammalian cells 53–62
- yeast cells 135, 140

Centrifugation 109
Checkpoints 63–65, 72–73, 94, 113
Chromosome pulverization 93, 95
Cooperation 88
Counterstaining 11, 41
Cross-reactivity 13
Cryostat sections 7
Crystal structure 190–192
– cAMP-dependent kinase 190
– Cdk2 190, 191
– p13Suc1 193
– PTP1B 192
Cy3 42
Cyclin depent kinases (Cdks)
– activation by Cak 166
– activation by cyclins 110, 165
– anti-peptide antibodies 272–273
– Cdks in DNA replication assay 176–178
– chromosomal anomalies 31
– crystal strucutre 181–191
– expression in E. coli 163
– expression in insect cells 240–241
– fluorescence in situ hybridization 30–32
– kinase assay 111, 166, 167, 241
– peptide mapping 265–266
– purification of recombinant Cdk 159–164
– transfection 87–88
Cyclins
– cyclin box 188–189
– cyclins in DNA replication assay 176–178
– destruction box 189
– expression in E. coli 162
– expression in insect cells 240–241
– fluorescence in situ hybridization 30–32
– immunofluorescence 15–16
– immunohistochemical detection 4–5, 11–13
– purification 162, 163
– synthesis 106
– transfection 87–88
Cytogenetic 30
Cytoplasmic mRNA 32

DAPI 42
Databases
– OWL 188
– PRINTS 188
– PROSITE 187, 188, 194
– SWISS-Prot 188
DAE dextran 90, 91
Diaminobenzidine 9
Digital cameras 34
– color cooled CCD 34
– cooled CCD digital camera 34
Dimethyl pimelidate 208, 279
6-dimethylaminopurine 93
DNA binding domain
– GAL4 146, 147, 148, 149, 150, 211–213
– LexA 147, 149, 150, 212, 213, 216
DNA content 46
DNA damage 49
– agents 72–73
DNA replication 47, 49, 113, 175–185
– in vitro assay 183–185
DOTAP 92
Drug-resistance 88

E2F Bandshift assay 243
Electroporation 75, 77, 90, 91
– buffer 78
– parameters 83
Elutriation (see Centrifugal elutriation)
Ethanolamine 277
Expression of cell cycle proteins in
– E. coli 157–168
– eukaryotic cells 75–86, 87–92
– insect cells 231–243
– Pichia pastoris 169–174
– Schizosaccharomyces pombe 119–131

Fetch 187
Fine needle aspiration 31
Fission yeast 132
FITC 42
Fixation 18
Fixatives 6–7, 18
– organic solvents 18
– cross-linking agents 18
FKB1 148, 149
FKBP-12 145, 146, 147, 148, 150–153
FKBP-12/rapamycin complex 153
FK-506 145, 147, 148, 153
Flow cytometry 46, 49, 51, 61
Fluorescence in situ hybridization (FISH) 29
– visualization 29
– sensitivity 29
Fluorescence microscopy 15, 75, 84

Fluorochromes 23
- AMCA 23
- filter sets 23
- FITC 23
- Texas Red 23
Footprinting 246
Foreign genes 29
Freund's adjuvant 275
- complete 275
- incomplete 275
FTP
- servers 187
- site 187

GAL1
- promoter 148
- UAS 146
Gelvatol 10
Genomic probes 31, 39
Giemsa 96, 98
Glutaraldehyde 274
ghopher 187
GVBD 201

Heat shock synchrony 140
Histone H1 kinase assay 111, 206
Hoechst 33258 22, 84, 95–97
Hydroxyurea 95, 96, 138

Image analysis 77
Immunoblotting 251, 259–261
- blocking solution 259
- Ponceau S solution 260
- transfer buffer 259
Immunofluorescence microscopy 15, 76
- analysis at a single cell level 16
- basic protocol 26
- controls 21
- direct immunofluorescence 20
- image analysis 77
- indirect immunofluorescence 20
- multiple labeling 15
Immunohistochemistry 3
- immunodetection 4
Immunoprecipitation 251
- pre-clear 255
Induction synchrony 138–140
Insect cells 182, 233–234
In situ fractionation 19
Insoluble cellular components 19
INTERNET 186
IP-Western 258
- blocking solution 259
- Ponceau S solution 260
- transfer buffer 259
Isoleucine 66
Isopentane 6

Keyhole limpet hemocyanin (KLH) 273
Kinase assay 111, 166, 167, 241

Large T antigen (SV-40-LT) 179–180
- expression in insect cells 182
- immunoaffinity 182
Libraries
- polyA+-primed 150
- random-primed 150
Ligands
- small molecule 145
Lipofection 92
Lovastatin 67–68
Lysis buffer
- NP-40 LB 254
- RIPA LB 255

MacPattern 194
m-Maleimidobenzol-N-hydroxysuccinimide ester (MBS) 273
Media
- bacterial minimal 151
- MB 121
- SP1 121
- SP2 122
- SP3 122
- SP4 122
- SP5 122
- stocks 123
- yeast extract medium 123
- yeast drop-out 151
- YPD 153
Meiotic maturation 201
Metabolic labeling
- monolayer cells 252
- suspension cells 253
Methanol-Acetone 7
Metharcan 7
Methyladenine 201
Methylation interference 246
Microinjection 75
- automatic injection system (AIS) 80
- bubble pressure 79
- coinjection markers 76
- double injection 82
- glass capillaries 78

Microinjection
- luciferase 82
- micro-biochemistry 82
- microlader 80
- micromanipulator 80
- pulling micropipettes 79
- sample volume 81
Microscope 32
- multiple band-pass filter 32
Microwave treatment 8
Mimosine 68
Mitosis
- normal 94, 97
- premature 94–96
Mitotic shake-off 70–72
Mounting 22
Mowiol 10
Multiple labeling 22
Nick translation 37

Nocodazole 70–72, 94–96
Nongenomic probes 39
Nuclear assembly 113
Nuclear envelope breakdown 94
Nuclear mRNA 32

Okadaic acid 93

p13Suc1
- affinity chromatography 206–207
- crystal structure 193
- p13Suc1 beads 206
p53 65
- chromosomal anomalies 31
- fluorescence in situ hybridization 30–32
- in DNA replication assay 176
Paraffin sections 4
Paraformaldehyde 6
Peptide coupling to a protein 273–275
Peptide coupling to a sepharose 276
- CNBr activated Sepharose 4B 276
Peptide mapping 266–269
Permeabilization 18
Peroxidase reagents 9
Phosphate anchor 189
Photography 32, 33
- color films 33
- enhance 33
- inexpensively 33
- PhotoCD 33
- pseudocoloring 33
- reciprocity failure 33

Pichia pastoris 169, 170
- expression of proteins in P. pastoris 169, 170
- secretion of proteins from P. pastoris 169, 170
p-NPP assay 193
Polyclonal antiserum 17
- signal strength of an antiserum 17
Polyglycine linker 150
Poly-L-lysine 17
Premature chromosome condensation 93, 95
Production of cell cycle proteins in E. coli 157–168
- insect cells 231–243
- Pichia pastoris 169–174
Pronase treatment 9
Propidium iodide 42, 48, 62
- DNA analysis 49
Protein phosphatase 110
Proteinase 42
Puling micropipettes 79

Quantitate copy number 31
Quenching enzyme activities 10

Rabbit immunization 275
Rapamycin 145, 146, 149
RAPT-1 154
Rat-6 cells 88
Reporter genes
- HIS3 146, 147, 149, 150
- lacZ 146, 147, 149, 150, 152, 212, 214–216, 218–222, 227
- LEU2 149, 213–215, 217, 218, 220–222
Required 42
Retinoblastoma protein (pRb)
- chromosomal anomalies 31
- fluorescence in situ hybridization 30–32
- transfection 87–88
Rnase 42
RP-A in DNA replication assay 176

S 100 lysate 179–180
S phase 24
- biotinylated dUTP 24
- bromo-2'-dexoyuridine (BrdU) 24
- Fluorescently labeled dUTP analogues 26
Saccharomyces cerevisiae 119, 120, 128, 169, 170

Schizosaccharomyces pombe 119, 120, 125, 128, 129, 170
Selectable markers
- ADE2 149
- LEU2 149, 150
- TRP1 149, 150
Sequence-specific binding 244
Serum starvation 65–66
Simultaneous mapping 35
Spodoptera frugiperda
- Sf9 182, 233
- Sf21 233
Stable transfection 88, 91, 92
Staphylococcus aureus V8 protease 268
Starfish homogenate 200
Staurosporine 93, 95
Structural motifs
- Cdc25 192
- cyclins 188
- cyclin-dependent kinases 189
- database search 194
Structural model
- Cdc2 191
Subcellular localization 15
Subnuclear compartments 16
Supershift 245
Sv-40 poliomavirus 175–185
Synchronization
- of mammalian cells 63–73, 53–61
- of human lymphocytes 35
- yeast synchronous culture 132–142

T-cells
- activation 145
- receptor 145
- synchronization 35
T-loop 190
Temperature sensitive (ts) mutants 119, 126
- premissive temperature 120, 124, 126
- restrictive temperature 120, 125, 126
Threonine 161, 200, 203
Thymidine block 69–70
^{3}H-Thymidine uptake 47–48
- monitoring progression through S phase 47
- monitoring proliferation state 47
Tissue preservation 5
Tissue processing 3
Tissue resecton 5
TOR 154
Transactivation
- activator domain 211–212
- ligand-dependent 146, 147
Transformations
- bacterial 151
- lithium acetate 152
- yeast 152
Transient transfection 87, 88, 90, 91, 92
Translocated 31
Trypsin treatment 8
Two-hybrid screen 152–153
Two-hybrid system, 145, 147, 148, 149, 150, 211–213, 227

Velocity gradient 133

Wee1
- expression in E. coli 164
- purification 164, 165
Western blotting 259–261
- blocking solution 259
- Ponceau S solution 260
- transfer buffer 159

X-GAL 148, 151, 152, 153
Xenopus laevis
- anesthesia 114
- egg extracts 106, 109
- eggs 103–105, 108, 207
- homogenate 207
- injection 106
- maintenance 105
- MO15 200, 205, 209
- oocyte maturation 103
- oocytes 103
- sperm 114

Z buffer 152–153
Zonal Rotor 137

Springer-Verlag and the Environment

We at Springer-Verlag firmly believe that an international science publisher has a special obligation to the environment, and our corporate policies consistently reflect this conviction.

We also expect our business partners – paper mills, printers, packaging manufacturers, etc. – to commit themselves to using environmentally friendly materials and production processes.

The paper in this book is made from low- or no-chlorine pulp and is acid free, in conformance with international standards for paper permanency.